BRAIN-GUT PEPTIDES AND REPRODUCTIVE FUNCTION

BRAIN-GUT PEPTIDES AND REPRODUCTIVE FUNCTION

Edited by

CRAIG A. JOHNSTON   CHARLES D. BARNES

College of Pharmacy

and

Department of Veterinary and Comparative Anatomy,

Pharmacology and Physiology

Washington State University

Pullman, Washington

CRC Press

Boca Raton Ann Arbor Boston London

Library of Congress Cataloging-in-Publication Data

Brain-Gut peptides and reproductive function / edited by Craig A.
   Johnston, Charles D. Barnes.
         p.   cm.
      Based on an International conference held at Washington State
   University on June 18–19, 1989.
      Includes bibliograghical references and index.
      ISBN 0-8493-8848-1
      1. Neuropeptides—Congresses. 2. Gastrointestinal hormones —
   Congresses. 3. Human reproduction—Regulation—Congresses.
   4. Neuroendocrinology—Congresses. I. Johnston, Craig A.
   II. Barnes, Charles D., 1935–   .
         [DNLM: 1. Gastrointestinal Hormones—physiology—congresses.
      2. Neuropeptides—physiology —congresses. 3. Neuroregulators—
      physiology—congresses. 4. Reproduction—drug effects—congresses.
      5. Reproduction—physiology—congresses.   WK 185 B14 1989]
      QP552.N39B72   1991
      612.6—dc20
      DNLM/DLC   91-23468
      for Library of Congress   CIP

Developed by Telford Press

This book represents information obtained from authentic and highly regarded sources. Reprinted material is quoted with permission, and sources are indicated. A wide variety of references are listed. Every reasonable effort has been made to give reliable data and information, but the author and the publisher cannot assume responsibility for the validity of all materials or for the consequences of their use.

Direct all inquiries to CRC Press, Inc., 2000 Corporate Blvd., N.W., Boca Raton, Florida, 33431.

© 1991 by CRC Press, Inc.

International Standard Book Number 0-8493-8848-1

Printed in the United States 1 2 3 4 5 6 7 8 9 0

EDITORS

Craig A. Johnston, Ph.D., is Assistant Professor of Pharmacology/Toxicology in the School of Pharmacy and Allied Health Sciences with an adjunct appointment in the Division of Biological Sciences of the University of Montana at Missoula. He is also an adjunct Assistant Professor in the College of Pharmacy of Washington State University at Pullman.

Dr. Johnston received his B.S. degree from the Massachusetts Institute of Technology in 1977 in chemistry. He received Ph.D. degrees from Michigan State University in 1982 in pharmacology/toxicology and in neurosciences.

After a year and one-half as a postdoctoral fellow in the Department of Physiology at Southwestern Medical School of the University of Texas Health Science Center at Dallas, he moved to the National Institute of Environmental Health Sciences in Research Triangle Park, North Carolina, where he held the position of Senior Staff Fellow from 1983 to 1987 in the Laboratory of Reproductive and Developmental Toxicology. In November, 1987, he became an Assistant Professor of Pharmacology/ Toxicology in the College of Pharmacy at Washington State University where he remained until taking his present position in September, 1990.

Dr. Johnston is a member of the American Association for the Advancement of Science, American Association of Colleges of Pharmacy, International Brain Research Organization, International Society of Neuroendocrinology, Kappa Psi Pharmaceutical Fraternity, New York Academy of Sciences, Rho Chi National Pharmacy Honor Society, Sigma Xi Research Honor Society, Society for Neuroscience, The Endocrine Society, and the World Federation of Neuroscientists. He has received grants from the National Science Foundation, the National Institute of Child Health and Human Development, the Diabetes Research and Education Foundation, and currently holds a FIRST Award and an ADAMHA Award from the National Institute of Mental Health.

Dr. Johnston is the author of over 40 papers and has been the author or editor of 2 books. He serves as Regional Educational Counselor of Admissions for the Massachusetts Institute of Technology, Satrap for Province X of Kappa Psi Pharmaceutical Fraternity, and as a member of the Biotechnology Roundtable for the Missoula Economic Development Corporation. His current research focuses on elucidating the central neurochemical mechanisms controlling ovulation; characterizing the manner, mechanisms, and sites of action by which the neurointermediate lobe of the pituitary influences the secretion of hormones from the anterior pituitary lobe; and examining the central causes for the hyperadrenalcorticoidism associated with

Diabetes Mellitus and the possible relationship of those alterations to the development of depression which is also observed with Diabetes Mellitus.

Charles D. Barnes, Ph.D., is Professor and Chairman of the Department of Veterinary and Comparative Anatomy, Pharmacology and Physiology in the College of Veterinary Medicine of Washington State University at Pullman.

Dr. Barnes received his B.S. degree from Montana State University in 1958 with double majors in biology and physics. He received an M.S. degree from the University of Washington in 1961 in physiology and biophysics and a Ph.D. from the University of Iowa in 1962 in physiology.

After two years as a postdoctoral fellow in the Department of Pharmacology at the University of California at San Francisco, he became an Assistant Professor of Anatomy and Physiology at Indiana University in 1964. He advanced to Associate Professor in 1968 and in 1971 became Professor of Life Sciences at Indiana State University. In 1975, he became Chairman of the Department of Physiology at Texas Tech University College of Medicine, where he remained until taking up his present position in 1983.

Dr. Barnes is a member of the American Association for the Advancement of Science, American Association of Anatomists, American Institute of Biological Sciences, American
Physiological Society, American Association of Veterinary Anatomists, American Society of Pharmacology and Experimental Therapeutics, American Society of Veterinary Physiologists and Pharmacologists, Association of Anatomy Chairmen, Association of Veterinary Anatomy Chair-persons, Association of Chairmen of Departments of Physiology, International Brain Research Organization, Radiation Research Society, Society for Experimental Biology and Medicine, Society for Neuroscience, Society of General Physiologists, and the Western Pharmacological Society. He has been the recipient of many research grants from the National Institutes of Health and the National Science Foundation.

Dr. Barnes is the author of more than 150 papers and has been the author or editor of 15 books. His current research interests relate to the modulation of nervous system output by centers in the brainstem.

PREFACE

Many neuropeptides originally isolated from the gastrointestinal system are also synthesized in the brain. Several of these "Brain-Gut" peptides have been shown capable of influencing the secretion of reproductive hormones from the anterior pituitary. Indeed, a major focus at recent meetings dealing with the basic and clinical functions of these individual peptides has been their possible role in neuroendocrine regulation. Although the hypothesis that these centrally derived peptides may play an important role in neuroendocrine regulation is becoming widely accepted, the sites and mechanisms by which that regulation is exerted are poorly understood. The possible interactions which may be involved, between central peptidergic and aminergic neurotransmitters, in regulating anterior pituitary secretion are substantial. It is more than likely that the mechanisms and sites of interaction involved in this regulation are going to be extremely complex and difficult to elucidate. However, uncovering such information may be critical both for our understanding of how these peptides interact with the reproductive neuroendocrine axis and for using this information in the clinical setting to develop novel therapeutic agents for fertility or contraceptive therapies.

A two-day international conference was held at the campus of Washington State University on June 18-19, 1989, and was preceded by two social events including white-water rafting on the Lower Salmon River and a sunset cruise on Lake Coeur d'Alene. Our objective was to provide the first forum where international experts on the reproductive neuroendocrine functions of each individual "brain-gut" peptide could gather to present a summary of recent advances in their particular area dealing with their individual peptide. We hoped that sharing their insights/speculations/impressions/biases with others working with different peptides would help to formulate hypotheses about the participation and interactions of these peptides in reproductive neuroendocrine control. We believe these goals were successfully met as 88 scientists from Australia, Canada, France, Italy, Japan, Mexico, Spain, Switzerland, the United States, and West Germany gathered and openly interacted both at the scientific sessions (lectures and poster sessions) and the varied social activities. Novel insights and directions for future research resulted; the complexity of interactions between the endocrine system, feeding and satiety control, environmental cues such as light and stress, and the immune system became dominant discussion themes. Questions which were raised included: Are food intake and gastrointestinal function closely tied to neuroendocrine function and reproductive behaviors? If so, through what mechanisms? Can the peripheral release of brain-gut peptides influence anterior pituitary hormone secretion? How and where does information from the immune system integrate with environmental cues to exert their regulatory influence on reproductive hormone secretion? Many of the major findings and suggested areas for future research are found in this text. In addition, several other chapters dealing with related areas are also provided so that the reader can assimilate new directions for exploration into this rapidly expanding and truly exciting area of research. If, after reading this text, more questions for investigation are opened than answers are provided, then this adventure has been successful.

ACKNOWLEDGEMENTS

As for the conference itself, financial assistance from many sources provided the resources necessary for its successful completion. These include: Amersham Corporation; Burroughs Wellcome Company; College of Pharmacy at Washington State University; the Graduate School at Washington State University; ICI Pharmaceuticals Group, ICI Americas, Inc.; Jackson Immunoresearch Labs, Inc.; National Institute of Child Health and Human Development; National Institute of Diabetes and Digestive and Kidney Diseases; National Science Foundation, Division of Behavioral and Neural Sciences; Peninsula Laboratories, Inc.; Promega; Research Biochemicals, Inc.; and Searle Research and Development, Monsanto Company. The editors and the Symposium Organizing Committee are deeply indebted to these companies for their financial contributions.

We are personally indebted to the members of the organizing committee: W. Les Dees, Ph.D.; John M. Farah, Ph.D.; James I. Koenig, Ph.D.; John K. McDonald, Ph.D.; Felice Petraglia, M.D.; Willis K. Samson, Ph.D.; and Dipak K. Sarkar, Ph.D.; without their dedicated efforts this meeting would never have taken place. The dedication of Ms. Lisa Bennett who helped organize all aspects of this conference are also deeply appreciated. The present volume grew out of the conference rather than being simply proceedings of. To achieve this, a considerable amount of editorial skill and several updated copies were involved. For this, the editorial assistance of Catherine Smith, and word processing skills of Rebecca Thompson, Paula Perron-Bates and Dana Spear are also gratefully acknowledged.

C. A. Johnston, Ph.D.*
Organizing Secretariat

C. D. Barnes, Ph.D.

*Craig Johnston's present address: School of Pharmacy and Allied Health Sciences, PsP Bldg., Rm. 237, University of Montana, 32 Campus Drive, Missoula, Montana 59812-1075 USA

CONTENTS

Preface   *v*

CHAPTER 1

                                                                                                               1

Influence of Ovarian Hormones on the Regulation of Luteinizing Hormone and Prolactin Release by Angiotensin II
M.K. Steele, L.S. Myers, C.F. Deschepper, W.F. Ganong,
K.N. Stephenson, and R.L. Shackelford
Department of Physiology
University of California
San Francisco, California

CHAPTER 2

                                                                                                              21

Renin-Angiotensin System in Female Reproductive Organs
Robert C. Speth, Kevin L. Grove, and Cesario S. Zamora
Department of Veterinary and Comparative
Anatomy, Pharmacology and Physiology
Washington State University
Pullman, WA

CHAPTER 3

                                                                                                              43

Renin-Angiotensin System in Male Reproductive Organs
Kevin L. Grove, Robert C. Speth, and Cesario S. Zamora
Department of Veterinary and Comparative
Anatomy, Pharmacology and Physiology
Pullman, WA

CHAPTER 4

                                                                                                              61

Neurohypophyseal Hormones and Reproductive Hormone Secretion
Craig A. Johnston
College of Pharmacy
Washington State University
Pullman, WA

CHAPTER 5

83

Vasoactive Intestinal Peptide: A Neural Modulator of Endocrine Function
   Willis K. Samson
   Department of Anatomy and Neurobiology
   University of Missouri School of Medicine
   Columbia, Missouri
   and
   Marc E. Freeman
   Department of Biological Science
   Florida State University
   Tallahassee, Florida

CHAPTER 6

105

Reproductive Neuroendocrine Effects of Neuropeptide Y and Related Peptides
   John K. McDonald, Ph.D.
   Department of Anatomy and Cell Biology
   Emory University School of Medicine
   Atlanta, Georgia

CHAPTER 7

155

Opioids and Induction of Ovulation: Mediation by Neuropeptide Y
   S.P. Kalra, L.G. Allen*, A. Sahu, and W.R. Crowley**
   R.W. Johnson Medical School*
   Department of Anatomy
   Piscataway, NJ
   and
   Department of Pharmacology**
   University of Tennessee
   College of Medicine
   Memphis, TN

CHAPTER 8

179

The Role of Neurotensin in Control of Anterior Pituitary Hormone Secretion
  S.M. McCann[1] and E. Vijayan[2]
  [1]Department of Physiology, Neuropeptide Division
  The University of Texas Southwestern
  Medical Center at Dallas
  Dallas, Texas
  and
  [2]Department of Biological Sciences
  Pondicherry University
  Pondicherry, India

CHAPTER 9

193

Galanin: A Potentially Significant Neuroendocrine Modulator
  J.I. Koenig[1], S.M. Gabriel[2], and L.M. Kaplan[3]
  Departments of Neurology[1] and Medicine[3]
  Massachusetts General Hospital
  and Harvard Medical School
  Boston, MA
  and
  Department of Psychiatry[2]
  Mount Sinai Medical School
  New York, NY

CHAPTER 10

209

Cholecystokinin and Neuroendocrine Secretion
  Joseph G. Verbalis and Edward M. Stricker
  Departments of Medicine and
  Behavioral Neuroscience
  University of Pittsburgh
  Pittsburgh, PA

CHAPTER 11

239

Motilin and Gastric Inhibitory Polypeptide (GIP)
   C.H.S. McIntosh
   Regulatory Peptide Group
   Medical Research Council of Canada
   Department of Physiology
   University of British Columbia
   Vancouver, B.C., Canada

Index   277

INFLUENCE OF OVARIAN HORMONES ON THE REGULATION OF LUTEINIZING HORMONE AND PROLACTIN RELEASE BY ANGIOTENSIN II

M.K. Steele, L.S. Myers, C.F. Deschepper, W.F. Ganong,
K.N. Stephenson, and R.L. Shackelford
Department of Physiology
University of California
San Francisco, California

I.   Introduction

II.   The Brain Angiotensin II System

      A.   Components

      B.   Regulation of Luteinizing Hormone Release

      C.   Regulation of Prolactin Release

III. The Pituitary Angiotensin II System

      A.   Components

      B.   Regulation of Prolactin Release

IV.   Summary

V.   Speculations

Introduction

Angiotensin II (AII) is a peptide of eight amino acids formed by an enzymatic cascade in which the substrate, angiotensinogen, is cleaved by the enzyme renin to form the decapeptide, angiotensin I. Angiotensin-converting enzyme then further cleaves angiotensin I to form AII. This cascade occurs in the bloodstream to maintain the circulating renin-angiotensin system(s), (RAS), which classically has been implicated in the control of blood pressure, vascular reactivity, and fluid-electrolyte homeostasis.

In addition to the circulating AII system, research during the past ten years has identified putative "local," intra-tissue, RAS in a variety of organs including the brain, pituitary gland, adrenal gland, pineal, testis and ovary (for reviews, see Moffett *et al.*, 1987; Bumpus *et al.*, 1988; Deschepper and Ganong, 1988; Steele, 1989; Ganong *et al.*, 1989). The function of local tissue-generated AII may be similar to that of circulating AII, *i.e.*, regulation of vascular tone and fluid/electrolyte shifts. However, other functions are also possible. For example, in the brain, as discussed in this chapter, AII acts as a neurotransmitter, affecting the release of other brain peptides and classical neurotransmitters intracerebrally. In the pituitary gland, AII appears to act as a paracrine hormone, being synthesized and released from one cell type (the gonadotroph) and acting at receptors on another cell type (the lactotroph and/or the corticotroph).

This chapter will focus on the renin-AII systems of the brain and pituitary gland. We will briefly discuss the evidence for a local AII system in these tissues, the influence of ovarian hormones on the components of the AII-generating cascade and the functions of these systems as they relate to reproductive physiology. Finally, we will conclude by speculating on how AII may interact with other brain/pituitary systems in the regulation of luteinizing hormone (LH) and prolactin secretion in the rat.

The Brain Angiotensin II System

Components

All the proteins necessary for the synthesis of AII have been measured in the rat brain. In addition, the mRNAs for angiotensinogen and renin (Campbell and Habener, 1986; Dzau *et al.*, 1986) have been identified, suggesting that the brain has the capacity to generate AII. It appears, however, that the entire cascade to synthesize AII may not occur in the same cell. Angiotensinogen has been localized almost exclusively within glial cells (Stornetta *et al.*, 1988), while renin has been identified within neurons (Slater *et al.*, 1980). Converting enzyme, on the other hand, is predominantly found on the extracellular surface of both neuronal and glial cell membranes (Pickel *et al.*, 1986). The process whereby AII is actually generated, given the differential localization of the cascade components, is presently a source of much scientific inquiry (see Moffett *et al.*, 1987 for further discussion). By immunocytochemistry, AII is found exclusively in neuronal elements in the brain stem, hypothalamus and limbic regions (Lind *et al.*, 1985), usually in the same areas where high-affinity binding sites for the peptide have been identified (Healy *et al.*, l986; Mendelsohn *et al.*, 1984).

Ovarian hormones have effects upon the components, as well as on the function, of the brain AII system. When measured over the rat estrous cycle, levels of angiotensinogen were decreased at estrus, compared to diestrus, in parts of the hypothalamus and limbic system (Printz *et al.*, 1983). Levels in other brain regions were unchanged, while concentrations in the subfornical organ were increased at estrus. Ovariectomy resulted in angiotensinogen levels similar to those seen on diestrus, while estrogen treatment decreased concentrations, resembling those on estrus (Printz *et al.*, 1983). The number of AII binding sites in certain brain areas also varies over the estrous cycle, such that preoptic area sites showed a tendency to decrease on estrus compared to diestrus (Chen *et al.*, 1982). These data were confirmed by Jonklaas and Buggy (1985) who showed that estradiol treatment decreased AII receptor binding in the preoptic area of female rats but not in male animals. Interestingly, Phillips *et al.* (1986) reported preliminary data suggesting that levels of AII in the hypothalamus, but not in brain stem or in cortex, varied over the estrous cycle. The highest concentrations were seen on proestrus, coincident with the time of the LH surge. To date, there have been no reports on the effects of ovarian steroids on levels of renin or converting enzyme in the brain.

Brain AII has long been known to participate in the control of a number of physiological functions, including blood pressure regulation, fluid and electrolyte balance, neuronal activity and secretion of hormones from the pituitary gland (Phillips, 1987). A functional significance for the steroid-related alterations in brain AII components is suggested by studies showing that both spontaneous and AII-induced water drinking vary over the rat estrous cycle (Findlay *et al.*, 1979, Kucharczyk, 1984). Fluid intake is lowest on proestrus and estrus, in concert with decreased binding sites for AII in the anterior hypothalamic-preoptic area. In addition, estrogen treatment to female rats reduced *ad lib* water intake 24 h after treatment and attenuated AII-induced drinking (Jonklass and Buggy, 1984). A similar suppressive response of estrogens on AII-induced pressor responses has been reported (Jonklass and Buggy, 1984). The modulatory effects of ovarian steroids on the effects of exogenous AII, as well as the functioning of endogenous AII, on LH and prolactin secretion will be discussed below.

A.

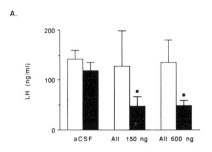

B.

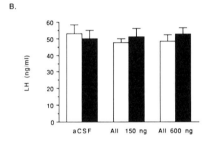

C.

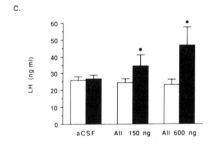

Fig. 1. Whole blood levels of LH (mean ± SEM) prior to (white bars) and during (black bars) intracerebroventricular infusion of artificial cerebrospinal fluid (aCSF, 25 μl/h) or AII at 150 or 600 ng/25 μl/h. A: ovariectomized rats; B: ovariectomized rats treated with estradiol (20 μg/rat) 48 h prior to blood collection; C: ovariectomized rats treated with estradiol (50 μg/rat) and progesterone (25 mg/rat) 72 h prior to blood collection. n = 6-9 rats per group. *p < 0.01 compared to preinfusion values.

Regulation of Luteinizing Hormone Release

Basal levels of LH in the blood are profoundly affected by the presence or absence of ovarian steroids. In their absence, (*e.g.*, after ovariectomy), blood LH levels are elevated and usually show a pulsatile profile. Following estrogen, with or without progesterone treatment, blood LH concentrations are reduced and show pulses of reduced amplitude. These effects are, of course, due to the negative feedback of the ovarian steroids. The sites of action for this negative feedback include the pituitary gland and several brain areas. With longer exposure to steroids, a positive feedback effect can be observed, such that a surge of LH occurs, similar to the proestrous preovulatory release of LH. In the rat, this action is likely due to an effect of the steroid(s) on the pituitary gland as well as on the anterior hypothalamic preoptic area, a region implicated in the phasic release of LH. As is suggested by the theme of this symposium, a constellation of brain peptides and classical neurotransmitters participates in maintaining both the basal and phasic secretion of LH. Ovarian steroids exert their effects upon virtually every brain system involved in LH control, ranging from effects upon synthesis and the sensitivity of release dynamics to receptor number and affinity. As has been described above, the AII system is no exception.

The effects of exogenous AII, administered into the third cerebral ventricle (ICV), on LH release are ovarian-steroid dependent (Fig. 1; Steele *et al.*, 1985). In ovariectomized rats (A), ICV-infusion of AII produced dose-dependent decreases in blood levels of LH. In estrogen-alone-treated animals (B), these same doses of AII did not affect LH concentrations. However, following treatment with both estradiol and progesterone (C), ICV-AII produced dose-related increases in LH. The brain site where AII is acting to stimulate LH release in ovarian steroid-treated rats is the anterior hypothalamus-preoptic area (Steele, 1987). Administration of AII directly into this area elicited dose-dependent elevations in blood LH, while equivalent concentrations of the peptide injected into the posterior hypothalamus were ineffective in modifying LH levels.

The physiological relevance of the stimulatory effects of AII on LH in steroid-treated animals was assessed by shifting our attention to the intact female animal on the day of proestrus, a time when the brain has been "primed" by endogenous ovarian steroids (Steele *et al.*, 1983). In this preparation, ICV-AII, administered on the morning of proestrus, stimulated LH release. Conversely, ICV administration of an AII-receptor blocker or synthesis inhibitor on the afternoon of proestrus blocked the naturally occurring LH surge. These data are compatible with those described earlier reporting increased levels of AII within the hypothalamus on the afternoon of proestrus (Phillips *et al.*, 1986). The decreased number of AII receptors in the hypothalamus on the day of estrus (Chen *et al.*, 1982), the day following proestrus, may be related to the suppressed levels of LH, as well as with the reduction in water intake observed on that day.

The brain circuitry involved in stimulating LH release in steroid-primed rats or in animals on the day of proestrus is obviously complex. We know that the AII effect in steroid-treated animals is mediated via the release of norepinephrine (Steele and Ganong, 1986). Inhibition of norepinephrine, but not epinephrine, synthesis within the hypothalamus abolished the facilitatory effects of ICV-AII on LH release. Furthermore, blocking α_2-adrenergic receptors also abolished the AII effect, while α_1 or β-receptor blockade had no effect. The stimulatory effects of ICV-norepinephrine were also abolished by α_2, but not α_1 or β-receptor antagonists. Finally, administration of an AII receptor blocker prevented the rise in LH due to ICV-AII but not the rise due to norepinephrine. Taken together, these data suggest that the stimulatory effects of ICV-AII on LH release are mediated by hypothalamic norepinephrine, acting at facilitatory α_2 receptors.

Ultimately, the effects of AII on LH are likely mediated by the release of LH-releasing hormone (LHRH). However, the brain circuitry distal to the LHRH neuron still needs to be unraveled. It appears to be clear that AII acts via norepinephrine, but the pattern of interaction of AII with other peptide systems, such as the opioids, oxytocin, or neuropeptide Y, is open to speculation and further experimentation.

Regulation of Prolactin Release

Prolactin secretion from the anterior pituitary gland is under tonic inhibitory control by dopamine, released from nerve terminals in the median eminence. Dopamine reaches the adenohypophysis via the hypophyseal portal vessels to act directly on the lactotrophs to inhibit prolactin secretion. Other brain peptides, as well as peptides in the pituitary gland itself and substances in the bloodstream (like ovarian hormones), interact with dopamine to influence basal and stimulated prolactin levels (Ben-Jonathan, 1985; Fink, 1988).

Ovarian hormones, especially estrogen, affect prolactin levels in the blood. Following exogenous estrogen treatment in ovariectomized rats or after the rise in endogenous estrogen in intact female animals, circulating prolactin levels are elevated. Some of the stimulatory effects of estrogen are directly at the level of the pituitary, such that prolactin synthesis, content, and basal release are increased. There are central nervous system effects of ovarian hormones also, which probably involve activation of prolactin releasing systems and/or inhibition of inhibitory ones.

The central action of AII to influence prolactin release appears to be inhibitory in nature. Central administration of the AII-generating enzyme, renin, or a high dose of AII inhibited prolactin release in male rats, as well as in ovariectomized animals, with or without estrogen treatment (Andersson *et al.*, 1982; Steele *et al.*, 1981, 1982b). The effect of renin was due to AII, since treatment with a converting enzyme inhibitor blocked the suppressive effects of the enzyme on prolactin secretion. It is likely that AII is affecting prolactin release through the activation of brain dopamine systems. For example, the administration of renin resulted in an increase in dopamine utilization (turnover) in the median eminence which was correlated with the reduction in plasma prolactin levels in male rats (Andersson *et al.*, 1982). In addition, the inhibitory effects of ICV-AII on prolactin release in female rats could be blocked by intravenous injection of a dopamine receptor blocker (Steele *et al.*, 1982b). This drug presumably acted directly at the pituitary lactotroph to antagonize the endogenous dopamine released by ICV-AII.

We have recently completed experiments designed to evaluate the sensitivity of the prolactin-lowering effects of ICV-AII in different animal models and also to determine the involvement of the endogenous brain AII system in tonically inhibiting prolactin release (Myers and Steele, 1989; Myers and Steele, unpublished data). In intact male rats, ovariectomized animals and ovariectomized rats treated with estradiol and progesterone, ICV-injection of AII, in doses as low as 50 ng, suppressed plasma prolactin levels. These levels remained suppressed for at least 15 min following ICV-AII and up to 60 min in male animals and in female rats treated with ovarian steroids. The duration of this effect suggests that other brain systems (*e.g.*, dopamine) are activated by AII to maintain the reduction in prolactin levels, since the half-life of ICV-AII in the brain is in the order of 20 to 30 sec (Harding *et al.*, 1986).

To investigate the role of endogenous brain AII in tonically inhibiting prolactin release, rats were administered either AII receptor blockers or an AII synthesis inhibitor, *e.g.*, a converting enzyme inhibitor. In male rats and in ovariectomized animals that had not been treated with estrogen and progesterone, neither of these drug regimens modified the low circulating levels of prolactin observed in these animal models (see Fig. 2A). In contrast, in ovariectomized rats that had been treated with ovarian hormones, both AII receptor blockade (Fig. 2B) and AII synthesis inhibition increased basal prolactin concentrations.

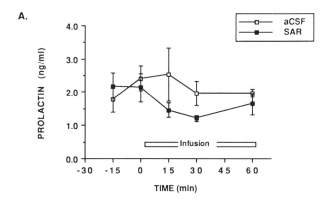

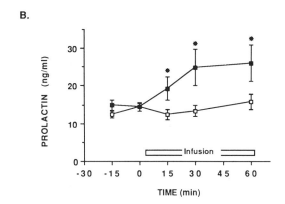

Fig. 2. Plasma prolactin levels (mean ± SEM) prior to and during intracerebroventricular infusion of aCSF (25 µl/h) or saralasin (15 µg/25 µl/h) in (A) ovariectomized rats and (B) ovariectomized rats treated with estradiol and progesterone. n = 6-10 rats per group.
*p < 0.05 compared to values from aCSF-infused rats at the same time point.
Modified and reproduced, with permission, from Myers and Steele, 1989.

The effect of receptor blockade was apparent within 15 min after the start of the drug infusion and continued for the duration of the drug treatment. Following administration of the converting enzyme inhibitor, prolactin levels did not rise significantly until 90 min and then remained elevated up to 120 min post-treatment. The rapid onset of the prolactin rise due to AII receptor blockade was not surprising since the drug is a competitive antagonist to AII at the receptor level. The delayed effect of the converting enzyme inhibitor was somewhat unexpected, but may reflect the time necessary to penetrate brain tissue and achieve complete inhibition of brain AII synthesis.

Taken together, the data demonstrate that, regardless of the steroid background of the animal, ICV-AII is capable of inhibiting prolactin release. This effect is quite sensitive, down to 50 ng, and, in fact, we have data demonstrating that even 10 ng of ICV-AII suppresses prolactin levels in male rats. However, the sensitivity to exogenous AII does not predict activity of endogenous brain AII. Our data using AII receptor antagonists or synthesis inhibitors show that only following exposure to ovarian steroids does endogenous brain AII participate in tonically inhibiting basal prolactin release. The involvement of brain AII in controlling patterns of prolactin release during the estrous cycle, gestation and/or lactation--times when ovarian steroids are fluctuating--is presently under investigation.

The Pituitary Angiotensin II System

Components

There is accumulating evidence for an intrapituitary AII system (Ganong *et al.*, 1989). Both renin and converting enzyme have been measured in the anterior pituitary gland (Celio *et al.*, 1980; Yang and Neff, 1973) and the mRNAs for angiotensinogen and renin have been identified (Helmann *et al.*, 1988; Deschepper *et al.*, 1986b). In addition, AII itself can be measured in extracts of fresh pituitaries, and AII immunoreactivity persists in pituitaries maintained in organ culture for 14 days in serum-free medium (Deschepper *et al.*, 1986a).

In the rat, the gonadotrophs appear to be the site of AII synthesis since renin, converting enzyme, and AII have been found in this cell type by immunocytochemistry (Naruse *et al.*, 1981; Steele *et al.*, 1982a; Deschepper *et al.*, 1985; Ganong *et al.*, 1989). Angiotensinogen, however, was not localized to gonadotrophs, corticotrophs, somatotrophs, thyrotrophs, lactotrophs, or with folliculo-stellate cells that stain with the S-100 protein. The nature of the unique cell type which contains angiotensinogen immunoreactivity remains to be established. Binding sites for AII are found on lactotrophs and corticotrophs (Aguilera *et al.*, 1982; Paglin *et al.*, 1984) and AII added directly to pituitary cells *in vitro* stimulates the release of prolactin and adrenocorticotropin (Steele *et al.*, 1981; Aguilera *et al.*, 1982; Sobel and Vagnucci, 1982). Taken together, these data suggest that AII is synthesized and stored within gonadotrophs. The mechanism by which angiotensinogen is made available for this synthesis is still unclear. The AII stored in gonadotrophs then has the potential for release and subsequent action on lactotrophs and corticotrophs to elicit prolactin and adrenocorticotropin secretion.

The influence of ovarian hormones on the components of the pituitary AII system or its function have not been extensively studied. To date, only the effects of estrogen on AII binding sites have been reported. Preliminary studies showed a reduction in AII binding on estrus compared to diestrus (Chen *et al.*, 1982). Subsequently, other reports have shown that both acute (24 h) and chronic (4 day) estrogen treatment down-regulate AII receptors, without decreasing the responsiveness of the tissue to AII stimulation (Carrière *et al.*, 1986; Platia *et al.*, 1986). In fact, the net release of prolactin in estradiol-treated cells was greater than that from controls.

Regulation of Prolactin Release

The function of AII within the pituitary gland has yet to be established. Exogenous AII stimulates prolactin release following intravenous administration *in vivo* or *in vitro* after direct application to either fresh tissue or cells that have been in culture for 2 to 4 days. Both the *in vivo* and *in vitro* effects display specificity since LH secretion is unaffected under these conditions and the prolactin release induced by AII can be abolished by prior administration of an AII-receptor blocker. However, an effect of the exogenous ligand does not automatically imply that the endogenous substance is functional. To assess the importance of endogenous pituitary AII in affecting basal prolactin release, we incubated fresh pituitaries from ovariectomized female rats with either AII synthesis inhibitors or with AII receptor blockers (Steele and Deschepper, unpublished data). Neither of these drug treatments, at doses as high as 10^{-7} M for 3 h, affected the release of prolactin (or LH) compared to control values. It is possible, given the results showing an activation of the brain AII system by estrogen and progesterone, that the pituitary system is also activated by ovarian hormones. However, at this time there are no data on the effects of AII receptor blockade or synthesis inhibition on basal prolactin release from pituitaries following treatment with estrogen and/or progesterone.

On the other hand, pituitary AII may be involved in controlling stimulated prolactin release. Co-localization of AII in LH secretory granules has been identified using electron microscopy (Deschepper *et al.*, 1986a). Therefore, co-release of AII and LH following LHRH stimulation is feasible. In addition to the well-documented effects of LHRH on LH release, a number of *in vivo* studies using human subjects have reported increases in prolactin secretion following LHRH (Mais *et al.*, 1988). Also, in reaggregates of anterior pituitary cells *in vitro*, Denef and Andries (1983) have demonstrated that a substance released from the gonadotrophs by LHRH stimulated prolactin release. Jones *et al.* (1988) recently provided evidence that this substance may be AII and that the LHRH-induced prolactin response from cultured rat pituitary cells was blocked by saralasin, an AII receptor blocker.

Table I. Levels of LH and prolactin (ng/ml plasma), expressed as the mean ± standard error of the mean, 10 min after intravenous injection of saline (100 ml) or LHRH (100 ng/100 ml saline) in male and female rats (n = 4 to 6/group).

		SALINE	LHRH
MALES	LH	0.53 ± 0.31	2.15 ± 0.30**
	Prolactin	3.40 ± 0.45	3.40 ± 0.36
FEMALES	LH	0.32 ± 0.05	5.95 ± 0.92**
	Prolactin	19.85 ± 2.31	32.45 ± 4.24**

**$p < 0.05$ versus saline value

We have investigated this phenomenon *in vivo*. In initial experiments, we compared the effects of intravenous LHRH (100 ng in 100 ml saline) on LH and prolactin release in intact male animals versus ovariectomized female rats that had been treated with estradiol and progesterone (Table 1; Steele, Shackelford and Ganong, unpublished data). Ten min following LHRH administration, plasma LH levels were maximally stimulated in both male and female rats. However, an increase in prolactin release was seen only in female animals, suggesting a facilitatory role for ovarian hormones in sensitizing the lactotrophs to stimulation. A similar effect has been seen in humans where estrogen treatment enhanced the prolactin-releasing effect of LHRH (Gooren *et al.*, 1985). To assess whether pituitary AII mediates the effect of LHRH on prolactin release, we infused ovarian steroid-treated rats with Sarthran, an AII-receptor blocker (Fig. 3). In saline-infused rats, LHRH induced a significant rise in

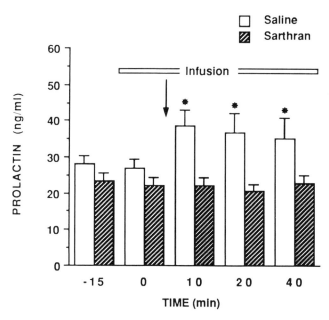

Fig. 3. Effect of intravenous infusion of Sarthran (5 mg/kg/min, n = 15) or saline (1.0 μl/h, n = 13) on the plasma prolactin response to intravenous LHRH (100 ng/100 μl saline). Infusions begun immediately following the -15 min blood sample and continued to the end of the sampling period. LHRH was injected immediately after the zero-time blood sample (arrow).
*p < 0.05 versus saline at the same time point.
Hormone values given as mean ± SEM.

prolactin levels; however, in Sarthran-infused animals, the prolactin increase was totally abolished. When considered with the data of Denef and Andries (1983) and Jones *et al.* (1988), these data are consistent with the hypothesis that LHRH stimulates the release of LH and AII from pituitary gonadotrophs. The AII, then, acts on the nearby lactotrophs to stimulate prolactin secretion.

To determine the physiological relevance of this pharmacologic effect (Steele and Myers, unpublished data) we investigated two conditions where plasma levels of both LH and prolactin were elevated; the LH rise due presumably to LHRH, and the prolactin rise hypothetically due to the release of AII from the gonadotroph. The first condition was an acute restraint stress that stimulated both LH and prolactin release. Although there is general agreement that chronic stress results in a decline in circulating LH levels, other studies have described temporary elevations in plasma LH (and prolactin) after acute stress (Briski and Sylvester, 1988). Levels of LHRH in hypophyseal portal blood have not been measured during such a stress; however, the increase in LH during this procedure is presumed to be due to a short-term increase in LHRH release. The second condition investigated was the surge in prolactin on the day of proestrus in intact female rats. The midcycle prolactin surge actually precedes that of LH by several hours. However, endogenous LHRH release displays a biphasic pattern of secretion over the day of proestrus, with the first phase (the self-priming phase for LH release) coincident with the start of the prolactin surge (Levine and Ramirez, 1982; Sarkar *et al.*, 1976).

To determine whether AII was involved in the prolactin rise in these two conditions, we infused rats with intravenous saralasin to block pituitary AII receptors during acute stress and during the day of proestrus. Compared to saline-infused animals, infusion of the AII-receptor blocker did not affect the stress-induced increase in prolactin nor the midcycle surge of prolactin. Levels of LH in plasma were also unaffected by saralasin administration. We know that the dose of saralasin used was sufficient to block pituitary AII receptors because the stimulation of prolactin induced by exogenous, intravenous AII was abolished by this drug treatment.

Taken together, the results of these experiments suggest that, first, the stimulatory effect of intravenous LHRH on prolactin release is mediated by AII, probably secreted from the pituitary gonadotroph. These data may prove useful in understanding and preventing, if necessary, one of the "paradoxical" effects of administering clinically high doses of LHRH. Second, the data do not support the hypothesis that endogenous pituitary AII participates in the prolactin rise due to acute restraint stress or during the afternoon of proestrus. Endogenous LHRH is likely to be elevated under these two conditions since LH is also stimulated at this time. However, thus far, we have no evidence that either AII or LHRH is involved in eliciting the prolactin increase due to stress or at midcycle. It is possible that pituitary AII may participate in regulating pulsatile prolactin release and future experiments will explore this hypothesis.

Summary

The data presented in this chapter strongly support a role for brain-derived AII in the control of LH and prolactin release. This control is particularly apparent following exposure of the brain to the ovarian steroids, estrogen, and progesterone. It appears that these hormones "activate" the brain AII system, at the synthesis and/or at the receptor level. Perhaps the clearest evidence for a role for endogenous brain AII in regulating anterior pituitary hormone release comes from experiments using receptor blockers or synthesis inhibitors. Using this approach, we have shown that the proestrous surge of LH is abolished and ovulation prevented when these interventions are restricted to the central nervous system. Similarly, we have shown a role for endogenous brain AII in influencing basal prolactin levels in rats that had been treated with estradiol and progesterone. These latter data suggest that the interaction of brain AII and dopamine in controlling prolactin release is functional following ovarian steroid "priming," either from exogenous or, possibly, from endogenous sources.

We have less evidence for a significant role of pituitary-derived AII in modulating prolactin secretion. Although peripheral AII receptor blockade abolished the prolactin-releasing effects of exogenous LHRH, we have no data suggesting a physiological condition where pituitary AII affects prolactin output. Neither AII synthesis inhibition nor receptor blockade affected basal prolactin release *in vitro*. In addition, *in vivo*, peripheral AII receptor blockade did not affect the prolactin rise due to stress or the increase on the day of proestrus. Further experiments are necessary to identify the conditions where pituitary-derived AII participates in controlling basal or stimulated prolactin secretion.

Speculations

As can be seen by the wealth of information presented in this volume, the control of reproductive hormone release is well studied, complex, and not totally understood. A constellation of brain and pituitary peptides interact with classical neurotransmitter systems to stimulate and/or inhibit the secretion of LH and prolactin. Coming from the brain side, the final effectors are, most likely, LHRH for LH and dopamine for prolactin. Yet, the circuitry of brain substances in a variety of brain areas proximal to these effectors still needs to be mapped.

In a recent review, Fink (1988) proposed a number of models whereby ovarian steroids could affect opioid, noradrenergic and dopaminergic systems in the hypothalamus to influence basal and phasic LH and prolactin release. Regarding LH, steroids are thought to decrease the inhibitory influence of the opioids (disinhibition) as well as stimulate (activate) facilitatory adrenergic inputs to LHRH neurons.

The ovarian hormones also affect the activity of other brain peptides, like AII, neuropeptide Y, neurotensin, corticotropin releasing hormone, and oxytocin, which have been demonstrated to influence LH release. The "stimulatory" peptides may impinge upon the noradrenergic inputs to the LHRH neuron. Disrupting any one of these inputs, at least on an acute basis, will interfere with the LH surge and ovulation. However, reproductive success is a powerful physiological and behavioral drive. It is likely that, on a chronic basis, elimination of any one of these inputs would still be compatible with normal estrous cyclicity. The brain circuitry responsible for reproductive hormone release is probably redundant and can overcome disruption of any one "facilitatory" element.

Inhibitory inputs to LHRH, on the other hand, may have more long-lasting effects. Activation of the opioid or corticotropin releasing hormone system by stressful situations appears to have chronic deleterious effects upon reproductive function. It may be more important to "turn off" the reproductive system when inhibitory inputs are maximally stimulated than when a single facilitory input is compromised.

In terms of LH, it is clear that the facilitatory effects of AII are mediated by norepinephrine which then acts, probably directly, at LHRH neurons. The AII-norepinephrine interaction apparently occurs within the anterior hypothalamus. Future experiments are necessary to determine where, for example, neurotensin and oxytocin fit into this scheme. Do they act via norepinephrine, the opioids, or directly on the LHRH neuron? Do they act in parallel or in sequence with AII to affect LHRH release? Do they act at the anterior hypothalamus or at the median eminence? Similar questions can be raised regarding AII and prolactin. Although we have evidence that the suppressive effects of AII on prolactin are mediated by dopamine, we do not know where this interaction occurs or whether AII acts directly on dopamine neurons (cell bodies or terminals), or via an intervening substance? Hopefully, the work presented both in this chapter and this volume will facilitate answers to these questions and ultimately provide a map to understand brain and pituitary peptide control over anterior pituitary hormone secretion.

ACKNOWLEDGMENTS

Supported by USPHS Grants HL29714, HDl8020, DK07265, and HL38774. We would like to thank Janet Fasbinder for typing this manuscript.

REFERENCES

Aguilera, G., C. Hyde and K.J. Catt. 1982. Angiotensin II receptors and prolactin release in pituitary lactotrophs. *Endocrinology* 111: 1045-1050.

Andersson, K., K. Fuxe, L.F. Agnati, D. Ganten, I. Zini, P. Eneroth, F. Mascagni and F. Infantillina. 1982. Intraventricular injections of renin increase amine turnover in the tuberoinfundibular dopamine neurons and reduce the secretion of prolactin in the male rat. *Acta Physiol. Scand.* 116: 317-320.

Ben-Jonathan, N. 1985. Dopamine: A prolactin inhibiting hormone. *Endocr. Rev.* 6: 564-589.

Briski, K.P. and P.W. Sylvester. 1988. Effect of specific acute stressors on luteinizing hormone release in ovariectomized and ovariectomized estrogen-treated female rats. *Neuroendocrinology* 47: 194-202.

Bumpus, F.M., A.G. Pucell, A.I. Daud and A. Husain. 1988. Angiotensin II: An intraovarian regulatory peptide. *Am. J. Med. Sci.* 31: 406-408.

Campbell, D.J. and J.F. Habener. 1986. Angiotensinogen gene is expressed and differentially regulated in multiple tissues of the rat. *J. Clin. Invest.* 78: 31-39.

Carrière, P.D., A. DeLean, J. Gutkowski and M. Cantin. 1986. Estrogens directly down-regulate receptors for angiotensin II in anterior pituitary cell culture without decreasing target cell responsiveness. *Endocrinology* 119: 429-431.

Celio, M., D. Clemens and T. Inagami. 1980. Renin in anterior pituitary, pineal and neuronal cells of mouse brain, immunohistochemical localization. *Biomed. Res.* 1: 427-431.

Chen, M.F., R. Hawkins and M.P. Printz. 1982. Evidence for a functional, independent brain-angiotensin system: Correlation between regional distribution of brain AII receptors, brain angiotensinogen and drinking during the estrous cycle of rats. In *The Renin-Angiotensin System in the Brain*, eds. D. Ganten, M. Printz, M.I. Phillips and B.A. Schölkens, 157-168. Raven Press, New York.

Denef, C. and M. Andries. 1983. Evidence for paracrine interaction between gonadotrops and lactotrops in pituitary cell aggregates. *Endocrinology* 112: 813-822.

Deschepper, C.F., D.A. Crumrine and W.F. Ganong. 1986a. Evidence that the gonadotrops are the likely site of production of AII in the anterior pituitary gland of the rat. *Endocrinology* 119: 36-43.

Deschepper, C.F. and W.F. Ganong. 1988. Renin and angiotensin in endocrine glands. In *Frontiers in Neuroendocrinology*, Vol. 10, eds. L. Martini and W.F. Ganong, 79-98. Raven Press, New York.

Deschepper, C.F., S.H. Mellon, F. Cumin, J.D. Baxter and W.F. Ganong. 1986b. Analysis by immunocytochemistry and *in situ* hybridization of renin and its mRNA in kidney, testis, adrenal and pituitary of the rat. *Proc. Natl. Acad. Sci. U.S.A.* 83: 7552-7556.

Deschepper, C.F., C.D. Seidler, M.K. Steele and W.F. Ganong. 1985. Further studies on the localization of angiotensin-like immunoreactivity in the anterior pituitary gland of the rat, using various antisera to pituitary hormones. *Neuroendocrinology* 40: 471-475.

Dzau, V., J. Ingelfinger, R.E. Pratt and K.E. Ellison. 1986. Identification of renin and angiotensinogen mRNA sequence in mouse and rat brains. *Hypertension* 8: 544-548.

Findlay, A.L., J.T. Fitzsimons and J. Kucharczyk. 1979. Dependence of spontaneous and angiotensin-induced drinking in the rat upon the oestrus cycle and gonadal hormones. *J. Endocrinol.* 82: 215-225.

Fink, G. 1988. Oestrogen and progesterone interactions in the control of gonadotropin and prolactin secretion. *J. Steroid Biochem.* 30: 169-178.

Ganong, W.F., C.F. Deschepper, M.K. Steele and A. Intebi. 1989. Renin-angiotensin system in the anterior pituitary of the rat. *Am. J. Hypertens.* 2: 320-322.

Gooren, L.J.G., W. Harmsen-Louman, L. van Bergeyk and H. van Kessel. 1985. Studies on the prolactin-releasing capacity of luteinizing hormone releasing hormone in male subjects. *Exp. Clin. Endocrinol.* 86: 300-304.

Harding, J.W., M.S. Yoshida, R.P. Diltz, T.M. Woods and J.W. Wright. 1986. Cerebroventricular and intravascular metabolism of [^{125}I] angiotensins in the rat. *J. Neurochem.* 46: 1292-1297.

Healy, D.P., A.R. Maciejewski and M.P. Printz. 1986. Localization of central AII receptors with [^{125}I]-sar^1,ile^8-angiotensin II: Periventricular sites of the anterior third ventricle. *Neuroendocrinology* 44: 15-21.

Helmann, W., F. Suzuki, H. Ohkubo, S. Nakanishi, G. Ludwig and D. Ganten. 1988. Angiotensinogen gene expression in extrahepatic rat tissue: Application of a solution hybridization assay. *Naunyn Schmiedebergs Arch. Pharmacol.* 338: 327-331.

Jones, T.H., B.L. Brown and P.R.M. Dobson. 1988. Evidence that angiotensin II is a paracrine agent mediating gonadotropin releasing hormone stimulated inositol phosphate production and prolactin secretion in the rat. *J. Endocrinol.* 116: 367-371.

Jonklaas, J. and J. Buggy. 1984. Angiotensin-estrogen interaction in female brain reduces drinking and pressor responses. *Am. J. Physiol.* 247: R167-R172.

Jonklaas, J. and J. Buggy. 1985. Angiotensin-estrogen central interaction: Localization and mechanism. *Brain Res.* 326: 239-249.

Kucharczyk, J. 1984. Localization of central nervous system structures mediating extracellular thirst in the female rat. *J. Endocrinol.* 10: 183-188.

Levine, J.E. and V.D. Ramirez. 1982. Luteinizing hormone releasing hormone release during the rat estrous cycle and after ovariectomy, as estimated with push-pull cannulae. *Endocrinology* 111: 1439-1448.

Lind, R.W., L.W. Swanson and D. Ganten. 1985. Organization of angiotensin II immunoreactive cells and fibers in the rat central nervous system. *Neuroendocrinology* 40: 2-24.

Mais, V., G.B. Melis, M. Gambacciani, A.M. Paoletti, D. Antinori and P. Fioretti. 1988. Gonadotropin-releasing hormone stimulates intrapituitary paracrine interactions: *In vivo* and *in vitro* evidence. In *Cell to Cell Communication in Endocrinology*, Vol. 49, eds. F. Piva, C.W. Bardin, G. Forti and M. Motta, 65-77. Raven Press, New York.

Mendelsohn, F.A.O., R. Quirion, J.M. Saavedra, G. Aguilera and K.J. Catt. 1984. Autoradiographic localization of angiotensin II receptors in rat brain. *Proc. Natl. Acad. Sci. U.S.A.* 81: 1575-1579.

Moffett, R.B., F.M. Bumpus and A. Husain. 1987. Minireview: Cellular organization of the brain-renin angiotensin system. *Life Sci.* 41: 1867-1879.

Myers, L.S. and M.K. Steele. 1989. The brain-renin angiotensin system and the regulation of prolactin secretion in female rats: Influence of ovarian hormones. *J. Neuroendocrinol.* 114: 299-303.

Naruse, K., Y. Takii and T. Inagami. 1981. Immunohistochemical localization of renin in luteinizing hormone containing cells in rat pituitary. *Proc. Natl. Acad. Sci. U.S.A.* 78: 7579-7583.

Paglin, S., H. Stukenbrok and J.D. Jamieson. 1984. Interaction of AII with dispersed cells from the anterior pituitary of male rats. *Endocrinology* 114: 2284-2292.

Phillips, M.I. 1987. Functions of angiotensin in the central nervous system. *Ann. Rev. Physiol.* 49: 413-435.

Phillips, M.I., E.M. Richards and A. Van Eekelen. 1986. Central and peripheral actions of angiotensin II. In *Central and Peripheral Mechanisms of Cardiovascular Regulation*, ed. A. Magro, W. Osswald, D. Reis and P. Vanhoutte, 385-441. Plenum Press, New York.

Pickel, V.M., J. Chan and D. Ganten. 1986. Dual peroxidase and colloidal gold-labeling of angiotensin converting enzyme and angiotensin-like immunoreactivity in the rat subfornical organ. *J. Neurosci.* 6: 2457-2469.

Platia, M.P., K.J. Catt and G. Aguilera. 1986. Effects of 17β-estradiol on angiotensin II receptors and prolactin release in cultured pituitary cells. *Endocrinology* 119: 2768-2772.

Printz, M.P., R.L. Hawkins, C.J. Wallis and F.M. Chen. 1983. Steroid hormones as feedback regulators of brain angiotensinogen and catecholamines. *Chest* 835: 3085-3115.

Sarkar, D.K., S.A. Chiappa, G. Fink and N.M. Sherwood. 1976. Gonadotropin-releasing hormone surge in proestrous rats. *Nature* 264: 461-463.

Slater, E.E., R. Defendini and E.A. Zimmerman. 1980. Wide distribution of immunoreactive renin in nerve cells of human brain. *Proc. Natl. Acad. Sci. U.S.A.* 77: 5458-5460.

Sobel, D. and A. Vagnucci. 1982. Angiotensin II mediated ACTH release in rat pituitary cell culture. *Life Sci.* 30: 1281-1286.

Steele, M.K. 1987. Effects of angiotensins injected into various brain areas on LH release in female rats. *Neuroendocrinology* 46: 401-405.

Steele, M.K. 1991. The brain renin-angiotensin system and the secretion of anterior pituitary hormones in the rat. In *Potentials of Molecular Biology in Fertility Regulation: Growth Regulatory Factors*, ed. F. Hazelton. WHO, Geneva, in press.

Steele, M.K., M.S. Brownfield and W.F. Ganong. 1982a. Immunocytochemical localization of angiotensin immunoreactivity in gonadotrops and lactotrops of the rat anterior pituitary gland. *Neuroendocrinology* 35: 155-158.

Steele, M.K., R.V. Gallo and W.F. Ganong. 1983. A possible role for the brain renin-angiotensin system in the regulation of LH secretion. *Am. J. Physiol.* 245: R805-R810.

Steele, M.K., R.V. Gallo and W.F. Ganong. 1985. Stimulatory or inhibitory effects of angiotensin II upon LH secretion in ovariectomized rats: A function of gonadal steroids. *Neuroendocrinology* 40: 210-216.

Steele, M.K. and W.F. Ganong. 1986. Effects of catecholamine depleting agents and receptor blockers on basal and angiotensin II or norepinephrine stimulated LH release in female rats. *Endocrinology* 119: 2728-2736.

Steele, M.K., S.M. McCann and A. Negro-Vilar. 1982b. Modulation by dopamine and estradiol of the central effects of angiotensin II on anterior pituitary hormone release. *Endocrinology* 111: 722-729.

Steele, M.K., A. Negro-Vilar and S.M. McCann. 1981. Effect of angiotensin II on *in vivo* and *in vitro* release of anterior pituitary hormones in the female rat. *Endocrinology* 109: 893-899.

Stornetta, R.L., C.L. Hawelu-Johnson, P.G. Guyenet and K.R. Lynch. 1988. Astrocytes synthesize angiotensinogen in brain. *Science* 242: 1444-1446.

Yang, H-Y. and N.H. Neff. 1973. Differential distribution of angiotensin converting enzyme in the anterior and posterior pituitary of the rat. *J. Neurochem.* 21: 1035-1036.

RENIN-ANGIOTENSIN SYSTEM IN FEMALE REPRODUCTIVE ORGANS

Robert C. Speth, Kevin L. Grove, and Cesario S. Zamora
Department of Veterinary and Comparative
Anatomy, Pharmacology and Physiology
Washington State University
Pullman, WA

I.   Introduction

II.   Angiotensinergic Activity in Female Reproductive Structures

        B.   Uterus

        C.   Other Structures

III.   Conclusions and Future Directions

Introduction

Angiotensin II (Ang II), initially named hypertensin (Braun-Menendez *et al.*, 1940) was first recognized for its vasoconstrictor and pressor actions. Subsequent studies demonstrated it to be a potent dipsogenic agent (Fitzsimons, 1978), and Ang II is now known to have a primary role in regulating fluid and electrolyte balance in the body. Ang II has also been shown to affect learning, respiration and thermoregulation (see review, Speth *et al.*, 1988). Regulation of reproductive function now appears to be a third major function for Ang II. The neuroendocrine role of Ang II in the regulation of luteinizing hormone and prolactin release are well established and are described in detail elsewhere in this volume (Steele *et al.*, Chapter 1). This chapter focuses on the actions of Ang II on female reproductive organs, while an accompanying chapter (Grove *et al.*, Chapter 3) describes renin-angiotensin system (RAS) activity in male reproductive organs.

Angiotensinergic activity in female reproductive structures

Ovary

All of the components of the RAS [angiotensinogen, renin, angiotensin I (Ang I), angiotensin converting enzyme (ACE), Ang II and Ang II receptor binding sites] have been identified in mammalian ovaries. Thus, Ang II may serve as an ovarian hormone or it may exert autocrine, paracrine and/or endocrine actions in the ovary.

Angiotensin II synthesis. The liver is the primary source of the Ang II precursor angiotensinogen (also known as renin substrate), although many other tissues (*e.g.*, the adrenal, adipose tissue and brain) also synthesize angiotensinogen (Campbell *et al.*, 1984; Ohkubo *et al.*, 1986; Cassis *et al.*, 1988; see also review, Campbell, 1987). Angiotensinogen has been identified in human follicular fluid (Glorioso *et al.*, 1986); however, the concentrations are similar to those found in plasma and may represent angiotensinogen taken up and sequestered from the bloodstream. While knowing the source of angiotensinogen is not critical to questions regarding ovarian Ang II synthesis, it is of interest to determine if the ovary generates each component of the renin-angiotensin system cascade. The definitive test for local generation of a substance is the demonstration of messenger ribonucleic acid (mRNA) for that substance in the tissue. Ohkubo *et al.* (1986) demonstrated that mRNA for angiotensinogen is present in the rat ovary at a concentration approximately one-thirtieth that found in liver. Blood angiotensinogen levels are increased by estrogens (Helmer and Griffith, 1952), and Ohkubo *et al.* (1986) demonstrated that estrogens increase the expression of angiotensinogen mRNA in the liver--which implied a similar mechanism in other tissues. Thus, it is possible that the ovary uses estrogen to increase Ang II production, either in the bloodstream or within the ovary itself, by increasing the amount of angiotensinogen precursor. The exact cellular localization of angiotensinogen in the ovary is as yet undetermined. If the ovary makes Ang II within a single type of cell, then all of the components should be present in those cells; however, if Ang II is made extracellularly, as in the bloodstream, the source of each component of the RAS within the ovary may differ.

Renin, the initial enzyme of the renin-angiotensin cascade, is derived from prorenin, a precursor of higher molecular weight. The primary source of renin is the kidney, although there are a number of alternative tissue sources (see review, Campbell, 1987). Glorioso *et al.* (1986) reported high concentrations of prorenin in follicular fluid of humans. The levels of prorenin and the ratio of prorenin to renin in the follicular fluid greatly exceed those in the blood suggesting that the ovary is a site of prorenin synthesis. Noting elevated blood prorenin levels (but not blood renin levels) at ovulation prompted Sealey *et al.* (1985) to suggest the ovary as a significant source of plasma prorenin in the bloodstream, but not as a source of plasma renin. Fernandez *et al.* (1985) not only showed that active renin is present in human follicular fluid at levels 15-fold higher than in plasma but also reported that follicular fluid renin activity was positively correlated with follicular fluid estradiol levels. Further evidence for renin involvement in ovarian function is the increase in follicular fluid renin activity associated with stimulation of the ovary by gonadotropins (Lightman *et al.*, 1987b), and the increase in ovarian renin activity during estrus in the rat (Howard *et al.*, 1988). Immunocytochemical data demonstrate that ovarian renin-like staining occurs primarily in the theca, stromal and luteal cells but is also present in gonadotropin-stimulated granulosa cells in humans (Palumbo *et al.*, 1989). *In situ* hybridization studies (Lightman *et al.*, 1987a) demonstrated renin mRNA in the corpus luteum, but not the theca or granulosa cells, suggesting that rat ovarian renin appears to be derived solely from luteal cells. Renin and prorenin have also been identified in bovine ovarian follicular fluid and are thought to be produced in theca cells in this species (Schultze *et al.*, 1989). Interestingly, Schultze *et al.* (1989) observed a negative correlation between follicular fluid prorenin and estradiol, suggesting that bovine ovarian prorenin was present primarily in atretic, rather than developing, follicles, as seen in humans (Fernandez *et al.*, 1985).

Both Ang I, the product of renin-hydrolyzed angiotensinogen, and Ang II have been identified in human follicular fluid (Glorioso *et al.*, 1986; Lightman *et al.*, 1987b); Ang II has also been identified in the rat ovary (Husain *et al.*, 1987). Immunocytochemical localization studies have demonstrated Ang II-like immunoreactivity in thecal, stromal, luteal, vascular smooth muscle and endothelial cells of ovaries from cycling women. Granulosa cells of gonadotropin-stimulated human ovaries also display Ang II-like immunoreactivity (Palumbo *et al.*, 1989). With the exception of the vascular elements, the distribution of Ang II was the same as that of renin-like immunoreactivity (Palumbo *et al.*, 1989); thus, Ang II may be produced in theca, granulosa, stromal or luteal cells of the ovary.

The other major enzyme of the renin-angiotensin cascade, ACE, is also present in the ovary. Its presence in the ovary was first demonstrated by Cushman and Cheung (1971), who found its concentration to be approximately 7% that of the lung (the major source of this enzyme). ACE-like immunoreactivity has been identified on the surface of follicular oocytes and zona pellucida of rabbit ova (Brentjens *et al.*, 1986). Recently, ACE localization in the rat ovary was demonstrated by enzyme autoradiographic techniques using the potent radiolabeled ACE inhibitor, [125]I-351A (Speth and Husain, 1988). ACE-like binding was observed in the germinal epithelium surrounding the corpora lutea, the granulosa cells of some, but not all follicles, stroma and vascular elements. ACE-like binding was lower in those portions of the autoradiograms corresponding to the theca cell layers. Thus, the localization of ACE in the rat ovary does not correspond to that of renin and AII in the human ovary. It is possible that there are species differences in the ovarian cells that produce Ang II, or that the conversion of Ang I to Ang II occurs extracellularly, after which Ang II is taken up by the renin-containing cells. ACE acts on a number of substrates other than Ang I (Thiele *et al.*, 1985) and only a small proportion of ovarian ACE may be involved in Ang II formation. Therefore, ACE is not an exclusive marker for Ang II synthesizing cells.

There is good evidence that the ovary releases components of the renin-angiotensin system which affect other tissues in the body, not unlike the mechanism whereby the kidney affects blood pressure as well as fluid and electrolyte balance. In addition, the ovary appears to be capable of generating the end product of the renin-angiotensin system, Ang II. In view of the short

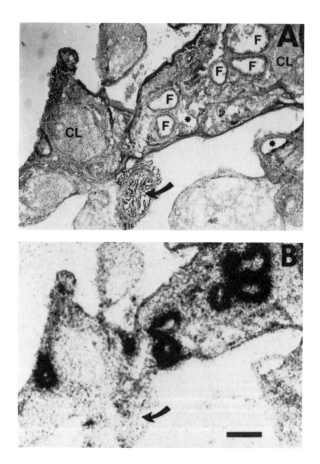

Fig. 1. Angiotensin Receptor Binding in Rat Ovary.
Panel A indicates a hematoxylin and eosin-stained section of a rat ovary at estrus. Panel B indicates the autoradiogram generated from the section shown in panel A. The tissue preparation and autoradiography for this and subsequent figures was done essentially as described previously (Husain *et al.*, 1987), except that the medium used for the incubations and rinses contained 150 mM NaCl, 1 mM Na$_2$EDTA, 0.1 mM bacitracin and 50 mM sodium phosphate buffer, pH 7.1, and the autoradiograms were visualized with SB-5, X-ray film (Eastman Kodak, Rochester, NY). Several antral follicles that display ^{125}I-SI Ang II binding sites are marked with an F while other antral follicles that do not display Ang II receptor binding sites are marked with an asterisk (*) in Panel A. Two corpora lutea (CL) are also denoted in panel A. The curved arrow in panels A and B points to the ampulla of the oviduct, which displays little Ang II receptor binding. The calibration bar in the lower right corner of panel B indicates 0.5 mm.

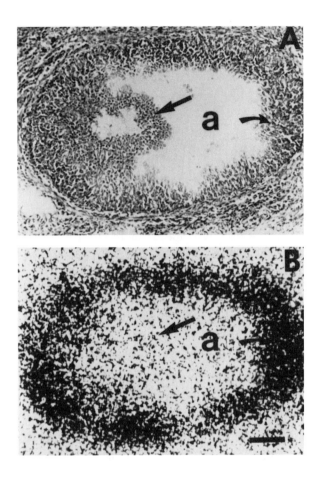

Fig. 2. Angiotensin Receptor Binding in an Antral Follicle with a Cumulus Mass. Panel A indicates a hematoxylin and eosin-stained section of a rat ovarian follicle at estrus. Panel B indicates the autoradiogram generated from the section shown in panel A. The antrum (a) and the corona radiata (indicated by straight arrow) display little ^{125}I-SI Ang II binding, while the stratum granulosa (indicated by the curved arrow) displays intense ^{125}I-SI Ang II binding. The calibration bar in the lower right corner of panel B indicates 0.1 mm.

half-life of Ang II in the body, it is likely that formation of Ang II in the ovary indicates a local, paracrine-type action on the ovary.

Responsiveness to Angiotensin II. The ovary appears to be a target organ of the renin-angiotensin system as well as a source of its active hormone. Ang II actions upon the ovary (Fernandez *et al.*, 1985) were predicated on the fact that Ang II acted on the vasculature to promote angiogenesis. To investigate this matter we determined whether the ovary contained Ang II receptor binding sites and characterized their cellular localization (Speth *et al.*, 1986). Radioligand binding studies indicated that the rat ovary contained specific (1 μmol, Ang II displaceable), saturable (3.3 fmol/mg initial wet weight), high-affinity (K_D = 643 pM) binding sites for ^{125}I-sarcosine[1], isoleucine[8] Ang II (^{125}I-SI Ang II) during diestrus. Displacement studies with Ang II analogs and fragments, and unrelated peptides, confirmed the Ang II receptor characteristics of this binding (Husain *et al.*, 1987). Additional studies with ^{125}I-Ang II in the presence and absence of guanosine triphosphate (GTP) analogs further established the characteristics of Ang II receptors in rat ovary (Pucell *et al.*, 1987).

   In vitro receptor autoradiographic studies demonstrated that the putative Ang II receptors were localized primarily to the granulosa and theca cells of some, but not all, follicles. Subsequently, Speth and Husain (1988) described lower numbers of ^{125}I-SI Ang II binding sites in ovarian stroma, blood vessels and discrete portions of a few corpora lutea. Harwood and Hausdorff (1987) also observed specific ^{125}I-SI Ang II binding in isolated luteal cells from rat ovaries. Zamora and Speth (1988) demonstrated that the distribution of ^{125}I-SI Ang II binding sites in the bovine ovary was similar to that of the rat. ^{125}I-SI Ang II binding sites are also abundant in the rat ovary during estrus and show a pattern similar to the diestrous rat ovary. Both secondary and tertiary (antral stage) follicles display intense Ang II binding (Fig. 1). As noted previously, not all follicles display Ang II binding sites. Figure 2 illustrates the binding of ^{125}I-SI Ang II in an antral follicle. While binding is clearly localized to the stratum granulosum, the corona radiata of the cumulus mass is devoid of specific Ang II binding sites. Due to the lack of tissue fixation (*in vitro* receptor autoradiography is done using fresh frozen sections) it is not possible to determine if the follicles displayed in these figures are atretic or preovulatory. The major significance of Ang II binding site observations is that the highest Ang II receptor binding occurs in a nonvascular region of the ovary (on the stratum granulosum); thus, the actions of Ang II on the ovary are not simply due to vascular actions of Ang II. It is unlikely that large quantities of blood-borne Ang II would enter the follicle, suggesting that Ang II produced in the ovary may be the primary stimulus for ovarian Ang II receptors.

   While it is clear that Ang II receptor binding sites are present in the ovary, the function of Ang II is less well defined. Pucell *et al.* (1987) examined Ang II receptor binding in immature ovaries of rats stimulated with pregnant mare's serum gonadotropin (PMSG). Although they observed no change in Ang II receptor binding site density, the size of the ovaries (hence the total number of Ang II receptor binding sites) increased substantially during PMSG-stimulated maturation. Pucell *et al.* (1987) also incubated portions of ovaries from immature

PMSG-treated rats in the absence and presence of 1 μM Ang II. They observed that ovaries treated with Ang II produced significantly more estrogen than the non-Ang II-exposed controls; Ang II did not affect *in vitro* progesterone synthesis in these experiments. Since granulosa cells are the primary source of estrogen in the female rat, the selective stimulation of estrogen production by Ang II is consistent with the localization of ^{125}I-SI Ang II binding sites to the granulosa cells. Harwood and Hausdorff (1987) observed that nanomolar concentrations of Ang II caused a rapid elevation of intracellular calcium in dispersed granulosa and luteal cells from rat ovaries, consistent with a receptor-mediated stimulation of cellular function. Further indication of a functional role for Ang II in the ovary came from Pellicer *et al.* (1988) who demonstrated that treatment of rats with the Ang II antagonist, SI Ang II, reduced the number of oocytes ovulated by immature rats treated with human chorionic gonadotropin (hCG). The specificity of this effect was demonstrated by the inability of SI Ang II to inhibit ovulation when administered more than 3 h after hCG or when given together with Ang II. Although this observation has recently been disputed (Daud *et al.*, 1989), an Ang II antagonist has also been shown to inhibit ovulation *in vitro* in a perfused rat ovary preparation that avoids the complications of systemic actions of the Ang II antagonist (Peterson *et al.*, 1989).

In contrast to the concept that Ang II acts on preovulatory follicles, a recent study by Daud *et al.* (1988) suggests that Ang II receptors in the rat ovary occur primarily on atretic follicles. Using histological analysis of the follicles they demonstrated that Ang II binding sites were present on atretic follicles. They also described a differential distribution of ^{125}I-SI Ang II binding sites and those for ^{125}I-follicle stimulating hormone (FSH) which were primarily observed on healthy preovulatory follicles. However, a small proportion (approximately 5%) of the follicles demonstrating Ang II receptor binding sites appeared to be healthy, and their development may be enhanced by Ang II.

A hypothesis that unifies the disparate observations of Daud *et al.* (1988) and Pellicer *et al.* (1988) is that Ang II may be capable of rescuing atretic follicles and returning them to a preovulatory state. Thus, follicles that begin to show signs of atresia may express Ang II receptors. If Ang II is present in sufficient quantities, it acts on the receptors to cause estrogen formation and restoration of the preovulatory state. If Ang II is not present, then the process of atresia continues. This would explain 1) why Ang II can stimulate estrogen formation despite its putative localization to atretic or early atretic follicles, 2) why an angiotensin antagonist can reduce the number of ovulating follicles, and 3) what renin and prorenin may be doing in healthy follicles.

Another hypothesis for the actions of Ang II at preatretic follicles is that it extends their estrogen-producing phase to supplement estrogen formation from preovulatory follicles. The potential benefits of such an action include: greater stimulation of LH release from the anterior pituitary, greater responsiveness of the uterus to contractile hormones, and increased estrous behavior.

To summarize, the ovary contains Ang II receptors that are not limited to vascular components. Possible functions and mechanisms of action of Ang II in the ovary include: stimulation of estrogen synthesis (Pucell *et al.*, 1987), promotion of ovulation (Pellicer *et al.*, 1988), stimulation of angiogenesis associated with the luteinization of follicles to form corpora lutea (Fernandez *et al.*, 1985), regulation of blood flow within the ovary (Lightman *et al.*, 1987b), regulation of atretic follicles (Daud *et al.*, 1988) and secretion of antral fluid by granulosa cells, contraction of granulosa cells or other cells leading to follicular rupture and movement of oocytes to the oviducts, as discussed by Speth and Husain (1988). Further studies of the components of the RAS and Ang II receptors at different stages of the cycle should help to refine our understanding of the physiologically relevant functions of the ovarian renin-angiotensin system in the cycling female rat.

Uterus

Synthesis of Angiotensin II. The uterus has long been known to contract in response to Ang II; however, it is also a source of components of the renin-angiotensin system. Renin is abundant in the nonpregnant rabbit uterus and is present in even higher concentrations in the pregnant uterus (Ferris *et al.*, 1967). It can serve as a source of renin in the blood of nephrectomized rabbits (Gorden *et al.*, 1967; Ferris *et al.*, 1972). The latter study suggested that uterine renin release can increase uterine blood flow via synthesis of Ang II, stimulation of adrenal medullary epinephrine secretion, and β_2-adrenergic-mediated relaxation of uterine arterioles. Capelli *et al.* (1968) suggested that the human uterus may also be a source of renin in the bloodstream following nephrectomy. We are not aware of any studies that have described angiotensinogen mRNA or immunocytochemical identification of angiotensinogen in the uterus. The uterus does contain ACE, at a level approximately 5% that of the lung (Cushman and Cheung, 1971). However, the isolated uterus is reported to be relatively insensitive to Ang I (see review, Khairallah, 1971), suggesting that there is little ACE activity. This discrepancy is difficult to resolve.

Responsiveness to Angiotensin II. The first demonstration of uterine responsiveness to Ang II dates back to 1940, when Luduena observed a contractile response of the uterus to Ang II (see Khairallah and Smeby, 1974). This ability of the estrogen-treated or estrous-phase uterus to respond to Ang II led to its application as a bioassay for Ang II (Dekanski, 1954; Khairallah and Smeby, 1974); it is still used to determine the myotropic actions of Ang II.

In the uterus of pregnant monkeys, Ang II stimulates prostaglandin formation (Franklin *et al.*, 1974), which may contribute to the contractile responses to Ang II in the uterus. Dubin and Ghodgaonkar (1980) examined the ability of Ang II to stimulate prostaglandin formation in the near-term (18 day) pregnant rat uterus, *in vitro*, and found no relationship between Ang II-stimulated prostaglandin formation and contractile responses; *e.g.*, SI-Ang II blocked the contractile actions of Ang II but did not inhibit prostaglandin formation. Furthermore, inhibition of prostaglandin formation did not affect the contractile response to Ang II in the uterus. This finding is consistent with an action of Ang II on smooth muscle of the rat uterus.

Indications that the actions of Ang II are mediated by specific Ang II receptor binding sites were initially determined in the rat by Lin and Goodfriend (1970), although the putative Ang II binding site they observed demonstrated low affinity for Ang II and did not demonstrate saturability. Subsequent studies verified the existence of specific Ang II receptor binding sites in the uteri of the rat (Devynck *et al.*, 1976), dog (Capponi and Catt, 1979) and monkey (Petersen *et al.*, 1985). Schirar *et al.* (1987) demonstrated that the density of Ang II receptors, but not the affinity for Ang II, varied during the estrous cycle of rats. The highest binding site density was seen at proestrus, when circulating estradiol levels were at their highest, and uterine contractile responses to Ang II were maximal (Baudouin-Legros *et al.*, 1974). Lowest uterine Ang II receptor density (one-fourth that of proestrus) was observed at diestrus I (metestrus) when blood estradiol levels were lowest (Schirar *et al.*, 1987) and uterine responsivity to Ang II was lowest (Baudouin-Legros *et al.*, 1974). These observations are consistent with those for other hormones, *i.e.*, oxytocin and prostaglandin $F_{2\alpha}$, that are capable of contracting the uterus, and probably reflect the trophic actions of estrogen on the uterine myometrium (Reeves, 1987).

Autoradiographic studies of ^{125}I-SI Ang II binding in the uterus of an estrous rat indicate a predominant localization of Ang II receptors within the myometrial layer (Fig. 3). The distribution of Ang II binding sites varies within the myometrium and appears to be highest in the innermost smooth muscle layer. A second band of binding sites occurs at the perimeter of the uterus and is localized either to the outermost smooth muscle layer, or is present on the perimetrium. Ang II binding in the endometrium is considerably less than that

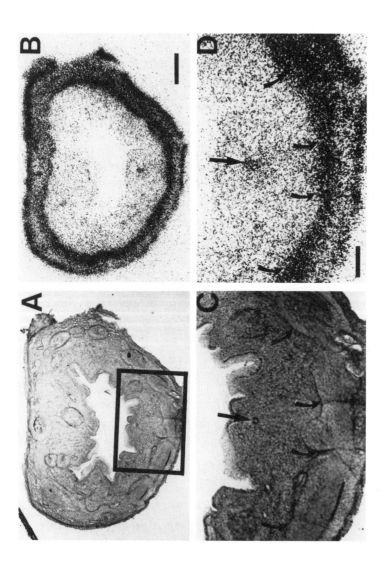

Fig. 3.   Angiotensin Receptor Binding in the Uterus of the Rat.
Panel A indicates a hematoxylin and eosin-stained section of a uterus from a rat at estrus. Panel B indicates the autoradiogram generated from the section shown in panel A. Panel C is a higher magnification of the area in the box in panel A. Panel D is the autoradiogram corresponding to panel C. The curved arrows in panels C and D denote the boundary between the inner endometrial layer and the outer myometrial layer of the uterus. The straight arrow in panels C and D points to a uterine gland. The calibration bar in the lower right corner of panel B indicates 0.5 mm.   The calibration bar in the lower left corner of panel D indicates 0.2 mm.

in the myometrium and is only slightly greater than the level of non-specific binding, except in uterine glands (Figs. 3C and 3D). This suggests that the primary function of Ang II in the uterus is to cause its contraction and secondarily to regulate secretion of uterine fluid. Interestingly, it has been reported that some species, in particular the bat, store spermatozoa in the uterine gland to allow fertilization up to several months following coitus (Harper, 1982); thus, Ang II might be involved in spermatozoal survival in the uterus as well. It

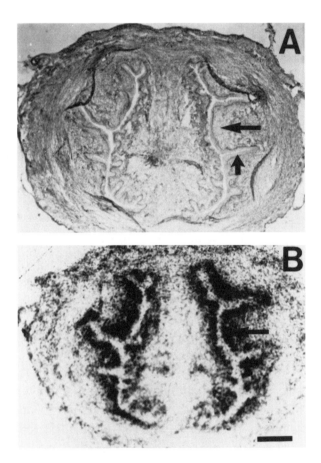

Fig. 4. Angiotensin Receptor Binding in the Cervix of the Rat.
Panel A indicates a hematoxylin and eosin-stained section of the cervix of a rat at estrus. Panel B indicates the autoradiogram generated from the section shown in panel A. The small vertical arrow in each panel points to the layer of epithelium lining the lumen of the cervical canals which is virtually devoid of ^{125}I-SI Ang II binding. The large horizontal arrow in each panel points to the lamina propria, at which intense ^{125}I-SI Ang II binding was present. The calibration bar in the lower right corner of panel B indicates 0.2 mm.

would be of interest to determine not only the density of Ang II receptors in the uterus at different times during pregnancy, but also whether Ang II contributes to parturition or spontaneous abortion.

The cervix, the most distal part of the uterus, also contains Ang II binding sites. Figure 4 demonstrates the presence of ^{125}I-SI Ang II binding sites in the lamina propria immediately beneath the epithelium of the two cervical canals. The lamina propria of the cervix corresponds to the endometrial stroma of the uterus; however, there are distinct anatomical differences between the cells of this layer of the cervix and the uterus. Our observation of Ang II receptor binding in the lamina propria, but not in the endometrium of the uterus proper (cf. Fig. 3), further confirms the differential characteristics of the cervix and the body of the uterus. In the cervix, the lamina propria is thought to be primarily connective tissue. However, the presence of Ang II receptor binding sites in such a high density suggests that it may have a functional importance. We speculate that the presence of these binding sites indicates possible contractile elements or fluid regulatory cells in this layer of the cervix, perhaps acting to narrow the lumen of the canal leading to retention of uterine fluids and/or exclusion of external fluids. Thus, Ang II released following coitus or in the seminal fluid (Grove et al., unpublished data) might cause contraction of the smooth muscle, closing the canal and preventing the loss of semen. Alternatively, Ang II may act tonically on the cervix during pregnancy to help prevent premature parturition. This action of Ang II may be antagonized by the hormone relaxin, which functions to relax the cervix at the time of parturition.

Other structures

The oviduct has not been studied to any great extent for its ability to form or respond to Ang II. There is immunocytochemical evidence for ACE in the oviduct of the rabbit (Brentjens et al., 1986); however, it appears to be limited to the capillary endothelium. Brentjens et al. also demonstrated that oviductal oocytes retain the plasma membrane ACE demonstrated in the follicle.

Since the oviduct secretes fluid and contains smooth muscle, it is reasonable to suggest that it may be responsive to Ang II in a manner similar to the uterus; however, this has not been established. Daud et al. (1988) noted a low amount of ^{125}I-SI Ang II binding in the superficial epithelium of the oviduct. Figure 1B indicates no ^{125}I-SI Ang II binding in the ampulla of the oviduct of a rat on estrus. Figure 5 indicates that the isthmus of the estrous rat oviduct also displays little ^{125}I-SI Ang II binding. Further studies are necessary to determine if there are functional oviductal Ang II receptors in the rat, *e.g.*, at different stages of the estrous cycle. Preliminary studies (Zamora and Speth, unpublished results) suggest that the pig oviduct is enriched in Ang II receptors.

The human placenta has been shown to contain renin-like material, localized to trophoblasts (Poisner *et al.*, 1981) and chorionic cells (Symonds *et al.*, 1968). Within the placental unit, chorionic laeve cells produced the greatest amounts of renin, while chorionic plate, placental and amniotic cells produced considerably smaller amounts of total renin (prorenin and renin). Prorenin, derived presumably from the chorionic cells, has been observed in human amniotic fluid

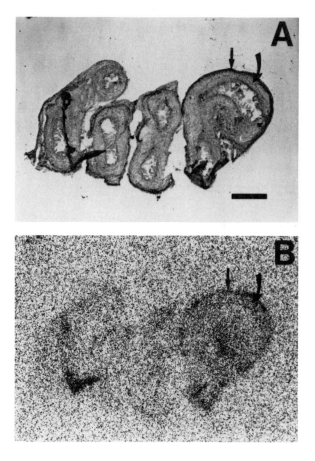

Fig. 5. Angiotensin Receptor Binding in the Oviduct of the Rat.
Panel A shows a hematoxylin and eosin-stained section of the oviduct of a rat at estrus. Panel B shows the autoradiogram generated from the section shown in panel A. The small straight arrow in each panel points to the tunica serosa of the oviduct, which shows moderate ^{125}I-SI Ang II binding. The curved arrow points to the oviduct's smooth muscle, which shows light ^{125}I-SI Ang II binding. The calibration bar in the lower right corner of panel B indicates 0.5 mm.

in concentrations nearly 100-fold higher than those in plasma (Skinner *et al.*, 1974). Human amniotic fluid also contains Ang I and Ang II at concentrations greater than those seen in plasma, suggesting local synthesis of angiotensins in the amniotic fluid (Skinner *et al.*, 1974). However, amniotic fluid angiotensinogen is present at concentrations at least ten times smaller than those found in plasma (Skinner *et al.*, 1974), suggesting that angiotensinogen is either produced in exceedingly low amounts in the placenta or that it is derived from the blood. Unlike uterine renin, placental renin does not seem to produce systemic effects, since nephrectomized pregnant dogs did not release significant quantities of renin in response to a variety of stimuli capable of releasing renal renin (Carretero *et al.*, 1972). Once again, species differences (rabbit and human versus dog) may also be involved.

The human placenta displays Ang II receptor binding sites. Cooke *et al.* (1981) demonstrated specific binding of ^{3}H-Ang II in the placenta, although the affinity of the higher of two binding sites observed was relatively low (K_D = 9 nM). Further, Cooke *et al.* (1981) observed little specific binding of ^{3}H-Ang II in the chorion and amnion. Ang II is capable of increasing the placental vascular resistance in the ewe (Parisi and Walsh, 1989), so the possibility exists that placental Ang II receptors are limited to its vasculature.

Conclusions and Future Directions

It is now clear that peripheral reproductive structures of the female are richly endowed in renin-angiotensin system components and receptor binding sites. The localization of these components is summarized in Table I. While functional responses to Ang II or its antagonists have been identified, it is still uncertain what the physiological significance of the RAS is to reproductive function. Given the widespread usage of ACE inhibitors for treatment of hypertension, it would be anticipated that some reproductive alterations would be observed in these patients if the RAS plays a key role in reproductive function. However, the access of ACE inhibitors into reproductive structures may be less than into structures regulating blood pressure. The reduction in the number of ovulated ova following Ang II antagonist administration suggests a role for endogenously produced Ang II in female reproductive function, at least in the rat. Many peptide hormones act on reproductive structures in addition to the sex steroids, and it will be necessary to consider the actions of Ang II within this hormonal milieu. Since synthesis of Ang II found in the bloodstream typically involves multiple organs, and synthetic components of the renin-angiotensin system occur in both male and female reproductive organs, it is possible that

Table I. Localization of renin-angiotensin system components in female reproductive organs.

Organ/Cell Type	Aogen.	Proren.	Renin	Ang I	ACE	Ang II	Recep.
Ovary	L	H	L	L	M	M	M
follicular fluid	M	H	L	M		M	
granulosa			M		H[a]	M	H[a]
theca			M		L[a]	M	H[a]
corpus luteum		M	M			M	L[b]
stroma			M		L	M	L
vasculature					M	M	M
germ. epithelium					H		
oocytes					M		
Oviduct							M[c]
superfic. epithel.							L
smooth muscle							
vasculature					L		
Uterus		M	H		L		H
endometrium							
uterine glands							M
myometrium							H
superfic. epithel.							
cervix							H
vasculature							
Placenta							L
chorionic cells		H	M				
trophoblasts			M				
amniotic fluid	L	H			M	M	
vasculature							L

[a] not present in all follicles
[b] not present in most corpora lutea
[c] pig, rat and cow show negligible binding

H, M and L indicate high, medium and low concentration and represent semiquantitative estimates of the presence of these components. Except as noted, these values represent a synopsis of all species studied. Blank values suggest the absence of the component, but may also reflect a lack of study of this issue. Germ. epithelium is germinal epithelium, superfic. epithel. is superficial epithelium, Aogen. is angiotensinogen, Proren. is prorenin, and Recep. is Ang II receptor. For more detail, please refer to the text.

synthesis of Ang II or the transport of Ang II to its receptor sites may occur, in part, in both sexes, *e.g.*, Ang I might be produced in semen in the testes and converted to Ang II in the uterus.

Future studies of the significance of angiotensinergic actions on reproductive function will need to be directed toward: 1) precise localization of the cells demonstrating Ang II receptor binding sites in reproductive structures, 2) determination of the biochemical and electrophysiological actions of Ang II on specific cell types in reproductive structures, 3) elucidation of the sources of Ang II acting on reproductive structures, 4) determination of the mechanisms whereby renin-angiotensin system components arising from the reproductive system act on other structures to promote reproductive success, and 5) continued demonstration of alterations in reproductive function by inhibitors and antagonists of the RAS.

ACKNOWLEDGMENTS

This work was supported by grants from NIH (NS-21305; BRSG 2 S07 RR05465-27), and Select Sires Inc.). The authors thank Drs. Philip L. Senger and Raymond W. Wright for reviewing this manuscript. Technical assistance was provided by Renee Andrea, Brenda Vavra and Machelle Schuster.

REFERENCES

Baudouin-Legros, M., P. Meyer and M. Worcel. 1974. Effects of prostaglandin inhibitors on angiotensin, oxytocin and prostaglandin $F_{2\alpha}$ contractile effects on the rat uterus during the oestrous cycle. *Br. J. Pharmacol.* 52: 393-399.

Braun-Menendez, E., J.C. Fasciolo, S.F. Leloir and J.M. Munoz. 1940. The substance causing renal hypertension. *J. Physiol.* 98: 283-298.

Brentjens, J.R., S. Matuso, G.A. Andres, P.R.B. Caldwell and L. Zamboni. 1986. Gametes contain angiotensin converting enzyme (kininase II). *Experientia* 42: 399-402.

Campbell, D.J., J. Bouhnik, J. Menard and P. Corvol. 1984. Identity of angiotensinogen precursors of rat brain and liver. *Nature* 308(5955): 206-208.

Campbell, D.J. 1987. Circulating and tissue angiotensin systems. *J. Clin. Invest.* 79: 1-6.

Capelli, J.P., L.G. Wesson, G.E. Aponte, C. Faraldo and E. Jaffe. 1968. Characterization and source of a renin-like enzyme in anephric humans. *J. Clin. Endocrinol.* 28: 221-230.

Capponi, A.M. and K.J. Catt. 1979. Angiotensin II receptors in adrenal cortex and uterus. *J. Biol. Chem.* 254: 5120-5127.

Carretero, O.A., C. Polomski, A. Piwonska, A. Afsari and C.P. Hodgkinson. 1972. Renin release and the uteroplacental-fetal complex. *Am. J. Physiol.* 223: 561-564.

Cassis, L.A., J. Saye and M.J. Peach. 1988. Location and regulation of rat angiotensinogen messenger RNA. *Hypertension* 11: 591-596.

Cooke, S.F., D.J. Craven and E.M. Symonds. 1981. A study of angiotensin II binding sites in human placenta, chorion, and amnion. *Am. J. Obstet. Gynecol.* 140: 689-692.

Cushman, D.W. and H.S. Cheung. 1971. Concentrations of angiotensin-converting enzyme in tissues of the rat. *Biochim. Biophys. Acta* 250: 261-265.

Daud, A.I., F.M. Bumpus and A. Husain. 1988. Evidence for selective expression of angiotensin II receptors on atretic follicles in the rat ovary: An autoradiographic study. *Endocrinology* 122: 2727-2734.

Daud, A.I., F.M. Bumpus and A. Husain. 1989. Angiotensin II: Does it have a direct obligate role in ovulation? *Science* 245: 870-871.

Dekanski, J. 1954. The quantitative assay of angiotonin. *Br. J. Pharmacol.* 9: 187-191.

Devynck, M.A., B. Rouzaire-Dubois, E. Chevillotte and P. Meyer. 1976. Variations in the number of uterine angiotensin receptors following changes in plasma angiotensin levels. *Eur. J. Pharmacol.* 40: 27-37.

Dubin, N.H. and R.B. Ghodgaonkar. 1980. Angiotensin II and [Sar^1,Ile^5,Ala^8]angiotensin II effect on contractile activity and prostaglandin production of *in vitro* pregnant rat uteri. *Endocrinology* 107: 1855-1860.

Fernandez, L.A., B.C. Tarlatzis, P.J. Rzasa, V.J. Caride, N. Laufer, A.F. Negro-Vilar, A.H. DeCherney and F. Naftolin. 1985. Renin-like activity in ovarian follicular fluid. *Fertil. Steril.* 44: 219-223.

Ferris, T.F., P. Gorden and P.J. Mulrow. 1967. Rabbit uterus as a source of renin. *Am. J. Physiol.* 212: 698-702.

Ferris, T.F., J.H. Stein and J. Kauffman. 1972. Uterine blood flow and uterine renin secretion. *J. Clin. Invest.* 51: 2827-2833.

Fitzsimons, J.T. 1978. Angiotensin, thirst, and sodium appetite: Retrospect and prospect. *Fed. Proc.* 37: 2669-2675.

Franklin, G.O., A.J. Dowd, B.V. Caldwell and L. Speroff. 1974. The effect of angiotensin II intravenous infusion on plasma renin activity and prostaglandins A, E and F levels in the uterine vein of the pregnant monkey. *Prostaglandins* 6: 271-280.

Glorioso, N., S.A. Atlas, J.H. Laragh, R. Jewelewicz and J.E. Sealey. 1986. Prorenin in high concentrations in human ovarian follicular fluid. *Science* 233: 1422-1424.

Gorden, P., T.F. Ferris and P.J. Mulrow. 1967. Rabbit uterus as a possible site of renin synthesis. *Am. J. Physiol.* 212: 703-706.

Harper, M.J.K. 1982. Sperm and egg transport. In *Reproduction in Mammals: 1. Germ Cells and Fertilization*, eds. C.R. Austin and R.V. Short, 102-127. Cambridge University Press, Cambridge.

Harwood, J.P. and W.P. Hausdorff. 1987. Angiotensin II receptors in the ovary: Characterization and linkage to calcium mobilization. *Endocrin. Soc. Abstr.* 227.

Helmer, O.M. and R.S. Griffith. 1952. The effect of the administration of estrogens on the renin-substrate (hypertensinogen) content of rat plasma. *Endocrinology* 51: 421-426.

Howard, R.B., A.G. Pucell, F.M. Bumpus and A. Husain. 1988. Rat ovarian renin: Characterization and changes during the estrous cycle. *Endocrinology* 123: 2331-2340.

Husain, A., F.M. Bumpus, P. De Silva and R.C. Speth. 1987. Localization of angiotensin II receptors in ovarian follicles and the identification of angiotensin II in rat ovaries. *Proc. Natl. Acad. Sci. U.S.A.* 84: 2489-2493.

Khairallah, P.A. 1971. Pharmacology of angiotensin. In *Kidney Hormones*, Vol. 1, ed. J.W. Fisher, 129-171. Academic Press, New York.

Khairallah, P.A. and R.R. Smeby. 1974. Bioassay of angiotensin. In *Handbook of Experimental Pharmacology, XXXVII*, eds. I.H. Page and F.M. Bumpus, 227-239. Springer-Verlag, New York.

Lightman, A., C.F. Deschepper, S.H. Mellon, W.F. Ganong and F. Naftolin. 1987a. *In situ* hybridization identifies renin mRNA in the rat corpus luteum. *Gynecol. Endocrinol.* 1: 237.

Lightman, A., B.C. Tarlatzis, P.J. Rzasa, M.D. Culler, V.J. Caride, A.F. Negro-Vilar, D. Lennard, A.H. DeCherney and F. Naftolin. 1987b. The ovarian renin-angiotensin system: Renin-like activity and angiotensin II/III immunoreactivity in gonadotropin-stimulated and unstimulated human follicular fluid. *Am. J. Obstet. Gynecol.* 156: 808-816.

Lin, S.-Y. and T.L. Goodfriend. 1970. Angiotensin receptors. *Am. J. Physiol.* 213: 1319-1328.

Ohkubo, H., K. Nakayama, T. Tanaka and S. Nakanishi. 1986. Tissue distribution of rat angiotensinogen mRNA and structural analysis of its heterogeneity. *J. Biol. Chem.* 261: 319-323.

Palumbo, A., C. Jones, A. Lightman, M.L. Carcangiu, A.H. DeCherney and F. Naftolin. 1989. Immunohistochemical localization of renin and angiotensin II in human ovaries. *Am. J. Obstet. Gynecol.* 160: 8-14.

Parisi, V.M. and S.W. Walsh. 1989. Fetal placental vascular responses to prostacyclin after angiotensin II-induced vasoconstriction. *Am. J. Physiol.* 257: E102-E107.

Pellicer, A., A. Palumbo, A.H. DeCherney and F. Naftolin. 1988. Blockage of ovulation by an angiotensin antagonist. *Science* 240: 1660-1661.

Petersen, E.P., J.W. Wright and J.W. Harding. 1985. Characterization of angiotensin II binding sites in African Green monkey uterus. *Life Sci.* 36: 177-182.

Peterson, C.M., C. Zhu, T. Mukaida and W.J. LeMaire. 1989. The angiotensin II antagonist, saralasin, inhibits ovulation in the perfused rat ovary. *Soc. Study Reprod.* 40: 56.

Poisner, A.M., G.W. Wood, R. Poisner and T. Inagami. 1981. Localization of renin in trophoblasts in human chorion laeve at term pregnancy. *Endo* 109: 1150.

Pucell, A.G., F.M. Bumpus and A. Husain. 1987. Rat ovarian angiotensin II receptors. *J. Biol. Chem.* 262: 7076-7080.

Reeves, J.J. 1987. Endocrinology of reproduction. In *Reproduction in Farm Animals*, ed. E.S.E. Hafez, 85-106. Lea and Fibiger, Philadelphia.

Schirar, A., A. Capponi and K. Catt. 1987. Regulation of uterine angiotensin II receptors by estrogen and progesterone. *Endocrinology* 106: 5-12.

Schultze, D., B. Brunswig and A.K. Mukhopadhyay. 1989. Renin and prorenin-like activities in bovine ovarian follicles. *Endo* 124: 1389-1398.

Sealey, J.E., S.A. Atlas, N. Glorioso, H. Manapat and J.H. Laragh. 1985. Cyclical secretion of prorenin during the menstrual cycle: Synchronization with luteinizing hormone and progesterone. *Proc. Natl. Acad. Sci. U.S.A.* 82: 8705-8709.

Skinner, S.L., E.J. Cran, R. Gibson, R. Taylor, W.A.W. Walters and K.J. Catt. 1974. Angiotensins I and II, active and inactive renin, renin substrate, renin activity, and angiotensinase in human liquor amnii and plasma. *Am. J. Obstet. Gynecol.* 121: 626-630.

Speth, R.C., F.M. Bumpus and A. Husain. 1986. Identification of angiotensin II receptors in the rat ovary. *Eur. J. Pharmacol.* 130: 351-352.

Speth, R.C. and A. Husain. 1988. Distribution of angiotensin converting enzyme and angiotensin II receptor binding sites in the rat ovary. *Biol. Reprod.* 38: 695-702.

Speth, R.C., J.W. Wright and J.W. Harding. 1988. Brain angiotensin receptors: Comparison of location and function. In *Angiotensin and Blood Pressure Regulation*, eds. J.W. Harding, J.W. Wright, R.C. Speth and C.D. Barnes, 1-34. Academic Press, New York, NY.

Symonds, E.M., M.A. Stanley and S.L. Skinner. 1968. Production of renin by *in vitro* cultures of human chorion and uterine muscle. *Nature* 217: 1152-1153.

Thiele, E.A., S.M. Strittmatter and S.H. Snyder. 1985. Substance K and substance P as possible endogenous substrates of angiotensin converting enzyme in the brain. *Biochem. Biophys. Res. Comm.* 128: 317-324.

Zamora, C.S. and R.C. Speth. 1988. Localization of angiotensin II binding sites in the bovine ovary. *Anat. Hist. Embryol.* 17: 380.

RENIN-ANGIOTENSIN SYSTEM IN MALE REPRODUCTIVE ORGANS

Kevin L. Grove, Robert C. Speth, and Cesario S. Zamora
Department of Veterinary and Comparative
Anatomy, Pharmacology and Physiology
Washington State University
Pullman, WA

I.   Introduction

II.   Angiotensinergic activity in male reproductive structures

      A.   Testis

      B.   Epididymis

      C.   Ductus Deferens

      D.   Prostate

      E.   Seminal Vesicles

      F.   Seminal Fluid

III.   Conclusion and future research directions

Introduction

The Renin-Angiotensin System (RAS) has been shown to affect the female and male reproductive systems of many different species including the human, rat, and bovine. The RAS acts both at the central nervous system (CNS) level, through stimulation of gonadotropin release (Steele *et al.*, see Chapter 1), as well as directly at the gonadal level, altering steroid release and gamete transport and survival. This chapter concentrates on the direct actions of Angiotensin II (Ang II), and other RAS components, on the male reproductive system; an accompanying chapter (Speth *et al.*, see Chapter 2) focuses on angiotensin in the female reproductive system.

The first indication that the RAS plays a role in the male reproductive system was the discovery that the epididymis and the testis contain high concentrations of angiotensin-converting enzyme (ACE) (Cushman and Cheung, 1971). ACE is a carboxypeptidase that cleaves the relatively inactive decapeptide, Ang I, to the active octapeptide, Ang II. According to Cushman and Cheung, (1971) the epididymis and testis both contain greater than twice the concentration of ACE found in the lung, which is thought to be the major source of ACE for circulating blood. The amount of ACE activity detected in the testis varied in parallel to testicular activities such as sperm and steroid production. These findings led to the studies describing the concentration, location, and function of the many RAS components in male reproductive structures reviewed in this chapter.

Angiotensinergic activity in male reproductive structures

Testis

Increasing evidence of multiple RAS components contained within the testis has led us to surmise about a possible testicular RAS, similar to that in the kidney (Parmentier *et al.*, 1983; Deschepper *et al.*, 1986; Pandey and Inagami, 1986; Sealey *et al.*, 1988; Strittmatter and Snyder, 1984; Vanha-Pertulla *et al.*, 1985; Pandey *et al.*, 1984; Ohkubo *et al.*, 1986; Cassis *et al.*, 1988). Using immunohistochemical staining Parmentier *et al.* (1983) demonstrated the localization of renin (an enzyme that cleaves angiotensinogen to Ang I) in the Leydig cells of the rat testis. This renin-like staining did not appear until the onset of puberty, after which the amount of staining increased with age. The concentration of renin-like staining in the Leydig cells was also shown to be directly proportional to the levels of gonadotropin released from the anterior pituitary, while estradiol-treated or hypophysectomized animals displayed no renin-like staining. Whereas renin staining was abolished by estradiol treatment or hypophysectomy, staining returned upon treatment with human chorionic gonadotropin (hCG). In support of locally produced renin in the testes, Pandey and Inagami, (1984) demonstrated the presence of renin mRNA in Leydig cells using a renin cDNA probe; the findings were later confirmed by Deschepper *et al.* (1986).

Pandey and Inagami, (1986) also reported a gonadotropin-dependent renin activity in the testes by demonstrating the effects of gonadotropins, bovine luteinizing hormone (bLH) and hCG on Ang I and Ang II synthesis in cultured murine Leydig tumor cells (MA-10). Ang I and Ang II were increased 150-fold and 40-fold, respectively, in response to these gonadotropins. The majority of the Ang I (85%) was found to be intracellular while 90% of the Ang II was extracellular, suggesting the Ang I was formed intracellularly and that the Ang I was cleaved into Ang II as it left the cell. Pandey and Inagami (1986) were uncertain if ACE was bound to the internal or external face of the cell membrane.

Questioning whether or not prorenin, renin's precursor, and renin were secreted from the testis into the blood stream Sealey *et al.* (1988) measured both components from blood collected from the spermatic vein of young men. Their results indicated that prorenin was secreted from the testis in high concentrations, but that the renin concentration was no different in spermatic vein blood than that found in spermatic artery blood. From this finding Sealey *et al.* (1988) proposed that extrarenal renin systems cannot process prorenin to renin. While this may be true, it is also possible that renin synthesized in the testis from locally produced prorenin is not secreted across the blood-testis barrier, or that testicular renin is controlled such that excess prorenin is secreted into the blood stream before it is converted to renin.

In further support of a testicular RAS, Pandey *et al.* (1984) reported that renin, ACE, Ang I, II, and III were contained in the Leydig cells of rats. However, Strittmatter and Snyder (1984), using autoradiographic visualization with ^{3}H-captopril, a specific radiolabeled ACE inhibitor, reported ACE was not localized to the Leydig cells. Strittmatter and Snyder (1984) reported ACE localized to spermatid heads and ACE unbound in the lumen of seminiferous tubules. Vanha-Pertulla *et al.* (1985) found similar results when they measured the amount of ACE in different components of the testis and epididymis, showing that germinal cells and residual bodies contained the greatest amount of ACE, while nongerminal cells contain negligible levels of ACE activity.

Even though Pandey *et al.* (1984) showed that the testis contained many of the components needed for a testicular RAS, it was not clear if all of these components were produced in, or transported to, the testis. Ohkubo *et al.* (1986) reported that at least angiotensinogen, the ultimate precursor to Ang II, was not produced in the testes as evidenced by the absence of the angiotensinogen mRNA. This report was later confirmed by Cassis *et al.* (1988), while at the same time they discovered that the epididymal fat pads contained high levels of adipocyte angiotensinogen mRNA. The perirenal fat pad was the only fat pad reported to contain higher concentrations of angiotensinogen mRNA. Therefore, it is possible that the Ang I found in the testis by Pandey *et al.* (1984) could come from conversion of the angiotensinogen produced in the epididymal fat pads and transported to the testis.

Characterization of functional Ang II receptors. Even though it had been confirmed that components of the RAS were present in the testis, the presence of functional Ang II receptor sites had yet to be confirmed. Two recent studies documented the presence and function of Ang II receptors in the testis. First, Millan and Aguilera (1988), using autoradiographic methods with ^{125}I-sarcosine[1], isoleucine[8] Ang II (^{125}I-SI Ang II), reported specific, high-affinity Ang II binding sites localized in the Leydig cells of the rat, rhesus monkey, cebus monkey, and human testis. Millan and Aguilera (1988) also showed the concentration of ^{125}I-SI Ang II binding sites to be age dependent, with the highest concentration of binding sites in the developing testis of prepubertal rats. In the younger rats some of the Ang II receptors were localized to the mesenchymal cells, suggesting that Ang II may play a role in the development of the testis. Millan and Aguilera (1988) commented that the approximate number of Ang II binding sites remained constant between prepubertal and mature testis.

In the second study, Khanum and Dufau (1988) described the ability of Ang II to inhibit the luteinizing hormone (LH) stimulation of testosterone production in rat Leydig cells. Khanum and Dufau (1988) reported that Ang II inhibited the production of cAMP, the second messenger of the Leydig LH receptors. These Ang II receptors also proved to be pertussis toxin sensitive, suggesting that the actions of Ang II are mediated via the G_i nucleotide binding protein.

The reports discussed thus far on the actions of the RAS in the testis suggest a possible feedback system between Ang II, LH and testosterone production: 1) LH stimulates Ang II production in the Leydig cells (Parmentier *et al.*, 1983; Pandey and Inagami, 1986); 2) Ang II inhibits testosterone secretion by Leydig cells (Khanum and Dufau, 1988); 3) Estrogens inhibit testis renin production (Parmentier *et al.*, 1983), thus decreasing Ang II production indirectly. In addition it is known that: 1) LH stimulates and increases estrogen and testosterone secretion by the testis (de Kretser, 1985); 2) Estrogens and testosterone inhibit LH secretion from the anterior pituitary (Karsh, 1985). Figure 1 displays a possible feedback system that could explain the RAS role in the cyclic pattern of steroid production and secretion.

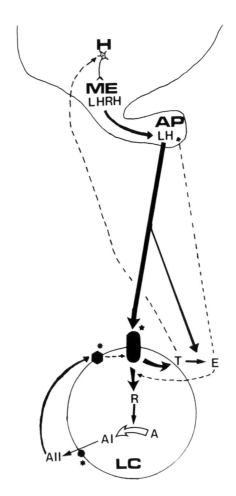

Fig. 1. Schematic Diagram Representing a Putative Feedback Loop in the Leydig Cells Between the RAS, Gonadotrophs, and Steroids. H = LHRH hypothalamic nuclei; ME = median eminence; AP = anterior pituitary; LC = Leydig cells; LHRH = luteinizing hormone releasing hormone; LH = luteinizing hormone; T = testosterone; E = estradiol; R = renin; A = angiotensinogen; AI = angiotensin I; AII = angiotensin II. Five-point star = LH receptor; eight-point star = AII receptor; six-point star = membrane-bound angiotensin converting enzyme. Dotted line represents inhibition; solid line represents stimulation.

Epididymis

Angiotensin II synthesis. Cushman and Cheung (1971) established the epididymis as containing the highest concentration of ACE out of all the major organs studied. However, little research on control or localization followed this report until Hohlbrugger *et al.* (1982) reported that the ACE activity in the epididymis, ductus deferens and testis was dependent on age and sexual stimulation of the rat. A significant increase in ACE activity with increased age was reported in all segments of the male reproductive tract studied (testis to ductus deferens). Male rats caged with female rats for two weeks also showed a significant increase in ACE activity in the testis, efferent ducts and caput epididymis.

Strittmatter and Snyder (1984), using autoradiography with ^{3}H-captopril, first reported ACE labeling localized in the epithelial wall of the rat epididymis, which decreased from the caput to the caudal regions. Little or no labeling was reported in the efferent ducts of the rat epididymis. Strittmatter and Snyder (1984) also reported ACE-like labeling on lumenal contents in the cauda epididymis. Vivet *et al.* (1987) confirmed Strittmatter and Snyder's report of ACE in the epithelial wall of the human epididymis, after using immunocytochemical staining for ACE. No ACE-like staining in the lumen of any region of the human epididymis was discovered. Species differences or differencee in research methods may explain the absence of the lumenal ACE staining in the human epididymis in contrast to that reported in the rat epididymis.

Responsiveness to Angiotensin II. Recently Grove and Speth (1989) reported that Ang II may play a direct role in the function of the epididymis. Radioligand binding studies were used to show a specific, saturable (63 fmol/mg protein), high-affinity (K_D < 1 nM) Ang II receptor binding site localized to the tubular wall of the epididymis of mature rats (Grove and Speth, 1989). Autoradiography, using ^{125}I-SI Ang II, displayed an uneven distribution of the putative Ang II binding sites, with the highest concentration in the proximal cauda epididymis (Fig. 2A and 2B). No specific ^{125}I-SI Ang II binding was observed on the spermatozoa present in the lumen of the epididymis of mature rats.

Using emulsion-coated coverslips we determined that binding was present on both the smooth muscle layer and the epithelial layer of the epididymal tubule. Again, negligible specific binding was found within the lumen on spermatozoa, or on the connective tissue between the convolutions of the tubule. Figures 2E

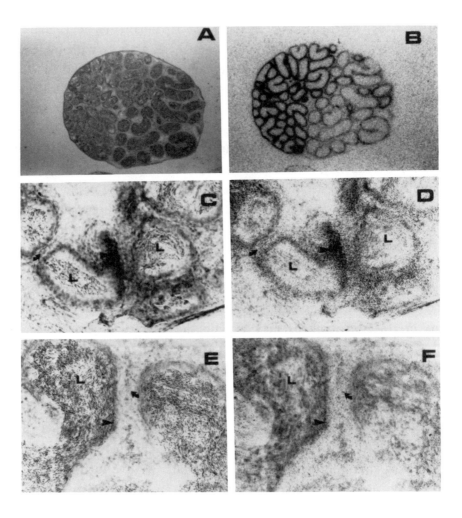

Fig. 2. Autoradiographic Localization of ^{125}I-SI Ang II Binding in the Rat Epididymis. A, C and E indicate hematoxlin and eosin-stained epididymal section. B is an autoradiogram of total ^{125}I-SI Ang II binding of a cross section of a whole epididymis. Nonspecific binding of ^{125}I-SI Ang II in the presence of 1 μM Ang II was minimal (see Fig. 2D, Grove and Speth, 1989). D and F indicate emulsion-coated coverslips with fixed overlays over C and E, respectively. (L) indicates the tubular lumen filled with spermatozoa. Black arrowhead indicates epithelial layer of tubule wall. Black curved arrow indicates smooth muscle layer of tubule wall.

and 2F represent a cross section of the distal cauda region of a rat epididymis, 40x magnification, where a thick band of smooth muscle lay between two tubules. The emulsion-coated coverslip displays a notable increase in concentration of exposed silver grains throughout the smooth muscle layer compared to the level of grains shown in the lumen, which is similar to the background exposure. Figures 2C and 2D show a cross section through the wall of a convoluted tubule in the proximal cauda epididymis, displaying a large area of epithelial cells. The ciliated epithelial cells play a role in the transport of spermatozoa and fluid and electrolyte transport across the tubule wall (Setchell and Brooks, 1988). The emulsion-coated coverslip displays a high concentration of grains over the epithelial cells and a low concentration of grains over the lumen. These findings support the hypothesis that Ang II receptors are present on both smooth muscle and epithelial cells of the epididymal wall, and play a role both in movement of spermatozoa and in intraluminal fluid and electrolyte content.

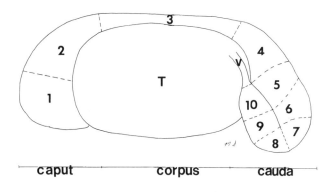

Fig. 3. Schematic Diagram of Epididymis. Lobes 1 and 2 in the caput (head), lobe 3 represents the corpus (body), lobes 4-7 are in the proximal cauda (tail), and lobes 8-10 are in the distal cauda.

To quantitate the uneven distribution of Ang II binding sites within the regions of the epididymis (Figs. 2A and 2B), we divided the epididymis into ten regions, the caput (regions 1 and 2), corpus (region 3), proximal cauda (regions 4-7), and distal cauda (regions 8-10) (Fig. 3) and performed ^{125}I-SI Ang II saturation isotherms on each of the regions. Rosenthal analysis (Rosenthal, 1967) of the

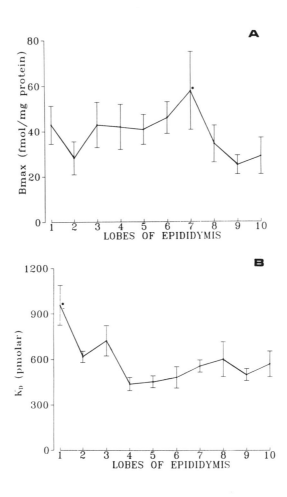

Fig. 4. Angiotensin II Receptor Binding in the Lobes of the Epididymis. A. Density (B_{max}) of specific ^{125}I-SI AII binding sites in ten lobes of the epididymis as indicated in figure 3. Values shown are mean ± SEM from five experiments. Randomized block analysis of variance indicated a significant variation between lobes (F = 2.17, df 9,34, p < 0.05). Duncan's multiple range test indicated that the binding site density in lobe 4 was significantly greater than that in lobes 1, 2, 3 and 9 (p < 0.05). B. Dissociation constant (K_D) of specific ^{125}I-SI AII binding sites in 10 lobes of the epididymis indicated in Fig. 3. Values shown are mean ± SEM from five experiments. Randomized block analysis of variance indicated a significant variation between lobes (F = 7.73, df 9,35, p < 0.0001). Duncan's multiple range test indicated that the K_D in lobe 10 was significantly greater (lower affinity) than lobes 1-9 (p < 0.05) and that the K_D in lobe 8 was significantly greater (lower affinity) than that in lobes 2, 5, 6 and 7 (p < 0.05).

saturation isotherms showed that lobe 4 contained a significantly higher concentration of putative Ang II receptors than lobes 1, 2, 3 and 9 (Fig. 4A). This supports the autoradiographic observations (Figs. 2A and 2B). However, the Rosenthal analysis of region 4 (B_{max} = 60 fmol/mg protein), which contained the greatest concentration of ^{125}SI-Ang II binding, was less than the concentration reported in the whole epididymis (B_{max} = 63 fmol/mg protein) (Grove and Speth, 1989). Possible explanations for this difference include variations in the preparation of the tissue used in the binding assay, including the additional handling required for dissection of the lobes, slight differences in the ages of the animals and/or seasonal differences. There was a large day-to-day variability of binding found in lobes 4 and 5 and this decreased the difference between these two lobes and the other eight lobes, compared to the difference seen in the autoradiograms. This may be due to the difficulty in precisely replicating the dissections such that the lobes may not be exactly the same every time. If the lobes with a high concentration of Ang II binding sites were mixed with the lobes possessing a lower concentration of Ang II binding sites, this would mask interlobal differences.

In this study we also calculated the dissociation constant (K_D) of each of these ten lobes (Fig. 3B) and observed a significant increase in the K_D (lower affinity) in lobe 10 in the caput epididymis. These data suggest that the caput may have a different receptor subtype than the cauda epididymis. However, there may be a higher level of metabolism of the radioligand in the caput epididymis, reducing the actual concentration of intact radioligand, with the appearance of less potent fragments and an artificially elevated K_D. A problem of high metabolism of the radioligand and some of the analogs was present in our previous study (Grove and Speth, 1989).

In situ, short segments of cauda epididymis displayed a strong contractile response to 1 μM Ang II, followed by a strong expulsion of spermatozoa (Grove and Speth, 1989). Other regions of the epididymis, from the caput to the ductus deferens, also displayed a similar contractile response to 1 μM Ang II. No contractile response was obtained from an equal dose of the less potent analog, 3-8 hexapeptide Ang II. The functional response of the epididymal segments was blocked by the specific Ang II antagonist, SI Ang II, but not by the specific α-adrenergic antagonist, phentolamine. A similar functional response was obtained from the specific α-adrenergic agonist, phenylephrine, which was blocked by phentolamine, but was not blocked by SI Ang II. This suggests that Ang II can directly contract the epididymis; indeed, the contractile response of the epididymis to Ang II suggests that the ^{125}I-SI Ang II binding sites localized to the smooth muscle layer, as shown by emulsion-coated coverslip autoradiography, (Figs. 2E and 2F) are functional receptors and also reaffirms the concept that the octapeptide may play a direct role in the controlled transport of spermatozoa during their final maturation process. This also leads to the question of what effects changes in plasma Ang II levels would have on spermatozoal transport and maturity. Is it possible that renovascular hypertension could be associated with fertility problems due to high levels of Ang II, or that the treatment of high blood pressure with ACE inhibitors could adversely affect spermatozoal maturity?

The distribution pattern of putative Ang II receptors in the epididymis does not correlate with the increase of smooth muscle cells from the caput epididymis to the distal cauda epididymis. In addition, the emulsion autoradiography method showed binding sites on the epithelial layer (Figs. 2C and 2D), which is responsible for fluid and electrolyte balance during the final maturation process of the spermatozoa. Work done on osmotic diffusion (Crabo and Gustafsson, 1964; Levine and Marsh, 1971) shows that the site of the lowest concentration of Ang II receptors, the caput epididymis, is the site of the least osmotic diffusion. While the area with the greatest osmotic diffusion, the proximal cauda epididymis, is the area that contains the greatest concentration of Ang II binding sites. Therefore, it is possible that the high amount of Ang II binding that we see in the proximal cauda epididymis may be primarily on epithelial cells responsible for fluid and electrolyte balance. More work needs to be done, however, to confirm this hypothesis. Again, since the fluid and electrolyte environment is very important to the maturation process of the spermatozoa, any alteration of the Ang II concentrations could potentially affect spermatozoal maturation and viability in the epididymis.

Preliminary studies (Grove *et al.*, unpublished observation) in radioimmunoassays done on bovine semen suggest that there is a sufficient concentration of Ang II in semen to elicit a physiological response from the epididymis. However, it is as yet undetermined whether this Ang II is produced by the epididymis or by one of the accessory glands.

Ductus Deferens

Angiotensin II synthesis. The ductus deferens also contains several components of the RAS. Strittmatter and Snyder (1984) demonstrated captopril binding to ACE localized on components inside the lumen of the ductus deferens of male rats, with little labeling on the tubular epithelium. Vivet *et al.* (1987), using immunocytochemical methods, found ACE localized to the tubular epithelium of human ductus deferens, while also demonstrating intense staining in the luminal fluid. The conflicting results reported in these two studies may have resulted from species or methodological differences.

Responsiveness to Angiotensin II. Magnan and Regoli (1979) first reported that the ductus deferens contained Ang II receptor binding sites; later it was discovered that the ductus deferens displayed a functional response to Ang II and Ang III (Freer *et al.*, 1980; Saye *et al.*, 1986; Trachte *et al.*, 1987; Bin Zhang, *et al.* 1988; Trachte, 1988). This series of papers reported that Ang III inhibited the electrically stimulated contraction (ESC) of the ductus deferens, while Ang II increased the ESC by potentiating the release of norepinephrine (NE) from adrenergic nerve terminals. While both Ang II and Ang III were also shown to stimulate the production of PGE_2, Ang II was more potent. Thus Ang II, by stimulating the release of norepinephrine and PGE_2, may increase the contraction of the ductus deferens and facilitate ejaculation. We recently determined the concentration (B_{max}) of specific (1 μM Ang II displaceable) ^{125}I-SI Ang II binding sites in the ductus deferens to be 450 fmol/g wet weight, with a dissociation constant (K_D) of 810 pM (Fig. 5). This concentration is approximately double that of the whole cauda epididymis.

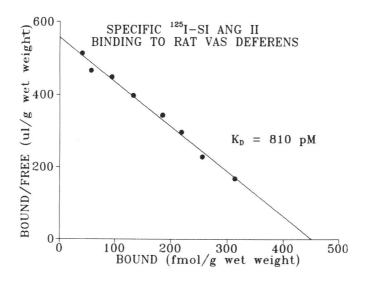

Fig. 5. Rosenthal Plot of Specific ^{125}I-SI Ang II Binding to Rat Ductus Deferens. ^{125}I-SI Ang II was present at concentrations ranging from 0.1-2.5 nM, in the presence and absence of 1 μM Ang II to define specific binding. Tissue concentrations were approximately 50 mg/ml with a total assay volume of 0.2 ml. Tissue preparation as in Grove and Speth (1989).

Prostate

The prostate is also known to contain ACE that has been localized to the endothelial cells of the vessels of the rat (Vivet *et al.*, 1987). The ACE concentration in the prostate of humans is low when compared to its concentration in the testis and epididymis (Mizutani and Schill, 1985). However, Lieberman and Sastre (1983) found the prostate of postmortem men to contain a higher level of ACE activity than that of the testis, 10.4 units/mg protein versus 7.8 units/mg protein, respectively. Measurements on the epididymis were not performed. Yokoyama *et al.* (1980) found a lower ACE activity (0.44 units/g of tissue) in the human male prostate than Lieberman and Sastre (1983). We observed little ^{125}I-SI Ang II binding in the rat prostate. Therefore, it is likely that the ACE produced by the prostate is secreted into the ductus deferens during ejaculation, for subsequent actions in the female reproductive system (Krassnigg *et al.*, 1986; Yokoyama *et al.*, 1980).

Seminal Vesicles

The seminal vesicles have also been shown to contain ACE on the free surface of the epithelial cells (Vivet *et al.*, 1987). To date, we have been unable to identify Ang II binding sites in tissue homogenates of the seminal vesicles of the rat.

Seminal Fluid

Along with measuring ACE in the epididymis, testis, and other reproductive structures, Cushman and Cheung (1971) also determined that the amount of ACE in human seminal plasma (47 munits/mg) is comparable to that found in the rat lung. The concentration of ACE present in the seminal plasma of the epididymis has also been reported to correlate directly with the quality of semen (spermatozoal density and motility) (Kanenko and Moriwaki, 1981). It is also noteworthy to mention that the ACE purified from the seminal fluid is similar to the blood plasma ACE rather than testicular ACE (El-Dorry *et al.*, 1983), which implies that it is synthesized outside of the testis (Vanha-Pertulla *et al.*, 1985). There are also several reports showing that ACE is contained within the spermatozoa (Vanha-Pertulla *et al.*, 1985; Brentjens *et al.*, 1986; Yotsumoto *et al.*, 1984). ACE has an optimal pH of 8.0 (Hohlbrugger *et al.*, 1982) which is the pH of cervical fluid in the female, providing the ACE its optimal environment for converting Ang I to Ang II at this site. Mizutani and Schill (1985) also showed that other components of the RAS (Ang I and II) increase sperm velocity. These findings suggest that components of the RAS increase fertility of the male. The ACE present in the seminal fluid could possibly function to convert the relatively inactive Ang I to the potent Ang II in the female reproductive tract after ejaculation. The Ang II produced from the ACE present in an ejaculate could also increase the fertility rate by acting on the female reproductive system to increase uterine motility following ejaculation. Ang II may also enter the bloodstream of the female following entrance to the uterus and arrive at the ovary via a countercurrent mechanism, which would then allow it to affect ovulation as suggested in the accompanying chapter (Speth *et al.*, this volume).

Conclusion and Future Research Directions

The increasing number of reports on RAS involvement in male reproductive physiology support the concept that Ang II and other components of the RAS play important roles affecting the delicate balance needed to keep the reproductive system fertile by: 1) regulating steroid production in the testis (Fig. 1); 2) stimulating contraction of the smooth muscle in the epididymal tubule wall to assist in spermatozoal transport through the epididymis and by causing expulsion of spermatozoa (Grove and Speth, 1989); 3) affecting fluid and electrolyte transport, a function that Ang II subserves in other tissues of the body; 4) potentiating the release of norepinephrine from adrenergic nerve terminals in the ductus deferens to potentiate the contraction of the smooth muscle responsible for ejaculation (Freer *et al.*, 1980; Magnan and Regoli, 1979); 5) increasing sperm velocity (Mizutani and Schill, 1985); and 6) playing a possible role in the female by acting on the cervix, uterus, and ovary (Speth *et al.*, this volume).

It is still unknown to what extent the RAS regulates reproductive function and what effects the alteration of the concentrations of the various components would produce. The present data suggest that Ang II and other components of the RAS may be necessary for normal reproductive function. If the RAS does play an important role in regulating reproductive function, the elucidation of how increased concentrations of blood Ang II in high blood pressure, or other conditions, might affect reproductive function and the determination on how the use of ACE inhibitors may affect reproduction then become important areas for future research.

REFERENCES

Bin Zhang, S., B. Dattatreyamurty and L.E. Reichert Jr. 1988. Regulation of follicle-stimulating hormone binding to receptors on bovine calf testis membranes by cholera toxin-sensitive guanine nucleotide binding protein. *Molec. Endocrinol.* 2: 148-158.

Brentjens, J.R., S. Matuso, G.A. Andres, P.R.B. Caldwell and L. Zamboni. 1986. Gametes contain angiotensin converting enzyme (kininase II). *Experientia* 42: 399-402.

Cassis, L.A., J. Saye and M.J. Peach. 1988. Location and regulation of rat angiotensinogen messenger RNA. *Hypertension* 11: 591-596.

Crabo, B. and B. Gustafsson. 1964. Distribution of sodium and potassium and its relation to sperm concentration in epididymal plasma of the bull. *J. Reprod. Fert.* 7: 337-345.

Cushman, D.W. and H.S. Cheung. 1971. Concentrations of angiotensin-converting enzyme in tissues of the rat. *Biochim. Biophys. Acta* 250: 261-265.

de Kretser, D.M. 1985. The testis. In *Hormonal Control of Reproduction* (second edition), Vol. 3, 76-90. Cambridge University Press, Cambridge.

Deschepper, C.F., S.H. Mellon, F. Cumin and J.D. Baxter. 1986. Analysis by immunocytochemistry and *in situ* hybridization of renin and its mRNA in kidney, testis, adrenal and pituitary of the rat. *Proc. Natl. Acad. Sci. U.S.A.* 83: 7552-7556.

El-Dorry, H.A., J.S. MacGregor and R.L. Soffer. 1983. Dipeptidyl carboxypeptidase from seminal fluid resembles the pulmonary rather than the testicular isoenzyme. *Biochem. Biophys. Res. Comm.* 115: 1096-1100.

Freer, R.J., J.C. Sutherland Jr. and A.R. Day. 1980. Organ selective angiotensin antagonists: Sarcosyl[1]-cysteinyl(S-methyl)[8]-angiotensin I. *Eur. J. Pharmacol.* 65: 349-354.

Grove, K.L. and R.C. Speth. 1989. Rat epididymis contains functional angiotensin II receptors. *Endocrinology.* 125: 223-230.

Hohlbrugger, G., H. Schweisfurth and H. Dahlheim. 1982. Angiotensin I converting enzyme in rat testis, epididymis, and vas deferens under different conditions. *J. Reprod. Fert.* 65: 97-103.

Kanenko, S. and C. Moriwaki. 1981. Studies on dipeptidyl carboxypeptidase in the male reproductive organs; its biological and pathological status. *J. Pharm. Dyn.* 4: 175-183.

Karsh, F.J. 1985. The hypothalamus and anterior pituitary gland. In *Hormonal Control of Reproduction*, Vol. 3, 1-20. Cambridge University Press, Cambridge.

Khanum, A. and M.L. Dufau. 1988. Angiotensin II receptors and inhibitory actions in Leydig cells. *J. Biol. Chem.* 263: 5070-5074.

Krassnigg, F., H. Niederhauser, R. Placzek, J. Frick and W.B Schill. 1986. Investigations on the functional role of angiotensin converting enzyme in human seminal plasma. *Adv. Exp. Biol.* 198: 477-485.

Levine, N. and D.J. Marsh. 1971. Micropuncture studies of the electrochemical aspects of fluid and electrolyte transport in individual seminiferous tubules, the epididymis and the vas deferens in the rat. *J. Physiol.* 213: 557-570.

Lieberman, J. and A. Sastre. 1983. Angiotensin-converting enzyme activity in postmortem human tissue. *Lab. Invest.* 48: 711-717.

Magnan, J. and D. Regoli. 1979. Characterization of receptors for angiotensin in the rat vas deferens. *Can. J. Physiol. Pharmacol.* 57: 417-423.

Millan, M.A. and G. Aguilera. 1988. Angiotensin II receptors in testes. *Endocrinology* 122: 1984-1990.

Mizutani, T. and W.B. Schill. 1985. Motility of seminal plasma-free spermatozoa in the presence of several physiological compounds. *Andrologia* 17: 150-156.

Ohkubo, H., K. Nakayama, T. Tanaka and S. Nakanishi. 1986. Tissue distribution of rat angiotensinogen mRNA and structural analysis of its heterogeneity. *J. Biol. Chem.* 261: 319-323.

Pandey, K.N. and T. Inagami. 1984. Detection of renin mRNA in mouse testis by hybridization with renin cDNA probe. *Biochem. Biophys. Res. Comm.* 125: 662-667.

Pandey, K.N. and T. Inagami. 1986. Regulation of renin angiotensins by gonadotropic hormones in cultured murine Leydig tumor cells. *J. Biol. Chem.* 261: 3934-3938.

Pandey, K.N., K.S. Misono and T. Inagami. 1984. Evidence for intracellular formation of angiotensins: Coexistence of renin and angiotensin-converting enzyme in Leydig cells of rat testis. *Biochem. Biophys. Res. Comm.* 122: 1337-1343.

Parmentier, M., T. Inagami, R. Pochet and J.C. Desclin. 1983. Pituitary dependent renin-like immunoreactivity in rat testis. *Endocrinology* 112: 1318-1323.

Rosenthal, H.E. 1967. A graphic model for the determination and presentation of binding parameters in complex systems. *Anal. Biochem.* 20: 525

Saye, J., S.B. Binder, G.J. Trachte and M.J. Peach. 1986. Angiotensin peptides and prostaglandin E_2 synthesis: Modulation of neurogenic responses in the rabbit vas deferens. *Endocrinology* 119: 1895-1903.

Sealey, J.E., M. Goldstein, T. Pitarresi, T.T. Kudlak, N. Glorioso, S.A Fiamengo and J.H. Laragh. 1988. Prorenin secretion from human testis: No evidence for secretion of active renin or angiotensinogen. *J. Clin. Endocrinol. Metab.* 66: 974-978.

Setchell, B.P. and D.E. Brooks. 1988. Anatomy, vasculature, innervation, and fluids of the male reproductive tract. In *The Physiology of Reproduction*, Vol. 1, eds. Knobil and Neill, 753-835. Raven Press, New York.

Strittmatter, S.M. and S.H. Snyder. 1984. Angiotensin-converting enzyme in the male rat reproductive system: Autoradiographic visualization with [^{3}H]captopril. *Endocrinology* 115: 2332-2341.

Trachte, G.J. 1988. Prostaglandins do not mediate the inhibitory effects of angiotensins II and III on autonomic neurotransmission in the rabbit vas deferens. *Prostaglandins* 36: 215-227.

Trachte, G.J., E. Stein and M.J. Peach. 1987. Alpha-adrenergic receptors mediate angiotensin-induced prostaglandin production in the rabbit isolated vas deferens. *J. Pharmacol. Exp. Ther.* 240: 433-440.

Vanha-Pertulla, T., J.P. Mather, C.W. Bardin, S.B. Moss and A.R. Bellve. 1985. Localization of the angiotensin-converting enzyme activity in testis and epididymis. *Biol. Reprod.* 33: 870-877.

Vivet, F., P. Callard and A. Gamoudi. 1987. Immunolocalization of angiotensin I converting enzyme in the human male genital tract by the avidin-biotin-complex method. *Histochemistry* 86: 499-502.

Yokoyama, M., K. Hiwada, T. Kokubu, M. Takaha and M. Takeuchi. 1980. Angiotensin-converting enzyme in human prostate. *Clin. Chim. Acta* 100: 253-258.

Yotsumoto, H., S. Shoichiro and M. Shibuya. 1984. Localization of angiotensin converting enzyme in swine sperm by immunofluorescence. *Life Sci.* 35: 1257-1261.

NEUROHYPOPHYSEAL HORMONES AND REPRODUCTIVE HORMONE SECRETION

Craig A. Johnston
College of Pharmacy
Washington State University
Pullman, WA

I.   Introduction

II.   Anatomy

III.   Influence of OXY and AVP Neurons on LH Release from the Anterior Pituit

        A.   OXY

        B.   AVP

IV.   Influence of OXY and AVP Neurons on FSH Release from the Anterior Pitu

        A.   OXY

        B.   AVP

V.   Influence of OXY and AVP Neurons on Prolactin Release from the Anterior Pituitary

        A.   OXY

        B.   AVP

VI.   Summary

Introduction

It is increasingly obvious that many of the brain-gut peptides discussed at this conference are not only found in the posterior pituitary, but are also synthesized there (Johnston *et al.*, 1988). Therefore, when considering the mechanisms by which brain-gut peptides might influence reproductive hormone secretion, the possibility that some of these effects might involve the neurohypophysis must be considered. In addition, reproductive hormone secretion is greatly influenced by the two major neurohypophyseal peptide hormones, oxytocin (OXY) and arginine vasopressin (AVP). It is this influence which will represent the major focus of this chapter, *i.e.*, the direct and indirect influence of the neurohypophyseal hormones, OXY and AVP, on the secretion of the anterior pituitary reproductive hormones, luteinizing hormone (LH), follicle stimulating hormone (FSH) and prolactin (PRL).

Anatomy

The distribution of central OXY- and AVP-containing neurons has been reviewed in detail elsewhere (Zimmerman *et al.*, 1984; Swaab *et al.*, 1975; Hou-Yu *et al.*, 1986; Dierickx, 1980). These neuronal systems share many characteristics including: regional location of major cell body- and terminal-containing fields; structure, synthesis and transport of the mature nonapeptides; and neurochemical controls. In addition, there are many features of these cyclic nonapeptide pathways unique to either OXY or AVP neurons. For example, although very similar in structure and distribution, to date, the two nonapeptides have never been demonstrated to be produced and/or colocalized in the same neuron (Dierickx, 1980). Furthermore, despite similarities in their general distribution, there is an identifiable topographical separation of OXY and AVP innervation to various brain areas (Swaab *et al.*, 1975; Sokol *et al.*, 1976; Swanson *et al.*, 1981; Sawchenko and Swanson, 1982). Even within particular brain nuclei containing OXY and AVP cell bodies, a distinct anatomical localization for the two peptides exists (Sawchenko and Swanson, 1985). For example, OXY cell bodies are concentrated in the anterior portion of the paraventricular nucleus (PVN) and supraoptic nucleus (SON), while AVP cell bodies are localized more posteriorly and caudally. The major projection of OXY- and AVP-containing neurons is to the posterior pituitary, although smaller projections provide innervation to the median eminence (ME) and other hypothalamic and limbic areas, with another substantial projection of OXY neurons to the spinal cord and areas of the pons-medulla which are intimately involved in autonomic control

(Zimmerman *et al.*, 1984; Hou-Yu *et al.*, 1986; Swanson *et al.*, 1981; Sawchenko and Swanson, 1985). These two nonapeptides also differ in their primary known physiologic functions (promoting milk ejection in response to a nursing stimulus and contracting the uterus at parturition associated with OXY, and water/electrolyte homeostasis and cardiovascular function with AVP) as well as how physiological stimuli and surgical manipulations affect their release (nursing, parturition, kappa receptor opiates and surgical removal of the posterior pituitary all influencing OXY release, while dehydration and hemorrhage stimulate AVP release) (Beleslin *et al.*, 1967; Gartan *et al.*, 1964; Summy-Long *et al.*, 1984, 1990; Wakerley *et al.*, 1973; Weitzman *et al.*, 1978; Johnston *et al.*, 1990a).

OXY and AVP have also been implicated in the neuroendocrine regulation of anterior pituitary hormone secretion, including the gonadotropins (Johnston and Negro-Vilar, 1988; Lumpkin *et al.*, 1983; Froehlich *et al.*, 1985; Robinson and Evans, 1990; Fagin and Neill, 1982). As these studies indicate, much more evidence presently supports a role for OXY in the regulation of gonadotropin secretion than supports a role for AVP. Anatomical support for an effect of both OXY and AVP on gonadotropin secretory control is present. Both OXY and AVP neurons send projections which terminate in the ME (Zimmerman *et al.*, 1984), and evidence indicates that both nonapeptides are released directly into the hypophyseal vasculature where they are carried to the anterior pituitary (Sarkar and Gibbs, 1984; Rhodes *et al.*, 1982). Thus, the possibility exists that OXY and/or AVP could directly influence reproductive hormone secretion at the level of the anterior pituitary.

In addition, anatomical evidence for an interaction of central OXY and/or AVP neurons with central LH-releasing hormone (LHRH) neurons is substantial. OXY and AVP neurons send projections which terminate in many areas of the brain containing LHRH cell bodies or terminals such as the medial preoptic nucleus (MPO), the organum vasculosum of the lamina terminalis (OVLT) and ME (Zimmerman *et al.*, 1984; Swaab *et al.*, 1975; Hou-Yu *et al.*, 1986; Dierickx, 1980; Sokol *et al.*, 1976; Swanson *et al.*, 1981; Sawchenko and Swanson, 1982). Besides this possible indirect action of OXY and AVP neurons to influence gonadotropin secretion by altering the release of LHRH, there are also several possible anatomical sites where central OXY or AVP neurons could influence the secretion of other neuropeptide or catecholaminergic LHRH secretagogues including angiotensin II (ANG II), neuropeptide Y (NPY), or norepinephrine (NE). These possibilities will be discussed in detail later in this chapter.

Finally, it has been known for a long time that the secretion of reproductive hormones from the anterior pituitary is greatly influenced by changes in the plasma levels of gonadal hormones, and this influence may be mediated, in part, by OXY and/or AVP neurons. For example, although LHRH neurons are greatly influenced by plasma estradiol concentrations, no binding of estradiol on LHRH neurons has yet been identified. On the other hand, subpopulations of OXY neurons demonstrate nuclear receptors for estradiol (Sar and Stumpf, 1980; Dufau and Catt, 1978), and may represent the anatomical loci for such an important link. The influence of estradiol on these systems and the potential importance of the gonadal hormones to these interactions will be expanded throughout this chapter.

Influence of OXY and AVP Neurons on LH Release from the Anterior Pituitary

The important physiological role which LH plays in the process of ovulation, and in the growth and maintenance of normal ovarian and testicular function is universally recognized (Dufau and Catt, 1978; Kulin and Reiter, 1973; Richards, 1980; Vaitukaitis *et al.*, 1976). In fact, the severe clinical consequences resulting from the aberrant secretion of LH are well-documented and include hypogonadism, anovulatory menstrual cycles with resulting infertility, and several other physical and behavioral alterations (Illig *et al.*, 1980; Skarin *et al.*, 1982; Meldrum *et al.*, 1982; Leyendecker *et al.*, 1980; Liu *et al.*, 1983; Marshall and Kelch, 1979). Furthermore, recognition of the important role that LH plays in regulating ovarian function and ovulation has resulted in a concentrated effort, by those developing successful oral contraceptive agents, to focus on inhibition of the preovulatory surge of LH as their endpoint in determining which therapies may represent potentially valuable contraceptive regimens (Diczfalusy, 1968). The magnitude and seriousness of clinical manifestations resulting from the aberrant secretion of LH, the frequency with which they occur in the normal population, and the proven therapeutic usefulness of altering plasma levels of LH to treat infertility or achieve contraception, make the elucidation of the neuroendocrine mechanisms controlling the secretion of LH very important.

In the past two decades, substantial advances have been made toward identifying and characterizing those neuropeptide/neurotransmitter substances contained within the brain which may play a significant role in the neuroendocrine regulation of LH secretion from the anterior pituitary (see reviews: Negro-Vilar and Ojeda, 1981; McCann and Ojeda, 1976; McCann, 1982). Furthermore, significant insight into some of the complex mechanisms and interactions which these individual substances undergo to achieve that neuroendocrine control, and the recognition that the gonadal steroids represent important modulating influences on those effects and interactions, has been provided by ensuing studies (Knobil, 1980). Evidence supporting direct and indirect actions for OXY and/or AVP on LH release have been reported and will be summarized below.

OXY

An increasing body of evidence supports an important role for OXY on LH secretion. Several factors involved in this regulation remain controversial, including: the qualitative nature (inhibitory or stimulatory), the possible sites of interaction (directly at the pituitary or indirectly in the CNS) and the physiological significance of this influence. Gambacciani *et al.* (1986) reported evidence which provided the ground work to support the hypothesis that the actions of OXY on LH secretion are primarily negative in nature, and are mediated through actions on central LHRH neurons. Those studies provided *in vitro* evidence demonstrating an inhibitory effect of OXY on LHRH release from ME tissues or mediobasal hypothalamic fragments obtained from male rats. Two years later, Franci *et al.* (1988) reported that the action of OXY on LH secretion might be minor, at best. They demonstrated that the intracerebroventricular administration of OXY antisera to chronically ovariectomized female rats did not alter basal plasma levels of LH, suggesting that central OXY neurons may not exert a physiologically relevant effect on plasma LH secretion. In the same year, our laboratory provided evidence that OXY neurons may exert a physiologically relevant stimulatory influence on the release of LHRH and LH which occurs just prior to, and throughout, the afternoon of proestrus in cycling female rats (Johnston and Negro-Vilar, 1988). Those experiments demonstrated a dose-dependent ability of two different specific OXY pharmacological antagonists to inhibit the preovulatory surge of LH normally seen on the afternoon of proestrus in cycling female rats. The specificity of this effect was confirmed by the fact that both OXY antagonists blocked the surge (even though they differed considerably in terms of their vasopressin antagonist profiles), and that administration of a potent AVP pharmacological antagonist, in a dose which had been shown to selectively block the vasopressor action of AVP without antagonizing OXY activity, did not affect the surge of LH on the afternoon of proestrus in cycling female rats. This apparent controversy may be easily explained by the difference in plasma estradiol concentrations associated with the experimental paradigms used. Both the experimental paradigms of 1) tissues obtained from male rats for *in vitro* analysis and of 2) *in vivo* results obtained from chronically ovariectomized female rats would be associated with very low levels of plasma estradiol, whereas the cycling female rat on the afternoon of proestrus would be associated with physiologically high levels of plasma estradiol.

Evidence supporting a dramatic influence of plasma estradiol on the distribution, content, synthesis, and release of OXY in the CNS has been provided. Treatment with estradiol has been shown not only to enhance the OXY-like immunoreactivity contained in discrete regions of the rat brain (Caldwell *et al.*, 1986), but also to allow additional immunoreactive OXY perikarya and processes to be observed in areas in which no OXY-like immunoreactivity was observed prior to estradiol treatment (Jirikowski *et al.*, 1986; Jirikowski *et al.*, 1988; Rhodes *et al.*, 1981). The administration of estradiol *in vitro* to magnocellular hypothalamic neurons has been shown to stimulate axonal sprouting (Thoran-Allerand, 1980). Estradiol has also been shown to selectively enhance the firing rate of OXY neurons in the PVN (Akaishi and Sakuma, 1985). *In vivo* administration of estradiol stimulates the release of OXY into the blood and cerebral spinal fluid (Unger and Schwarzenberg, 1970). Nuclear receptors for estradiol have been demonstrated on subpopulations of OXY neurons (Rhodes *et al.*, 1982; Sar and Stumpf, 1980). Finally, measurements of OXY in plasma of monkeys, human females, or in hypophysial portal blood of rats throughout the reproductive cycle indicate that the highest levels of OXY secretion occur when plasma levels of estradiol are highest (Falconer *et al.*, 1980; Mitchell *et al.*, 1980; Sarkar and Gibbs, 1984; Robinson *et al.*, 1976).

Recently, we have examined more closely the dependence of the OXY influence on plasma LH release on circulating levels of estradiol. The ability of a central injection of OXY to affect plasma LH on proestrus (when plasma levels of estradiol are physiologically high) was compared to the response observed on metestrus or diestrus (when estradiol levels are low). Stainless steel cannulae were placed into the third cerebral ventricle of cycling female rats using a stereotaxic apparatus and the atlas of König and Klippel (1967) as a guide. Rats were then monitored for regularity of cycles by examining the cytology in their daily vaginal lavage. Animals demonstrating two consecutive regular four-day cycles were then utilized for further experiments beginning with the next cycle. On diestrus of the experimental cycle, a cannula was placed in the right external jugular vein as previously described (Harms and Ojeda, 1974). Each animal was then treated on proestrus and either metestrus or diestrus as follows: at 1100 h intravenous cannulae were connected to collecting tubes. OXY (10 μg/5 μl) or vehicle (5 μl saline, saline group) was administered icv at 1500 h and plasma samples were obtained via the chronic jugular cannulae from the unanesthetized, freely moving animals at 10 min before and 5, 15, 30 and 60 min following icv injection. The results of these experiments are shown in Table I.

TABLE I.   EFFECT OF OXY (ICV) ON PLASMA LH

Day of Cycle	Treatment	Plasma LH (ng/ml)				
		Pre	5 min	15 min	30 min	60 min
PROESTRUS	OXY	6.0+0.4	23.5+3.2*A	42.3+4.7*A	33.9+3.0*A	8.7+0.7*A
	Saline	5.4+0.4	4.8+0.4	5.2+0.3	5.1+0.4	5.2+0.3
MET/DI	OXY	4.7+0.4	3.9+0.3	3.2+0.2*	3.5+0.3	3.9+0.4
	Saline	4.3+0.2	3.9+0.2	4.5+0.3	4.2+0.3	4.0+0.2A

Values represent mean ± SEM as determined from N = 7-8 rats. *Significantly different from saline control value on same day of cycle (p < 0.01). ASignificantly different from value for same treatment group on different day of cycle (p < 0.01).

The data clearly demonstrate that the central administration of OXY will stimulate LH secretion on proestrus, but not on metestrus or diestrus (metestrus and diestrus values were combined because the values on those days were not significantly different). The results support the hypothesis that in order for OXY to stimulate LH secretion, physiologically high levels of estradiol must be present.

Important questions remaining to be answered concerning the ability of centrally administered OXY to influence plasma LH include: a) Is this influence an OXY receptor-mediated effect, and b) Is the action of centrally injected OXY to influence plasma LH mediated by an interaction with LHRH neurons? To answer these questions, the ability of a pretreatment with an OXY antagonist ([1-(β-mercapto-β,β-cyclopentamethylene propanoic acid)2-0-methyltyrosine,8-ornithine]vasotocin; 45 μg/kg BW, iv, 20 min prior to icv injection) or a potent LHRH antagonist (D-Glu[1], D-Phe[2], D-Trp[3,6]) LHRH; 750 μg/rat, iv, 50 min prior to icv injection) to alter the stimulation of plasma LH in response to the central administration of OXY on proestrus was examined. The results of this experiment are shown in Table II.

TABLE II. EFFECT OF OXY OR LHRH ANTAGONISM ON OXY-INDUCED

CHANGES IN PLASMA LH ON PROESTRUS

Treatment	Control	OXY (10μg,20')	OXY-A/OXY	LHRH-A	LHRH-A/OXY
LH (ng/ml plasma)	4.7+0.4	28.4+3.1[*]	3.8+0.3	2.8+0.2[A]	3.1+0.3[A]

Values represent mean + SEM as determined from N = 7-8 rats. [*]Significantly different from all other groups (p<0.01). [A]Significantly different from control values (p<0.01). OXY-A = OXY antagonist pretreatment; LHRH-A = LHRH antagonist pretreatment.

Pretreatment with either an OXY antagonist or an LHRH antagonist blocked the increase in plasma LH resulting from a central injection of OXY. These results demonstrate that the OXY-induced increase in plasma LH is an OXY receptor-mediated event and is apparently mediated by influencing the release of LHRH. This evidence, that OXY may work primarily at a central (indirect) site to influence LH release, is supported by additional evidence from the literature. Two experiments have examined the effect of intravenous administration of an OXY antiserum on increases in LH associated with physiologically high levels of plasma estradiol. Samson *et al.* (1986) used the pharmacological model for the preovulatory LH surge of ovariectomy with estrogen replacement to evaluate the influence of OXY on that surge. These authors found very little effect of passive immunization versus OXY on the pharmacologically induced LH surge, and concluded that the influence of OXY on LH release was minimal, at best. In late 1988, another paper examined the ability of passive immunization against OXY (using the intravenous injection of OXY antisera) to alter the preovulatory LH surge seen on the afternoon of proestrus in cycling female rats (Sarkar, 1988). This latter study also demonstrated no significant effect of passive immunization against OXY on plasma LH concentrations measured at a limited number of timepoints on the afternoon of proestrus. Once again, it was concluded that OXY exerts little, if any, physiologically relevant control over LH secretion. Results of experiments being conducted in our laboratory at the same time indicated that the effect of OXY on plasma LH might be mediated at a central site which would not be affected by the peripheral injection of the OXY antisera. If true, then OXY antisera injected intravenously might not gain access to the central site(s) involved in such an effect. To check this possibility more directly, the ability of OXY antisera administered intracerebroventricularly to affect the preovulatory LH surge on the afternoon of proestrus was compared with the ability of this same antisera to affect the LH surge following intravenous administration (Johnston *et al.*, 1990b). The antisera utilized has been previously characterized (Samson *et al.*, 1985) and demonstrates no cross-reactivity with AVP, vasoactive intestinal peptide, secretin, peptide histidine isoleucine, β-endorphin, the enkephalins, neurotensin, thyrotropin releasing hormone, LH, or PRL. This antisera demonstrates a maximum binding capacity (MBC) in plasma of 4059 ng/ml, with an MBC at 2 h of 1930 ng/ml, at 3 h of 1562 ng/ml and at 6 h after administration of 685 ng/ml. Proestrous rats were either a) injected intravenously with 0.8 ml of OXY antisera (OXY AB) or normal rabbit serum (NRS) at 1300 h, or b) administered an icv injection of 5 μl of OXY antisera or NRS at 1400 h. Blood samples were collected hourly from 1400 h to 2100 h for LH analysis. The results are shown in Table III.

TABLE III.   CENTRAL VERSUS PERIPHERAL OXY ANTISERUM ON LH

Time (h) on Proestrus:	1400	1500	1600	1700	1800	1900	2000	2100
NRS (iv)	8.2+2.1	7.3+1.6	17.3+2.6	44.7+5.3	69.1+7.7	37.4+2.8	38.2+4.0	15.0+5.2
OXY AB (iv)	8.0+1.8	4.6+2.7	4.1+3.2[*]	12.6+2.4[*]	27.2+3.3[*]	34.6+7.0	32.3+1.8	7.9+2.6
NRS (icv)	7.4+1.5	5.7+1.7	6.4+1.3	41.4+4.7	73.6+8.4	32.5+3.8	36.6+5.1	17.4+3.8
OXY AB (icv)	7.5+0.8	4.2+0.7	3.6+0.2[*]	3.1+0.2[*A]	2.8+0.1[*A]	2.5+0.2[*A]	2.3+0.1[*A]	2.7+0.3[*A]

Values represent mean + SEM as determined from N = 8-10 animals. [*]Significantly different from appropriate NRS value (P<0.01). [A]Significantly different from OXY AB (iv) value at the same time (p<0.01).

The administration of just 5 µl of OXY antiserum into the third cerebral ventricle completely abolished the preovulatory LH surge observed in the NRS-treated proestrous rats. The profound inhibition of the LH surge which resulted from the icv injection of just 5 µl of OXY antisera, compared to the much weaker effect of a large injection (0.8 ml) of OXY antisera given intravenously, strongly suggests that a central site(s) is primarily involved in the physiologically relevant stimulatory action of OXY on LH secretion. The weaker, but still significant, effect which intravenous administration of OXY antisera produced on the preovulatory LH surge could have resulted from interference with some weaker, yet physiologically relevant, direct effect of OXY on LH secretion at the level of the anterior pituitary (see discussion below). However, it is more probable that following the iv injection of such a large amount of OXY antisera, some antisera may have either gained access to relevant central sites or that some of the actions of central OXY neurons on LHRH release are mediated at sites in circumventricular areas of the brain which would lack a strong blood brain barrier, and would thus be susceptible to the actions of a large dose of OXY antisera injected peripherally.

As mentioned previously, several other neuropeptides and monoaminergic neurotransmitters have been implicated in the neuroendocrine regulation of LH secretion, and primarily appear to influence that secretion by altering the net release of LHRH from the ME into the hypophyseal portal blood. Of the myriad of candidates, a few of these LHRH secretagogues or inhibitors appear to be highly dependent on the levels of plasma estradiol. These candidates include the monoaminergic neurotransmitters, norepinephrine (NE) and 5-hydroxytryptamine (5-HT or serotonin), the peptidergic LHRH secretagogues, NPY and ANG II (see chapters by MacDonald, Speth, and Steele in this volume), and the peptidergic LHRH inhibitor, β-endorphin (see Kalra, this volume). Anatomical support for an interaction between central OXY and noradrenergic, serotonergic, NPY, β-endorphin and/or ANG II neurons exists; however, direct evidence for those interactions and their significance in the neuroendocrine regulation of LHRH and LH release, as well as the modulatory influence of plasma estradiol levels on those possible interactions is sparse, controversial, and at present, highly speculative. Preliminary results from our laboratory suggest that ANG II and OXY neurons may work in concert to stimulate LHRH release (Johnston *et al.*, 1990c).

Evidence in support of a direct effect of OXY on LH secretion at the anterior pituitary was first supplied by Evans (Evans and Catt, 1989; Evans *et al.*, 1989) demonstrating that OXY could stimulate LH release from rat anterior pituitary cells in static culture.   Recently, Robinson and Evans (1990) have confirmed our results that an OXY antagonist could inhibit the preovulatory LH surge in a dose-dependent manner.   In this paper, the authors also demonstrated that the intraperitoneal administration of OXY in the morning of proestrus, but not metestrus, diestrus or estrus, could advance the onset of the LH surge occurring on the afternoon of proestrus in a dose-dependent and time-dependent manner without altering the descending phase or amplitude of the LH surge. Because these effects follow peripheral injections of OXY, the authors concluded that they resulted from LH-releasing activity mediated by OXY directly at the level of the anterior pituitary.   Further investigations into the possible site(s) and mechanisms by which OXY can influence LH secretion from the anterior pituitary are warranted.

AVP

Whether AVP exerts a physiologically important role on the neuroendocrine regulation of gonadotropin secretion is quite controversial.   In fact, most of the actions of AVP may be due to its structural similarity to vasotocin, a peptide synthesized in the pineal which has been hypothesized to represent a physiologically relevant antigonadotropic factor.   Nonetheless, systemically administered AVP can inhibit the preovulatory surge of LH on proestrus (Cheesman *et al*., 1977), or the estrogen-induced LH surge in ovariectomized female rats (Salisbury *et al*., 1980).   Two results bring into question the physiological significance of these data.   First, AVP is much weaker in its ability to suppress the LH surge than vasotocin (Salisbury *et al*., 1980; Blask *et al*., 1978).   Second, administration of a potent AVP antagonist throughout the afternoon of proestrus did not significantly affect the preovulatory LH surge observed in vehicle-treated controls (Johnston and Negro-Vilar, 1988).   In contrast, a greater release of LH in response to LHRH injection has been reported in castrated male rats who were provided with a simultaneous injection of AVP or vasotocin than in rats treated with LHRH alone (Vaughan *et al*., 1979a).   The intravenous injection of a high dose of AVP has been reported to increase basal presurge LH concentrations in ovariectomized female rats (Salisbury *et al*., 1980), whereas low doses do not affect basal LH levels in castrated or intact male rats (Vaughan *et al*., 1979b; Turkelson *et al*., 1981).   A similar ability of AVP to elevate basal LH levels during the luteal phase in female baboons has been reported (Koyama and Hagino, 1983).   Once again, because the doses of AVP needed to affect LH secretion are greater than those which would cause elevations in blood pressure, the specificity and physiological significance of the effect of AVP on LH secretion must be questioned. Physiological significance aside, the pharmacological ablity of AVP to alter gonadotropin release does not appear to be mediated directly at the level of the anterior pituitary. Neither basal nor LHRH-stimulated LH release from female rat pituitaries incubated *in vitro* is affected by AVP administration (Turkelson *et al.*, 1981; Cheung, 1983; Vaughan *et al.*, 1975). Whether AVP may interact as a neurotransmitter in a physiologically relevant manner to influence either static or dynamic LH secretion from the anterior pituitary is not known. However, at present, the importance of this possible influence, even if demonstrated, might be minimal.

Influence of OXY and AVP Neurons on FSH Release from the Anterior Pituitary

Very few studies have been performed specifically to examine the influence of OXY or AVP on FSH secretion. Oftentimes, LH secretion has been used as the endpoint in evaluating effects on gonadotropin secretion despite the fact that increasing evidence suggests that a specific FSH-releasing factor, separate from LHRH, may exist and, therefore, FSH could easily undergo differential control from LH secretion. In light of the explosive discovery of physiological effects for activin and inhibin, and the differential effects which these peptides appear to produce on LH and FSH secretion, as well as the recent findings emerging which demonstrate a differential ability of activin to influence AVP and OXY synthesis and release, a more careful evaluation of the influence which OXY and/or AVP may exert on FSH secretion is warranted.

OXY

Specific to FSH, Franci *et al.* (1988) reported that neither OXY antiserum nor AVP antiserum injected intracerebroventricularly altered plasma FSH concentrations in chronically ovariectomized female rats. Direct stimulation of FSH release *in vitro* from anterior pituitary cells following OXY application (Evans and Catt, 1989; Evans *et al.*, 1989) has been implicated as the possible cause for the accelerated follicular maturation and ovulation in mice given OXY (Robinson *et al.*, 1985). Recently, Robinson and Evans (1990) have shown that the administration of OXY at 900 h, 1000 h and 1100 h on proestrus advanced the FSH surges observed on proestrus as well as on estrus. The data suggest that OXY may influence the timing of the FSH surges seen during the reproductive cycle.

AVP

The effect of exogenously administered AVP on FSH secretion in response to estrogen treatment in ovariectomized female rats (Blask *et al.*, 1978) or in castrated male rats (Vaughan *et al.*, 1979b) is minimal.

In summary, at present very little data supporting a physiological role for either OXY or AVP on plasma FSH release exists; however, the question has not been carefully examined.

Influence of OXY and AVP Neurons on Prolactin Release from the Anterior Pituitary

Unlike the influence of OXY and AVP on LH and FSH secretion, most studies dealing with this subject have agreed on the influence of OXY and AVP on PRL secretion. What remains controversial is the physiological significance of that influence in various paradigms, and what neurochemical interactions comprise the neuronal networks mediating those effects.

OXY

Indirect evidence supporting a possible role of OXY on PRL secretion abounds. Exogenous administration of OXY can stimulate PRL secretion both *in vitro* and *in vivo* (Lumpkin *et al.*, 1983; McCann *et al.*, 1984). OXY receptors have been identified in the anterior pituitary (Antoni, 1986). OXY is secreted directly into the hypophysial portal blood (Gibbs, 1984) and OXY concentrations in the hypophysial portal blood vary throughout the estrous cycle, with the highest concentrations correlating temporally with the preovulatory surge of PRL (Sarkar and Gibbs, 1984). Furthermore, surgical removal of the neurointermediate pituitary (which would contain OXY terminals) has been shown to attenuate the increase in plasma PRL associated with suckling (Murai and Ben-Jonathan, 1987), 5-hydroxytryptophan administration (Johnston *et al.*, 1986) or the proestrous surge (Murai *et al.*, 1989). The applicability of these latter studies to OXY, alone, must be questioned in light of the recent evidence that a specific, peptidergic PRL releasing factor may be synthesized and released from the posterior pituitary (Hyde and Ben-Jonathan, 1988; Samson *et al.*, 1990) and could explain the neurointermediate lobectomy results.

In a series of experiments, we examined the ability of pharmacological antagonists of OXY to influence basal PRL release, as well as the dynamic *in vivo* increases in PRL associated with 5-hydroxytryptophan administration, acute ether stress or suckling stimuli, or the surge of PRL observed on the afternoon of proestrus in the cycling female rat (Johnston and Negro-Vilar, 1988). We also examined the *in vitro* release of PRL from dispersed anterior pituitary cells in response to OXY administration. The pharmacological antagonists of OXY blocked the *in vitro* release of PRL from dispersed anterior pituitary cells seen in response to OXY, and also inhibited the preovulatory surge of PRL observed on the afternoon of proestrus in cycling female rats. No effect of OXY antagonism on basal *in vivo* PRL levels or on the increase in PRL seen in response to 5-hydroxytryptophan, ether stress or suckling stimuli were observed. The data provide evidence that OXY plays a physiologically important role in the release of PRL seen on the afternoon of proestrus, but not in response to other secretory stimuli examined. The role for OXY in the proestrous PRL surge (Sarkar, 1988) or in the estrogen-induced PRL surge in ovariectomized rats (Samson *et al.*, 1986) has been confirmed in studies using passive immunization against OXY. The lack of effect of OXY antagonism on the acute increase in PRL occurring in response to suckling stimulus is supported by other studies (Nagai *et al.*, 1983; Grosvenor *et al.*, 1986), although there may be a role for OXY in the maintenance of high levels of plasma PRL observed in the continuously suckled lactating female rat (Samson *et al.*, 1986; Johnston and Negro-Vilar, 1988). In addition to direct actions of OXY on PRL release, there is good evidence that OXY may interact with other central neuropeptides, like vasoactive intestinal peptide, to influence PRL release (Samson *et al.*, 1989). Studies concerning the mechanisms and neuronal interactions involved with OXY to influence PRL secretion are sure to be the subject of considerable investigation in the near future.

AVP

Administration of high doses of AVP have been reported to stimulate PRL release in steroid-primed male rats (Valverde-R *et al.*, 1972; Boyd *et al.*, 1976; Szabo and Frohman, 1976). Although this effect is much more apparent in male rats pretreated with estrogen and progesterone, it has also been reported in untreated male rats (Blask *et al.*, 1984; Shin, 1982) as well as in ovariectomized, steroid-primed female rats (Salisbury *et al.*, 1980). Some augmentation of the PRL release observed following transplantation of anterior pituitaries under the kidney capsule by AVP administration has also been reported (Shin, 1982). Two important points should be noted. First, the high doses of AVP used in these studies may be interacting with OXY receptors which then, in turn, could affect PRL secretion. Therefore, the specificity of these effects must be questioned.

Second, the AVP-induced effects on PRL secretion do not appear to be mediated directly at the level of the anterior pituitary. AVP has been reported in most studies to be ineffective in releasing PRL from dispersed anterior pituitary cells obtained from rats (Ben-Jonathan, 1980; Hanew *et al.*, 1980; Lamberts *et al.*, 1981), although both a non-dose-dependent stimulation (Turkelson *et al.*, 1981) and inhibition (Lumpkin *et al.*, 1983) of PRL in vitro in response to AVP administration have also been reported. The possibilities that actions of AVP on PRL secretion may be due to non-specific stressful responses resulting from the peripheral administration of large doses of AVP, or may be due to its ability to affect OXY receptors or interact with vasotocinergic receptors all have to be considered. This, plus the lack of effect which AVP demonstrates on the in vitro release of PRL from dispersed anterior pituitary cells, makes the hypothesis that AVP exerts a physiologically relevant influence on PRL secretion highly unlikely. This conclusion is supported further by a general lack of evidence that AVP influences PRL secretion in other mammalian species including sheep (Drummond *et al.*, 1980), dogs (Zucker *et al.*, 1983) or humans (del Pozo *et al.*, 1980; Lamberts *et al.*, 1981).

Summary

Neurohypophyseal hormones apparently exert important regulatory control on reproductive hormone secretion from the anterior pituitary. Evidence for a physiologically relevant role for OXY on the secretion of LH, FSH and PRL is much more abundant and convincing than that for AVP. Specifically, OXY appears to exert a physiologically important stimulatory influence on the preovulatory surge of LH. This influence appears to be mediated primarily at a central site and, in addition to interacting with LHRH neurons, may involve interactions with other central peptidergic and monoaminergic neurotransmitters such as ANG II, NPY, endogenous opiates and NE. Furthermore, this central OXY influence requires physiologically high levels of plasma estradiol in order to be present. OXY may also influence LH secretion, in part, by a direct action at the level of the anterior pituitary. This activity also appears to require the presence of estradiol, may influence the responsiveness of the pituitary to LHRH, and may primarily influence the timing of the preovulatory LH surge.

Future studies aimed at elucidating the interaction, mechanisms and sites of influence of OXY on LHRH and LH secretion should not only advance our basic understanding concerning the nature, sites, mechanisms and physiological significance of OXY in regulating LH release, but should potentially permit the development of dynamic tests to characterize and diagnose the nature of central causes responsible for the aberrant secretion of LH, and provide insight into the development of possible novel therapeutic regimens to either alleviate clinical dysfunctions resulting from the abnormal secretion of LH or to cause contraception.

The ability of OXY to affect PRL secretion directly at the pituitary level is well documented. The major question concerning this influence is its physiological significance to a clinical situation. Indeed, the only PRL secreting paradigm in which OXY has been clearly demonstrated to exert a physiologically relevant influence is on proestrus--an experimental setting which has no clinical correlate. Whether OXY plays a physiological role in other clinically applicable paradigms associated with dynamic changes in PRL release or in hyperprolactinemic states is currently a matter of considerable research effort.

ACKNOWLEDGMENTS

The authors would like to thank Christopher Greer and Michael V. Templin, B.Ph., for their excellent technical assistance in the laboratory. We also thank Mrs. Gina Cox for her superb secretarial help in preparation of the manuscript and for her patience, good nature and friendship during this ordeal.

REFERENCES

Akaishi, T. and Y. Sakuma. 1985. Estrogen excites oxytocinergic, but not vasopressinergic cells in the paraventricular nucleus of female rat hypothalamus. *Brain Res.* 335: 302-305.

Antoni, F.A. 1986. Oxytocin receptors in rat adenohypophysis: Evidence from radioligand binding studies. *Endocrinology* 119: 2393-2395.

Beleslin, D., G.W. Bisset, J. Haldar and R.L. Polak. 1967. The release of vasopressin without oxytocin in response to hemorrhage. *Proc. R. Soc. Biol.* 166: 443-458.

Ben-Jonathan, N. 1980. Catecholamines and pituitary prolactin release. *J. Reprod. Fertil.* 58: 501-512.

Blask, D.E., M.K. Vaughan, R.J. Reiter and L.Y. Johnson. 1978. Influence of arginine vasotocin on the estrogen-induced surge of LH and FSH in adult ovariectomized rats. *Life Sci.* 23: 1035-1040.

Blask, D.E., M.K. Vaughan, T.H. Champney, L.Y. Johnson, G.M. Vaughan, R.A. Becker and R.J. Reiter. 1984. Opioid and dopamine involvement in prolactin release induced by arginine vasotocin and vasopressin in the male rat. *Neuroendocrinology* 38: 56-61.

Boyd, A.E., III, E. Spencer, I.M.D. Jackson and S. Reichlin. 1976. Prolactin-releasing factor (PRF) in porcine hypothalamic extract distinct from TRH. *Endocrinology* 99: 861-871.

Caldwell, J.D., E.R. Greer, M.F. Johnson, A.J. Prange, Jr. and C.A. Pederson. 1986. Oxytocin and vasopressin immunoreactivity in hypothalamic and extrahypothalamic sites in late pregnant and postpartum rats. *Neuroendocrinology* 46: 39-47.

Cheesman, D.W., R.B. Osland and P.H. Forsham. 1977. Suppression of the preovulatory surge of luteinizing hormone and subsequent ovulation in the rat by arginine vasotocin. *Endocrinology* 101: 1194-1202.

Cheung, C.Y. 1983. Prolactin suppresses luteinizing hormone secretion and pituitary responsiveness to luteinizing hormone-releasing hormone by a direct action at the anterior pituitary. *Endocrinology* 113: 632-638.

del Pozo, E., J. Kleinstein, R.B. del Re, F. Derrer and J. Martin-Perez. 1980. Failure of oxytocin and lysine-vasopressin to stimulate prolactin release in humans. *Horm. Metab. Res.* 12: 26-28.

Diczfalusy, E. 1968. Mode of action of contraceptive drugs. *Am. J. Obstet. Gynecol.* 100: 136-163.

Dierickx, K. 1980. Immunocytochemical localization of the vertebrate cyclic nonapeptide neurohypophysial hormones and neurophysins. *Int. Rev. Cytol.* 62: 119-185.

Drummond, W.H., A.M. Rudolph, L.C. Keil, P.D. Gluckman, A.A. MacDonald and M.A. Heymann. 1980. Arginine vasopressin and prolactin after hemorrhage and the fetal lamb. *Am. J. Physiol.* 238: E214-E219.

Dufau, M.L. and K.J. Catt. 1978. Gonadotropin receptors and regulation of steroidogenesis in the testis and ovary. *Vitam. Horm.* 36: 461-592.

Evans, J.J. and K.J. Catt. 1989. Gonadotropin-releasing activity of neurohypophysial hormones: II. The pituitary oxytocin receptor mediating gonadotropin release differs from that of corticotrophs. *J. Endocrinol.* 122: 107-116.

Evans, J.J., G. Robinson and K.J. Catt. 1989. Gonadotrophin-releasing activity of neurohypophysial hormones. I. Potential for modulation of pituitary hormone secretion in rats. *J. Endocrinol.* 122: 99-106.

Fagin, K.D. and J.D. Neill. 1982. Involvement of the neurointermediate lobe of the pituitary gland in the secretion of prolactin and luteinizing hormone in the rat. *Life Sci.* 30: 1135-1141.

Falconer, J., M.D. Mitchell, L.A. Mountford and J.S. Robinson. 1980. Plasma oxytocin concentrations during the menstrual cycle in the Rhesus monkey, Macaca mulatta. *J. Reprod. Fertil.* 59: 69-72.

Franci, C.F., J. Anselmo-Franci, G.P. Kozlowski and S.M. McCann. 1988. Effects of vasopressin or oxytocin antiserum on anterior pituitary secretion in the rat. *Soc. Neurosci. Abstr.* 14: 629.

Froehlich, J.C., M.A. Neill and N. Ben-Jonathan. 1985. Interaction between the posterior pituitary and LHRH in the control of LH secretion. *Peptides* 6(1): 127-131.

Gambacciani, M., S.S.C. Yen and D.D. Rasmussen. 1986. GnRH release from the mediobasal hypothalamus. In vitro regulation by oxytocin. *Neuroendocrinology* 42: 181-183.

Gartan, E., E. Cobo and M. Mizrachi. 1964. Evidence for the differential secretion of oxytocin and vasopressin in man. *J. Clin. Invest.* 43: 2310-2322.

Gibbs, D. 1984. High concentrations of oxytocin in hypophysial portal plasma. *Endocrinology* 114: 1216-1218.

Grosvenor, C.E., S.-W. Shyr, G.T. Goodman and F. Mena. 1986. Comparison of plasma profiles of oxytocin and prolactin following suckling in rat. *Neuroendocrinology* 43: 679-685.

Hanew, K., M. Shiino and E.G. Rennels. 1980. Effect of indoles, AVT, oxytocin, and AVP on prolactin secretion in rat pituitary clonal (2B8) cells (40858). *Proc. Soc. Expt'l. Biol. Med.* 164: 257-261.

Harms, P.G. and S.R. Ojeda. 1974. A rapid and simple procedure for chronic cannulation of the rat jugular vein. *J. Appl. Physiol.* 36: 391-392.

Hou-Yu, A., A.T. Lamme, E.A. Zimmerman and A.J. Silverman. 1986. Comparative distribution of vasopressin and oxytocin neurons in the rat brain using a double-label procedure. *Neuroendocrinology* 44: 235-246.

Hyde, J.F. and N. Ben-Jonathan. 1988. Characterization of prolactin-releasing factor in the rat posterior pituitary. *Endocrinology* 122: 2533-2539.

Illig, R., T. Torresani, H. Bucher, M. Zachmann and A. Prader. 1980. Effect of intranasal LHRH therapy on plasma LH, FSH and testosterone and relation to clinical results in prepubertal boys with cryptorchidism. *Clin. Endocrinol. (Oxf)* 12: 91-97.

Jirikowski, G.F., J.D. Caldwell, W.E. Stumpf and C.A. Pedersen. 1986. Effects of estradiol on hypothalamic oxytocinergic neurons. *Soc. Neurosci. Abstr.* 12: 388.

Jirikowski, G.F., J.D. Caldwell, C.A. Pedersen and W.E. Stumpf. 1988. Estradiol influences oxytocin-immunoreactive brain systems. *Neuroscience* 25: 237-248.

Johnston, C.A., K.D. Fagin, R.H. Alper and A. Negro-Vilar. 1986. Prolactin release after 5-hydroxytryptophan treatment requires an intact neurointermediate pituitary lobe. *Endocrinology* 118: 805-810.

Johnston, C.A., K.D. Fagin and A. Negro-Vilar. 1988. Neuropeptide mediated hormone secretion: role of neurointermediate and anterior pituitary lobe interactions. In *Horizons in Endocrinology. Vol. 52*, eds. Maggi, M. and C.A. Johnston, 307-317. Serono Symposia Publications, Raven Press, New York.

Johnston, C.A. and A. Negro-Vilar. 1988. Role of oxytocin on prolactin secretion during proestrus and in different physiological or pharmacological paradigms. *Endocrinology* 122: 341-350.

Johnston, C.A., K.D. Fagin and A. Negro-Vilar. 1990a. Differential effect of neurointermediate lobectomy on central oxytocin and vasopressin. *Neurosci. Lett.* 113: 101-106.

Johnston, C.A., F. Lopez, W.K. Samson and A. Negro-Vilar. 1990b. Physiologically important role for central oxytocin in the preovulatory release of luteinizing hormone. *Neurosci. Lett.* 120: 256-258.

Johnston, C.A., J.B. Gelineau-vanWaes and M.V. Templin. 1990c. Oxytocin mediates the angiotensin-induced release of LH. *Soc. Neurosci. Abstr.* 16: 394.

Knobil, E. 1980. The neuroendocrine control of the menstrual cycle. *Recent Prog. Horm. Res.* 35: 53-88.

König, J.F.R. and R.A. Klippel. 1967. *The Rat Brain: A Stereotaxic Atlas.* Krieger, Huntington, New York.

Koyama, T. and N. Hagino. 1983. The effect of vasopressin on LH release in baboons. *Horm. Metab. Res.* 15: 184-186.

Kulin, H.E. and E.O. Reiter. 1973. Gonadotropins during childhood and adolescence: A review. *Pediatrics* 51: 260-271.

Lamberts, S.W.J., M. de Quijada and T.J. Visser. 1981. Regulation of prolactin secretion in patients with Cushing's disease. *Neuroendocrinology* 32: 150-154.

Leyendecker, G., L. Wildt and M. Hansmann. 1980. Pregnancies following chronic intermittent (pulsatile) administration of Gn-RH by means of a portable pump ('Zyklomat')--a new approach to the treatment of infertility in hypothalamic amenorrhea. *J. Clin. Endocrinol. Metab.* 51: 1214-1216.

Liu, J.H., R. Durfee, K. Muse and S.S.C. Yen. 1983. Induction of multiple ovulation by pulsatile administration of gonadotropin-releasing hormone. *Fertil. Steril.* 40: 18-22.

Lumpkin, M.D., W.K. Samson and S.M. McCann. 1983. Hypothalamic and pituitary sites of action of oxytocin to alter prolactin secretion in the rat. *Endocrinology* 112: 1711-1717.

Marshall, J.C. and R.P. Kelch. 1979. Low dose pulsatile gonadotropin-releasing hormone in anorexia nervosa: A model of human pubertal development. *J. Clin. Endocrinol. Metab.* 49: 712-718.

McCann, S.M. and S.R. Ojeda. 1976. Synaptic transmitters involved in the release of hypothalamic releasing and inhibiting hormones. In *Reviews of Neuroscience, Vol. 2*, eds. S. Ehrenpreis and I.J. Kopin, 91-148. Raven Press, New York, New York.

McCann, S.M. 1982. The role of brain peptides in the control of anterior pituitary hormone secretion. In *Neuroendocrine Perspectives, Vol. 1*, eds. E.E. Muller and R.M. MacLeod, 1-22. Elsevier Biomedical Press, Amsterdam.

McCann, S.M., M.D. Lumpkin, H. Mizunuma, O. Khorramy, A. Ottlecz and W.K. Samson. 1984. Peptidergic and dopaminergic control of prolactin release. *TINS* 7: 127-131.

Meldrum, D.R., R.J. Chang, J. Lu, W. Vale, J. Rivier and H.L. Judd. 1982. "Medical oophorectomy" using a long-acting GnRH agonist--a possible new approach to the treatment of endometriosis. *J. Clin. Endocrinol. Metab.* 54: 1081-1083.

Mitchell, M.D., P.J. Haynes, A.B.M. Anderson and A.C. Turbull. 1980. Oxytocin in human ovulation. *Lancet* 2: 704.

Murai, I. and N. Ben-Jonathan. 1987. Posterior pituitary lobectomy abolishes the suckling-induced rise in prolactin (PRL): Evidence for a PRL-releasing factor in the posterior pituitary. *Endocrinology* 121: 205-211.

Murai, I., S. Reichlin and N. Ben-Jonathan. 1989. The peak phase of the proestrous prolactin surge is blocked by either posterior pituitary lobectomy or antisera to vasoactive intestinal peptide. *Endocrinology* 124: 1050-1055.

Nagai, T., T. Makino, A. Nakayama, K. Nakazawa, H. Suzuki and R. Iizuke. 1983. Study on role of oxytocin in lactating rats by passive immunization. *Asia-Oceania J. Obstet. Gynecol.* 1: 109-115.

Negro-Vilar, A. and S.R. Ojeda. 1981. Hypophysiotropic hormones of the hypothalamus. In *Endocrine Physiology II, Vol. 24*, International Review of Physiology, ed. S.M. McCann, 97-156. University Park Press, Baltimore, Maryland.

Rhodes, C.H., J.I. Morrell and D.W. Pfaff. 1981. Changes in oxytocin content in the magnocellular neurons of the rat hypothalamus following water deprivation or estrogen treatment. Quantitative immunohistological studies. *Cell Tissue Res.* 216: 47-55.

Rhodes, C.H., J.I. Morrell and D.W. Pfaff. 1982. Estrogen-concentrating neurophysin-containing hypothalamic magnocellular neurons in the vasopressin-deficient (Brattleboro) rat: A study combining steroid autoradiography and immunocytochemistry. *J. Neurosci.* 2: 1718-1724.

Richards, J.S. 1980. Maturation of ovarian follicles: Actions and interactions of pituitary and ovarian hormones on follicular cell differentiation. *Physiol. Rev.* 60: 51-89.

Robinson, A.G., M. Ferin and E.A. Zimmerman. 1976. Plasma neurophysin levels in monkeys: Emphasis on the hypothalamic response to estrogen and ovarian events. *Endocrinology* 98: 468-475.

Robinson, G., J.J. Evans and M.E. Forster. 1985. Oxytocin can affect follicular development in the adult mouse. *Acta Endocrinol.* 108: 273-276.

Robinson, G. and J.J. Evans. 1990. Oxytocin has a role in gonadotropin regulation in rats. *J. Endocrinol.* 125: 425-432.

Salisbury, R.L., R.J. Krieg, Jr. and H.R. Seibel. 1980. Effects of arginine vasotocin, oxytocin, and arginine vasopressin on steroid-induced surges of luteinizing hormone and prolactin in ovariectomized rats. *Acta Endocrinol. (Copenh.)* 94: 166-173.

Samson, W.K., J.K. McDonald and M.D. Lumpkin. 1985. Naloxone-induced dissociation of oxytocin and prolactin releases. *Neuroendocrinology* 40: 68-71.

Samson, W.K., M.D. Lumpkin and S.M. McCann. 1986. Evidence for a physiological role for oxytocin in the control of prolactin secretion. *Endocrinology* 119: 554-560.

Samson, W.K., R. Bianchi, R.J. Mogg, J. Rivier, W. Vale and P. Melin. 1989. Oxytocin mediates the hypothalamic action of vasoactive intestinal peptide to stimulate prolactin secretion. *Endocrinology* 124: 812-819.

Samson, W.K., L. Martin, R.J. Mogg and R.J. Fulton. 1990. A non-oxytocinergic prolactin releasing factor and a non-dopaminergic prolactin inhibiting factor in bovine neurointermediate lobe extracts: In vivo and in vitro studies. *Endocrinology* 126: 1610-1617.

Sar, M. and W.E. Stumpf. 1980. Simultaneous localization of [3H] estradiol and neurophysin I or arginine vasopressin in hypothalamic neurons demonstrated by a combined technique of dry-mount autoradiography and immunohistochemistry. *Neurosci. Lett.* 17: 179-184.

Sarkar, D.K. and D.M. Gibbs. 1984. Cyclic variation of oxytocin in the blood of pituitary portal vessels of rats. *Neuroendocrinology* 39: 481-483.

Sarkar, D.K. 1988. Immunoneutralization of oxytocin attenuates preovulatory prolactin secretion during proestrus in the rat. *Neuroendocrinology* 48: 214-216.

Sawchenko, P.E. and L.W. Swanson. 1982. Immunohistochemical identification of neurons in the paraventricular nucleus of the hypothalamus that project to the medulla or to the spinal cord in the rat. *J. Comp. Neurol.* 205: 260-272.

Sawchenko, P.E. and L.W. Swanson. 1985. Relationship of oxytocin pathways to the control of neuroendocrine and autonomic function. In *Oxytocin. Clinical and Laboratory Studies.* International Congress Series 666, ed. J.A. Amico and A.G. Robinson, 87-103. Elsevier Science Publishers B.V. (Biomedical Division), Amsterdam.

Shin, S.H. 1982. Vasopressin has a direct effect on prolactin release in male rats. *Neuroendocrinology* 34: 55-58.

Skarin, G., S.J. Nillius, L. Wibell and L. Wide. 1982. Chronic pulsatile low dose GnRH therapy for induction of testosterone production and spermatogenesis in a man with secondary hypogonadotropic hypogonadism. *J. Clin. Endocrinol. Metab.* 55: 723-726.

Sokol, H.W., E.A. Zimmerman, W.H. Sawyer and A.G. Robinson. 1976. The hypothalamic-neurohypophysial system of the rat: Localization and quantitation of neurophysin by light microscopic immunocytochemistry in normal rats and in Brattleboro rats deficient in vasopressin and a neurophysin. *Endocrinology* 98: 1176-1188.

Summy-Long, J.Y., D.S. Miller, L.M. Rosella-Dampman, R.D. Hartman and S.E. Emmert. 1984. A functional role for opioid peptides in the differential secretion of vasopressin and oxytocin. *Brain Res.* 309: 3623-3660.

Summy-Long, J.Y., L.M. Rosella-Dampman, G.L. McLemore and E. Koehler. 1990. Kappa opiate receptors inhibit release of oxytocin from the magnocellular system during dehydration. *Neuroendocrinology* 51: 376-384.

Swaab, D.F., C.W. Pool and F. Nijveldt. 1975. Immunofluorescence of vasopressin and oxytocin in the rat hypothalamo-neurohypophysial system. *J. Neural Transm.* 36: 195-215.

Swanson, L.W., P.E. Sawchenko, A. Berod, B.K. Hartman, K.B. Helle and D.E. VanOrden. 1981. An immunohistochemical study of the organization of catecholaminergic cells and terminal fields in the paraventricular and supraoptic nuclei of the hypothalamus. *J. Comp. Neurol.* 196: 271-285.

Szabo, M. and L.A. Frohman. 1976. Dissociation of prolactin-releasing activity from thyrotropin-releasing hormone in porcine stalk median eminence. *Endocrinology* 98: 1451-1459.

Toran-Allerand, C.D. 1980. Sex steroids and the development of the newborn mouse hypothalamus and preoptic area in vitro. II. Morphological correlates and hormonal specificity. *Brain Res.* 189: 413-427.

Turkelson, C.M., A. Arimura, M.D. Culler, J.B. Fishback, K. Groot, M. Kanda, M. Luciano, C.R. Thomas, D. Chang, J.K. Chang and M. Shimizu. 1981. In vivo and in vitro release of ACTH by synthetic CRF. *Peptides* 2: 425-429.

Unger, H. and H. Schwarzenberg. 1970. Untersuchungen uber vorkommen und bedeutung von vasopressin und oxytocin im liquor cerebrospinalis und blut fur nervose funktionen. *Acta Biol. Med. Germ.* 25: 267-280.

Vaitukaitis, J.L., G.T. Ross, G.D. Braunstein and P.L. Rayford. 1976. Gonadotropins and their subunits: Basic and clinical studies. *Recent Prog. Horm. Res.* 32: 289-331.

Valverde-R.C., V. Chieffo and S. Reichlin. 1972. Prolactin-releasing factor in porcine and rat hypothalamic tissue. *Endocrinology* 91: 982-993.

Vaughan, M.K., D.E. Blask, L.Y. Johnson and R.J. Reiter. 1975. Prolactin-releasing activity of arginine vasotocin in vitro. *Horm. Res.* 6: 342-350.

Vaughan, M.K., C. Trakulrungsi, L.J. Petterborg, L.Y. Johnson, D.E. Blask, W. Trakulrungsi and R.J. Reiter. 1979a. Interaction of luteinizing hormone-releasing hormone, cyproterone acetate and arginine vasotocin on plasma levels of luteinizing hormone in intact and castrated adult male rats. *Mol. Cell. Endocrinol.* 14: 59-71.

Vaughan, M.K., D.E. Blask, L.Y. Johnson and R.J. Reiter. 1979b. The effect of subcutaneous injections of melatonin, arginine vasotocin, and related peptides on pituitary and plasma levels of luteinizing hormone, follicle-stimulating hormone, and prolactin in castrated adult male rats. *Endocrinology* 104: 212-217.

Wakerley, J.B., R.E.J. Dyball and D.W. Lincoln. 1973. Milk ejection in the rat: The result of a selective release of oxytocin. *J. Endocrinol.* 57: 557-558.

Weitzman, R.E., T.H. Glatz and D.A. Fisher. 1978. The effect of hemorrhage and hypertonic saline upon plasma oxytocin and arginine vasopressin in conscious dogs. *Endocrinology* 103: 2154-2160.

Zimmerman, E.A., G. Nilaver, A. Itou-Yi and A.J. Silverman. 1984. Vasopressinergic and oxytocinergic pathways in the central nervous system. *Fed. Proc.* 43: 91-96.

Zucker, I.H., A.J. Gorman, K.G. Cornish, L.J. Huffman and J.P. Gilmore. 1983. Influence of left ventricular receptor stimulation on plasma vasopressin in conscious dogs. *Am. J. Physiol.* 245: R792-R799.

VASOACTIVE INTESTINAL PEPTIDE: A NEURAL MODULATOR OF ENDOCRINE FUNCTION

Willis K. Samson
Department of Anatomy and Neurobiology
University of Missouri School of Medicine
Columbia, Missouri
and
Marc E. Freeman
Department of Biological Science
Florida State University
Tallahassee, Florida

I.   Vasoactive Intestinal Peptide (VIP) and Structurally Related Peptides

     A.   Characterization

     B.   Sites of Production and Release

II.   Neuroendocrine Actions of Vasoactive Intestinal Peptide

     A.   Presence in Hypophyseal Portal Plasma

     B.   VIP Receptors in the Anterior Pituitary Gland

     C.   Releasing Factor Activity of VIP *in vivo*

     D.   Releasing Factor Activity of VIP *in vivo*

III.   A Dual Site Hypothesis for the Reproductive Neuroendocrine Effects of VI

Introduction: Vasoactive Intestinal Peptide and Structurally Related Peptides

As early as 1902 (Bayliss and Starling), descriptions of acid-extractable substances in gut tissue possessing a wide spectra of bioactivity hinted at the presence of peptides now known to comprise the secretin/vasoactive intestinal peptide (VIP) family of hormones. Originally called the Secretin Family, perhaps because the first of the two peptides characterized was the 27-amino acid secretin (Jorpes *et al.*, 1962), it has become popular to refer to this group of structural homologs as the VIP Family. No doubt this reflects the explosion of literature on VIP which occurred from 1975 to the present (Said and Mutt, 1988).

Characterization

Porcine intestinal tissue proved to be an excellent source for the isolation of these structurally related peptides (Jorpes *et al.*, 1962; Brown and Dryburgh, 1971; Said and Mutt, 1970, 1972).   The sequencing of peptide histidine isoleucine-27 (PHI-27) in 1981 (Tatemoto and Mutt) completed the structural characterization of this family which includes, in addition to secretin and VIP, PHI-27 (PHM-27 in humans; Itoh *et al.*, 1983; Nishizawa *et al.*, 1985), gastric inhibitory polypeptide (GIP; Brown and Dryburgh, 1971), and the previously identified hormone glucagon (Bromer *et al.*, 1957).   Some recent reports have included human growth hormone releasing factor (hpGRF) in this family due to the relative homology of that peptide's N-terminus to VIP (Itoh *et al.*, 1983); however, the level of homology within the first 28 amino acids of the N-terminus of hpGRF is not striking when compared to GIP (4/28), glucagon (4/28), or secretin (5/28). Indeed, considerable homology exists between hpGRF and VIP (9/28) or PHI-27 (11/28), yet it clearly has bioactivity not expressed by any of the VIP family proper. The strongest homology within the family exists between VIP and PHI-27, peptides derived from a common precursor (Itoh *et al.*, 1983; Nishizawa *et al.*, 1985), since 13 of the 27 amino acids in PHI are positioned similarly to those in VIP (Table 1).   It is not surprising then that many of VIP's actions are mimicked by PHI (Bataille et al., 1981; Samson *et al.*, 1983; Werner et al., 1983).

Sites of Production and Release

Numerous, extensive reviews on this subject have been published (Larsson, 1982;   Hakanson *et al.*, 1982; Hökfelt *et al.*, 1982; Polak and Bloom, 1982;

Table 1. Comparison of the Amino Acid Composition of Members of the Secretion/VIP Family of Hormones.

PHI	HADGVFTSDFSRLLGQLSAKKYLESL I[a]
VIP	HSDAVFTDNYTRLRKQMAVKKYLNSIL N[a]
Secretin	HSDGTFTSELSRLRDSARLQRLLQGL V[a]
Glucagon	HSGGTFTSDYSKYLDSRRAQDFVQWLMNT
GIP	YAEGTFISDYSIAMDKIRQQDFVNWLLAQKGKKSDWKHNITQ
GRF	YAOAIFTNSYCKVLGQLSARKLLGDIM

A, alanine; D, aspartic acid; E, glutamic acid; F, phenylalanine; G, glycine; H, histidine; I, isoleucine; K, lycine; L, leucine; M, methionine; N, asparagine; Q, glutamine; R, arginine; S, serine; T, threonine; V, valine; W, tryptophane, Y, tyrosine; GIP, gastric inhibitory peptide. [a] = Amidated COOH terminus.

Samson, 1982; Beinfeld *et al.*, 1988; Gozes, 1988; Yamagami *et al.*, 1988). In the interest of brevity, only the distribution of VIP is discussed here.

Gastrointestinal and Other Tissues. Although it was originally thought that VIP was produced in endocrine cells of the gut, numerous immunocytochemical studies have now established the presence of VIP immunoreactivity in neural elements of all tissues in which the peptide has been detected. VIP-containing neurons in the large and small intestines appear to have diverse targets, including: neurons of the enteric plexi, vascular elements, and smooth muscle cells. Secretory motor fibers containing VIP are present in the pancreas. The anatomical distribution of VIP mirrors in its extent the wide variety of activities of the peptide within the gut. VIP is present as well in the urogenital tract, endocrine tissues and all exocrine glands that have been studied. Additionally, VIP neurons are present in the respiratory tract, peripheral ganglia, and blood vessels. These VIP elements have been suggested to play important neuromodulatory, vasomotor, sudomotor and secretory functions in the tissues described (Said and Mutt, 1988).

Central Nervous System. The highest concentrations of immunoreactive VIP in the CNS are found in the cerebral cortex (Samson *et al.*, 1979; Loren *et al.*, 1979). Both neo- and allocortex contain numerous VIP-positive neurons, most localized in layers II through IV (Hökfelt *et al.*, 1982; Marley and Emson, 1982). Abundant evidence exists for its role as a neuromodulatory agent within the cortex (Magistretti *et al.*, 1988; Marley and Emson, 1982) and brain, in general, on the basis of its presence in secretory neuronal structures, the existence of specific receptors on other neurons (Robberecht *et al.*, 1982; Besson, 1988), the presence of enzyme systems to terminate its actions (Strauss *et al.*, 1982), its ability to affect neural transmission (Haskins *et al.*, 1982; Phillis, 1982), to be released by depolarizing stimuli (Giachetti *et al.*, 1977, Emson *et al.*, 1978; Magistretti *et al.*, 1988), and to affect the biochemical activity of neurons within its receptive field (Magistretti *et al.*, 1988). In addition to its colocalization with acetylcholine (ACH) in cortical neurons (Eckenstein and Baughman, 1984), VIP potentiates phosphoinositide turnover in response to ACH (Raiteri *et al.*, 1987), suggesting a functional interaction of these neuroactive substances in the control of inter- and intracortical communication and organization.

Limbic structures contain the next highest concentrations of VIP (Samson, 1982) with the hippocampus, amygdala and hypothalamus all possessing abundant immunoreactivity. These semiquantitative measurements (Loren *et al.*, 1979; Samson *et al.*, 1979), obtained by microdissection and radioimmunoassay (RIA), are mirrored by results from immunocytochemical mapping studies which demonstrated (Loren *et al.*, 1979; Sims *et al.*, 1980; Hökfelt *et al.*, 1982) VIP-positive cell bodies in the subiculum, in the hippocampus close to the pyramidal cell layer in CA1 and CA3 regions, and in the dentate gyrus. Amygdaloid nuclei contain VIP-positive perikarya whose axons appear to exit via the stria terminalis to innervate the bed nucleus and hypothalamus (Loren *et al.*, 1979; Roberts *et al.*, 1980; Sims *et al.*, 1980). Within the hypothalamus the major aggregation of VIP-positive cell bodies is in the suprachiasmatic nucleus. Axons from these perikarya appear to innervate other hypothalamic structures including the paraventricular, dorsomedial, and ventromedial nuclei (Sims *et al.*, 1980). While levels of the peptide are low in the median eminence (ME), immunoreactive axon terminals have been identified in both the internal and external layers (Loren *et al.*, 1979) and more recently a second population of VIP-positive neurons originating in the parvocellular paraventricular nucleus and projecting to the ME has been revealed by immunocytochemistry (Hökfelt *et al.*, 1983; Mezey and Kiss, 1985).

Neuroendocrine Actions of Vasoactive Intestinal Peptide

The localization of VIP in brain structures known to be important in the control of anterior pituitary function suggested, to several groups, possible neuroendocrine actions of the peptide. The very first evidence for such an action (Kato *et al.*, 1978; Ruberg *et al.*, 1978) indicated a potential prolactin (PRL) releasing activity for VIP; it became clear within the next two years that VIP could be considered to have both pituitary and hypothalamic actions related to the control of PRL secretion (Vijayan *et al.*, 1979).

Presence in Hypophyseal Portal Plasma

For VIP to be considered a physiologically relevant PRL-releasing factor (PRF), several criteria had to be satisfied, not the least of which was the presence of abundant levels in the portal blood. Said and Porter (1979) detected VIP levels in hypophyseal portal plasma which were in excess of 40 times greater than those present in the peripheral circulation. Shimatsu *et al.* (1981, 1983) similarly described the gradient of VIP in portal versus peripheral plasma and reported that both serotonin (Shimatsu *et al.*, 1982) and prostaglandin D_2 (Shimatsu *et al.*, 1984) would stimulate its release into the portal circulation. Recently, Sarkar (1989) demonstrated that PRL could feed back at the hypothalamic level to regulate the release of VIP into the portal vessels.

VIP Receptors in the Anterior Pituitary Gland

The presence of VIP receptors in the anterior pituitary was deduced from the evidence that the peptide could not only stimulate the secretion of PRL, but also stimulate pituitary adenylate cyclase activity (Borghi *et al.*, 1979; Robberecht *et al.*, 1982; Rosselin et al., 1988). As in the brain, two classes of VIP receptors, based on binding affinities, were described (Besson, 1988): The high-affinity and low-capacity sites had an apparent dissociation constant (K_d) of about 1 nM, and a maximal binding capacity (B_{max}) of approximately 40 fmol/mg protein; the low-affinity displayed a K_d of 68 nM and B_{max} of 700 fmol/mg protein.

Releasing Factor Activity of VIP in vitro

The original descriptions of the PRF activity (Kato *et al.*, 1978; Ruberg *et al.*, 1978) of VIP have been confirmed by many laboratories (Samson *et al.*, 1980; Enjalbert *et al.*, 1980; Rotsztejn *et al.*, 1980; Nicosia *et al.*, 1982). In partially purified populations of lactotrophs (Rotsztejn *et al.*, 1980), VIP, in doses ranging from 0.1-100 nM, significantly and in a dose-related fashion stimulated PRL release with an ED_{50} of 2 nM. VIP has not been demonstrated to synergize with other PRFs and it tends to reverse the inhibitory effect of dopamine, although this action is not mediated via an interaction with the dopamine receptor (Enjalbert *et al.*, 1980). The effect is specific for PRL since the release of luteinizing hormone (LH), follicle stimulating hormone, growth hormone and thyroid stimulating hormone were not affected by the peptide (Samson *et al.*, 1980, Nicosia *et al.*, 1982). There is some controversy over possible direct effects of VIP on corticotrophs since, in dispersed cells, Rotsztejn *et al.* (1980) failed to observe any effect; however, others (Nicosia *et al.*, 1982) have observed significant stimulation of ACTH release from human pituitary adenomas. We have seen some effect of VIP on ACTH release from dispersed cells *in vitro* (Samson, unpublished observations), but the dose required exceeds that of corticotropin releasing factor (CRF) by four to five orders of magnitude.

Releasing Factor Activity of VIP in vivo

In vivo studies quickly identified the hypothalamus as well as the pituitary as a possible VIP site of action to stimulate PRL secretion. Imura's group (Kato *et al.*, 1978) and ours (Vijayan *et al.*, 1979) demonstrated that high doses of the peptide given i.v. would significantly stimulate PRL release in the conscious rat. Their group employed doses of 1 and 10 µg/100 g body weight and observed transient stimulation similar to that seen by us with 1 µg per rat. We were concerned that the PRF activity of these supraphysiologic doses (circulating VIP levels in the rat are in the range of 10 to 100 pg/ml plasma) of the peptide was not entirely due to a direct pituitary action, but, at least in part, to a generalized stress evoked by its profound hypotensive effect and the reflex sympathetic activation that ensued. Indeed, our dose lowered resting mean arterial blood pressure from 104 mmHg prior to infusion, to 82 and 86 mmHg 20 sec and 2 min later. There followed a baroreceptor-mediated reflex tachycardia which persisted for 15 min despite blood pressure normalizing during the 5 to 10 min after VIP administration. Still the *in vitro* data suggested a direct effect on the lactotroph. Both groups also reported a central action of the peptide to release PRL. Kato *et al.* (1978) observed the PRF action after administration of 5.0, 1.0, and 0.2 µg VIP into the lateral cerebroventricle. In our hands (Vijayan *et al.*, 1979) doses of 40 and 100 ng VIP given into the third cerebroventricle significantly elevated PRL in peripheral plasma. It was hypothesized that either the centrally administered VIP acted after leakage into the portal circulation or that the peptide was effectively modulating the release of hypothalamic PRL releasing and inhibiting factors. It would be ten years before the answer to that controversy was obtained (Arey and Freeman, 1989; Samson *et al.*, 1989).

The structural homologs PHI-27 and secretin are also active *in vivo*. PHI-27 is cosecreted with VIP into the portal vessels following serotonin treatment, and passive immunization with antisera directed against either VIP or PHI blunts the PRL rise seen in response to 5-hydroxy-1-tryptophan (Kaji *et al.*, 1985). Lateral ventricular infusion of PHI-27 results, as was the case with VIP, in a significant elevation of PRL levels, further suggesting a hypothalamic PRF action of these peptides (Kato *et al.*, 1985). In our hands, secretin, when administered i.v. to conscious rats, significantly affected PRL release only at the highest dose tested (10 μg) and then only doubled levels present in controls 60 min after injection (Samson *et al.*, 1984). Also, unlike the effects we had previously observed following VIP administration (Vijayan *et al.*, 1979), when injected into the third cerebroventricle, secretin significantly inhibited PRL secretion. Thus, although its action at the pituitary level under *in vivo* and *in vitro* conditions was stimulatory (albeit only at higher doses than required for VIP in both paradigms), the central action was the opposite.

In our original *in vivo* study (Vijayan *et al.*, 1979) central administration of VIP also resulted in a significant elevation of plasma LH levels. We later determined that VIP exerts this effect at the hypothalamic level by stimulating the release of LH-releasing hormone (LHRH) from nerve terminals in the ME (Samson *et al.*, 1981). The effect of VIP on LHRH release from synaptosomal preparations of ME tissue was dose-related (Fig. 1) and specific for the peptide

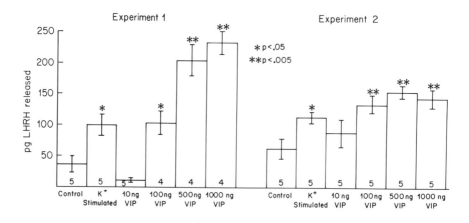

Fig. 1. Mean (± SEM) luteinizing hormone releasing hormone (LHRH) release from synaptosomal preparations of rat ME tissue in the presence of control incubation medium alone or medium containing 60 mM potassium or synthetic vasoactive intestinal peptide (VIP). Reprinted with the permission of Elsevier Scientific Publishers from the original manuscript (Samson *et al.*, 1981).

since the homolog glucagon was ineffective. These findings were corroborated by recent studies of Ohtsuka *et al.* (1988), in which VIP stimulated LHRH release from hypothalamic fragments perifused in series with anterior pituitaries. The effect was dose-related and relatively long-lasting, remaining significant for the 90 min sampling period after cessation of VIP exposure. Taken together, these data would seem to indicate that VIP plays a neuromodulatory role in the control of LH secretion by stimulating the release of LHRH into the portal vessels. Contradictory evidence was presented by Alexander *et al.* (1985) in studies employing long infusions of VIP into the third ventricle which lowered the mean LH levels in ovariectomized rats and significantly reduced LH pulse frequency, but not amplitude, in their animals. Recent results (Murai *et al.*, 1989) using passive immunoneutralization with antiVIP serum suggested no effect of VIP on LH secretion on proestrus, while our experience with VIP antagonists (Fig. 2) has been that pharmacologic blockade of VIP's action in steroid-primed, ovariectomized rats blocks not only the PRL but also the LH surge observed in controls. The long-lasting effect of the VIP antagonist [D-4-Cl-Phe^6,Leu^{17}]VIP (Pandol *et al.*, 1986) on both hormones suggests that the prolonged action of VIP on LHRH release, observed by Ohtsuka *et al.* (1988), has physiological significance.

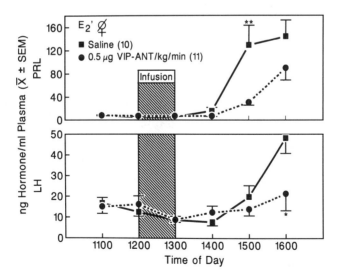

Fig. 2. Plasma prolactin (PRL) and luteinizing hormone (LH) levels in ovariectomized, estrogen-primed rats following i.v. infusion of 1 ml isotonic saline alone or containing the VIP antagonist, (4C^1-d-Phe^6,Leu^{17})VIP. *p < 0.05, **p, 0.01, versus control.

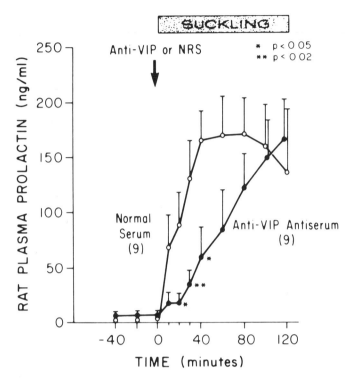

Fig. 3. Plasma prolactin levels in lactating rats exposed after three h of pup denial to a suckling stimulus following i.v. infusion of normal rabbit serum (NRS) or antiserum to VIP. Reprinted from the original manuscript (Abe *et al.*, 1985) with the permission of the author, S. Reichlin (Tufts University), and the Williams and Wilkins Publishing Company.

In terms of the physiological relevance of the PRF activity of VIP, there appears to be a consensus that VIP is at least one of a number of peptides which exert significant control over PRL secretion in the rat.   VIP antiserum infusionprevents the secretion of PRL in response to serotonin (Kaji *et al.*, 1985; Kato *et al.*, 1985), and i.v. infusions of antiserum to VIP significantly reduce the PRL response to suckling (Fig. 3) or suckling followed by ether exposure (Abe *et al.*, 1985).   Similarly the proestrous secretion of PRL is blunted by antiVIP (Murai *et al.*, 1989).   In our hands, i.v. infusion of a VIP antagonist not only altered steroid-induced PRL secretion but significantly reduced and delayed the secretory response to suckling (Samson *et al.*, 1989).   It was hypothesized in all of these reports that either the antiserum or the antagonist results were due primarily to prevention of a VIP action at the pituitary level.   Interestingly, the blockade induced by these maneuvers was incomplete, suggesting either the importance of dopamine withdrawal in the PRL secretory response, or the importance of other factors regulating PRL release.

It had been recognized for some time that one of the other putative PRFs of hypothalamic origin was thyrotropin releasing hormone (TRH). The tripeptide is active *in vitro* (Tashjian *et al.*, 1971) and *in vivo* (Rivier and Vale, 1974), and immunoneutralization of endogenous TRH results in a significant reduction and delay in the proestrous PRL surge (Horn *et al.*, 1983a). Another putative PRF is oxytocin (OT). The nonapeptide is present in hypophyseal portal blood in levels that exceed those in the periphery (Gibbs, 1984) and OT concentrations in portal plasma fluctuate in concert with prolactin secretion (Horn *et al.*, 1983b; Sarkar and Gibbs, 1984). Oxytocin stimulates PRL secretion when infused i.v. (Lumpkin *et al.*, 1983) at high doses, and exerts a significant, dose-related PRF action in cell culture (Samson *et al.*, 1986). Indeed, specific OT receptors are present in the pituitary (Antoni, 1986) and OT immunoreactivity has been detected on the plasma and secretory membranes of lactotrophs (Morel *et al.*, 1988). The physiological relevance of the PRF activity of OT was demonstrated in passive immunoneutralization studies (Samson *et al.*, 1986) which demonstrated that both the PRL surge to suckling stimulus (Fig. 4) and steroids could be

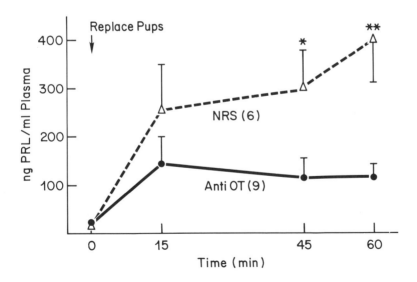

Fig. 4. Plasma prolactin levels in lactating rats exposed after 4 h of pup denial to a suckling stimulus following i.v. infusion of normal rabbit serum or antiserum to oxytocin. Reprinted from the original manuscript (Samson *et al.*, 1986) with the permission of the Williams and Wilkins Publishing Company.

significantly reduced by antiOT serum. Subsequent studies employing antagonists of OT supported those findings. Using the OT antagonist [deamino Cys^1,D-Trp^2,Val^4,Orn^8]OT (Melin *et al.*, 1986), we successfully reduced the PRL surge seen in ovariectomized steroid-primed rats (Samson *et al.*, 1989) and Johnston and Negro-Vilar (1988) obtained similar effects in cycling rats.

Finally, it is becoming increasingly clear that there exists at least one more physiologically relevant peptidergic PRF. Ben-Jonathan and colleagues have isolated a peptide from neurointermediate lobe extracts which possesses potent PRF activity *in vitro* (Hyde and Ben-Jonathan, 1988) and *in vivo* (Hyde, 1989). Removal of this tissue *in vivo* significantly attenuates the PRL surge to suckling (Murai and Ben-Jonathan, 1987) or 5-hydroxytryptophan administration (Johnston *et al.*, 1986). We too have been isolating this novel PRF from rat and cow neurointermediate lobes and have observed significant, dose-related, *in vivo* and *in vitro* bioactivity with chromatographically purified material which exceeds that for OT or VIP (Samson *et al.*, 1990; Samson and Mogg, 1990). These most recent findings suggest that, as in the case of the hypothalamic control of ACTH secretion (Vale *et al.*, 1983), there appear to be multiple releasers of PRL and only a well-coordinated interplay between a variety of stimulatory and inhibitory agents can eventuate the secretion of PRL seen in various physiologic states (Mogg and Samson, 1990).

A Dual-Site Hypothesis for the Reproductive Neuroendocrine Effects of VIP

As noted above, VIP has been shown to stimulate reproductive hormone secretion through neurohumoral and neuromodulatory mechanisms. Autocrine mechanisms have also been described whereby VIP produced within the pituitary regulates the secretion of PRL (Hagen *et al.*, 1986). While the classical neuroendocrine action of VIP, to stimulate PRL secretion, is now well established (Reichlin, 1988), there is abundant evidence that VIP may act as a neuromodulator which interacts with other hypothalamic PRFs. Effects of VIP on neural excitability have been demonstrated within the hypothalamus (Haskins *et al.*, 1982) and VIP immunopositive fibers originating in the suprachiasmatic nuclei (Sims *et al.*, 1980) terminate in the paraventricular nuclei adjacent to OT-positive cell bodies. Peripheral (Bardrum *et al.*, 1987) or central VIP administration (Samson *et al.*, 1989) results in the selective release of OT in a time course which mirrors that of the central effect on PRL secretion (Figs. 5 and 6). The central action of VIP to stimulate PRL secretion can be blocked by prior administration i.v. of either antiOT antiserum (Fig. 6) or an OT antagonist (Fig. 7). Prior i.v. administration of a selective VIP antagonist did not block the

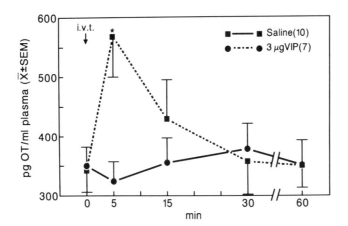

Fig. 5. Plasma oxytocin levels in ovariectomized rats before and after central administration of saline vehicle alone or containing vasoactive intestinal peptide intraventricular injection (i.c.v.). *p < 0.005. Reprinted from the original manuscript (Samson *et al.*, 1989) with the permission of the Williams and Wilkins Publishing Company.

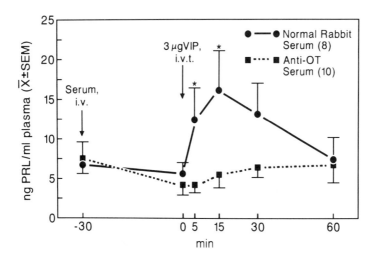

Fig. 6. Effects of pretreatment of ovariectomized rats with normal rabbit serum or antiserum to oxytocin on the prolactin response to central administration of vasoactive intestinal peptide. *p < 0.05. Reprinted from the original manuscript (Samson *et al.*, 1989) with the permission of the Williams and Wilkins Publishing Company.

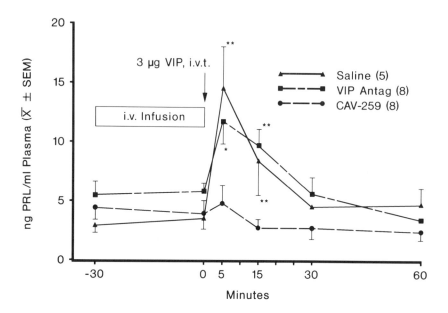

Fig. 7. Effects of pretreatment of ovariectomized rats by i.v. infusion of saline, or antagonist to oxytocin (CAV-259) or vasoactive intestinal peptide, on the prolactin response to central administration of vasoactive intestinal peptide. *p < 0.05, *p < 0.025. Reprinted from the original manuscript (Samson *et al.*, 1989) with the permission of the Williams and Wilkins Publishing Company.

central effect of VIP on PRL release (Fig. 7). These data suggest that VIP acts as a neuromodulatory agent to stimulate the release of OT into portal blood which, in turn, induces PRL release from the lactotroph.

We now have additional data to support such a relationship. Using a pharmacological model, we have uncovered an endogenous PRL-stimulatory rhythm involving VIP as a possible neuromodulatory agent (Arey and Freeman, 1989). When ovariectomized rats are injected i.v. with the dopamine receptor antagonist, domperidone (DOM), a large release of PRL occurs within 5 min. This PRL-secretory response differs depending upon the time-of-day the animal is challenged with DOM (Arey *et al.*, 1989); that is, the PRL-secretory response to DOM is twice as great at 0300 and 1700 h as that at 1200 h. This implies that a stimulatory rhythm which is unmasked by a lowering of dopaminergic tone may control PRL secretion at differing times of day. Similar differential time of day responses were not observed in male rats. To test the possibility that the sex-specific rhythm involved OT or VIP, we challenged ovariectomized rats with DOM superimposed upon the simultaneous infusion of synthetic antagonists of OT and VIP (as described above in the studies of Samson *et al.*, 1989).

Interestingly, the OT antagonist prevented the differential time-of-day response in ovariectomized rats; that is, the PRL-secretory response to DOM was the same at 0300, 1200, and 1700 h. On the other hand, only the differential PRL-secretory response to DOM at 0300 h was prevented by the VIP antagonist. Conversely, prior treatment with p-chlorophenylalanine, a serotonin synthesis inhibitor, blocked the differential PRL-secretory response to DOM at 1700 h but not 0300 h. Based on these data we have developed a model for the putative control of PRL secretion during dopamine withdrawal (Fig. 8). We have proposed two independent oscillators which are active at differing times of day: the 0300 h oscillator is VIPergic and the 1700 h oscillator is serotonergic. These oscillators synapse on OT neurons and affect OT release into the portal vessels at the appropriate time of day. This OT release only eventuates in PRL release if dopaminergic tone is lowered by exteroceptive stimuli such as suckling or cervical stimulation (CS). Indeed, intravenous infusion of OT is more effective in stimulating PRL secretion in conscious rats during dopamine withdrawal induced by DOM injection (Mogg and Samson, 1990).

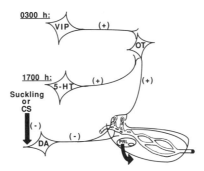

Fig. 8. Proposed mechanism for the roles of dopamine (DA), serotonin (5HT), vasoactive intestinal peptide (VIP), and oxytocin (OT) in the release of prolactin to suckling or mating-induced stimulation of the uterine cervix (CS). While the dopaminergic tone is constantly inhibitory in nature, two independent oscillators regulate oxytocin's stimulatory influence by causing the release of OT into the portal blood at 0300 h or 1700 h. The 0300 h (N) oscillator is VIP-ergic and the 1700 h (D) oscillator is serotonergic. The OT released by these neuromodulators only eventuates PRL release if suckling or the mating stimulus has lowered the inhibitory DA-ergic tone at the lactotroph. +, stimulation; -, inhibition. Reprinted from the original manuscript (Arey *et al.*, 1989) with the permission of the Williams and Wilkins Publishing Company.

Thus, VIP can act at two distinct loci to control PRL secretion. The apparent neuromodulatory actions of VIP suggest a positive feed-forward which magnifies the ability of the hypothalamus to stimulate anterior pituitary hormone secretion. In addition to possible multiple stimulatory influences acting at the level of the lactotroph, neuropeptides (VIP, in particular) possessing PRF activity themselves might act at more than one locus within the hypothalamo-pituitary axis to assure the coordinated control of PRL secretion. This phenomenon of recruitment of multiple PRFs could be hypothesized to play an important role in magnifying the pituitary response to dopamine withdrawal, thereby allowing rapid and appropriate responsiveness to stimuli such as suckling, mating and steroid milieu. In a more general sense, this dual-site hypothesis may provide a framework for the examination of multiple cascade systems regulating the secretion of a variety of pituitary hormones.

REFERENCES

Abe, H., D. Engler, M.E. Molitch, J. Bollinger-Gruber, and S. Reichlin. 1985. Vasoactive intestinal peptide is a physiological mediator of prolactin release in the rat. *Endocrinology* 116: 1383-1390.

Alexander, M.J., D.K. Clifton, and R.A. Steiner. 1985. Vasoactive intestinal polypeptide effects a central inhibition of pulsatile luteinizing hormone secretion in ovariectomized rats. *Endocrinology* 117: 2134-2139.

Antoni, F.A. 1986. Oxytocin receptors in the rat adenohypophysis: Evidence from radioligand binding studies. *Endocrinology* 119: 2393-2395.

Arey, B.J., R.L.W. Averill, and M.E. Freeman. 1989. A sex-specific endogenous stimulatory rhythm regulating prolactin secretion. *Endocrinology* 124: 119-124.

Arey, B., and M.E. Freeman. 1989. Hypothalamic factors involved in the endogenous stimulatory rhythm regulating prolactin secretion. *Endocrinology* 124: 878-884.

Bardrum, B., B. Ottesen, and A.-R. Fuchs. 1987. Preferential release of oxytocin in response to vasoactive intestinal polypeptide in rats. *Life Sci.* 40: 169-173.

Bataille, D., C. Gespach, M. Laburthe, B. Amiranoff, K. Tatemoto, N. Vauclin, V. Mutt, and G. Rosselin. 1981. Porcine peptide having N-terminal histidine and C-terminal isoleucine amide (PHI). Vasoactive intestinal peptide (VIP) and secretin-like effects in different tissues in the rat. *FEBS Lett.* 114: 240-242.

Bayliss, W.M., and E.H. Starling. 1902. On the causation of the so-called "peripheral reflex secretion" of the pancreas. *Proc. R. Soc.* 69: 352-353.

Beinfeld, M.C., P.L. Brick, A.C. Howlett, I.L. Holt, R.M. Pruss, J.P. Moskal, and L.E. Eiden. 1988. The regulation of vasoactive intestinal peptide synthesis in neuroblastoma and chromaffin cells. *Ann. N.Y. Acad. Sci.* 527: 68-76.

Besson, J. 1988. Distribution and pharmacology of vasoactive intestinal peptide receptors in brain and pituitary. *Ann. N.Y. Acad. Sci.* 527: 204-219.

Borghi, C., S. Nicosia, A. Giachetti, and S.I. Said. 1979. Vasoactive intestinal polypeptide (VIP) stimulates adenylate cyclase in selected areas of the rat brain. *Life Sci.* 24: 65-70.

Bromer, W.W., L.G. Sinn, and O.K. Behrens. 1957. The amino acid sequence of glucagon. V. Location of amide groups, acid degradation studies and sequential evidence. *J. Am. Chem. Soc.* 79: 2807-2810.

Brown, J.C., and J.R. Dryburgh. 1971. A gastric inhibitory polypeptide. II. The complete amino acid sequence. *Can. J. Biochem.* 49: 867-872.

Eckenstein, F., and R.W. Baughman. 1984. Two types of cholinergic innervation in cortex, one co-localized with vasoactive intestinal polypeptide. *Nature* 309: 153-155.

Emson, P.C., J. Fahrenkrug, O.B. Schaffalitzky de Muckadell, T.M. Jessell, and L.L. Iversen. 1978. Vasoactive intestinal peptide (VIP): Vesicular localization and potassium evoked release from rat hypothalamus. *Brain Res.* 143: 174-178.

Enjalbert, A., S. Arancibia, M. Ruberg, M. Priam, M.T. Blue-Pajot, W.H. Rotsztjen, and C. Kordon. 1980. Stimulation of *in vitro* prolactin release by vasoactive intestinal peptide. *Neuroendocrinology* 31: 200-204.

Giachetti, A., S.I. Said, R.C. Reynolds, and F.C. Koniges. 1977. Vasoactive intestinal polypeptide in brain: Localization in and release from isolated nerve terminals. *Proc. Natl. Acad. Sci. U.S.A.* 74: 3424-3428.

Gibbs, D.M. 1984. High concentrations of oxytocin in hypophyseal portal plasma. *Endocrinology* 114: 1216-1218.

Gozes, I. 1988. Biosynthesis and regulation of expression. The vasoactive intestinal peptide gene. *Ann. N.Y. Acad. Sci.* 527: 77-86.

Hagen, T.C., M.A. Arnaout, W.J. Scherzer, D.R. Martinson, and T.L. Garthwaite. 1986. Antisera to vasoactive intestinal polypeptide inhibits basal prolactin release from dispersed anterior pituitary cells. *Neuroendocrinology* 43: 641-645.

Hakanson, R., F. Sundler, and R. Uddman. 1982. Distribution and topography of peripheral VIP nerve fibers: Functional implications. In *Vasoactive Intestinal Peptide*, ed. S.I. Said, 121-144. Raven Press, New York.

Haskins, J.T., W.K. Samson, and R.L. Moss. 1982. Evidence for vasoactive intestinal polypeptide (VIP) altering the firing rate of preoptic, septal and midbrain central gray neurons. *Regul. Peptides* 3: 113-123.

Hökfelt, T., J. Fahrenkrug, K. Tatemoto, V. Mutt, S. Werner, A.L. Hulting, L. Terenius, and K.J. Chang. 1983. The PHI (PHI-27)/corticotropin releasing factor/enkephalin immunoreactive neuron: Possible morphological basis for integrated control of prolactin, corticotropin, and growth hormone secretion. *Proc. Natl. Acad. Sci. U.S.A.* 80: 895-898.

Hökfelt, T., M. Schultzberg, J.M. Lundberg, K. Fuxe, V. Mutt, J. Fahrenkrug, and S.I. Said. 1982. Distribution of vasoactive intestinal polypeptide in the central and peripheral nervous systems as revealed by immunocytochemistry. In *Vasoactive Intestinal Peptide*, ed. S.I. Said, 65-90. Raven Press, New York.

Horn, A.M., H.M. Fraser, and G. Fink. 1983a. Effects of antiserum to thyrotropin releasing hormone on concentrations of plasma prolactin, thyrotropin and luteinizing hormone in the pro-oestrous rat. *J. Endocrinol.* 104: 205-209.

Horn, A.M., I.C.A.F. Robinson, and G. Fink. 1983b. Vasopressin and oxytocin in hypophyseal portal blood: Experimental studies in normal and Brattleboro rats. *J. Endocrinol.* 104: 211-224.

Hyde, J.F. 1989. Posterior pituitary extracts specifically increase plasma prolactin (PRL) levels in ovariectomized rats. *71st Ann. Mtng. Endocrine Soc. Abstract*, #12, p. 25.

Hyde, J.F., and N. Ben-Jonathan. 1988. Characterization of prolactin-releasing factor in the rat posterior pituitary. *Endocrinology* 122: 2533-2539.

Itoh, N., K. Obata, N. Yanaihara, and H. Okamoto. 1983. Human preprovasoactive intestinal polypeptide contains a novel PHI-27-like peptide, PHM-27. *Nature* 304: 547-549.

Johnston, C.A., K.D. Fagin, R.H. Alper, and A. Negro-Vilar. 1986. Prolactin release after 5-hydroxytryptophan treatment requires an intact neurointermedial pituitary lobe. *Endocrinology* 118: 805-810.

Johnston, C.A., and A. Negro-Vilar. 1988. Role of oxytocin on prolactin secretion during proestrus and in different physiological or pharmacological paradigms. *Endocrinology* 122: 341-350.

Jorpes, J.E., V. Mutt, S. Magnusson, and B.B. Steele. 1962. Amino acid composition and N-terminal amino acid sequence of porcine secretin. *Biochem. Biophys. Res. Commun.* 9: 275-279.

Kaji, H., K. Chihara, H. Abe, T. Kita, Y. Kashio, Y. Okimura, and T. Fujita. 1985. Effect of passive immunization with antisera to vasoactive intestinal polypeptide and peptide histidine isoleucine amide on 5-hydroxy-1-tryptophan-induced prolactin release in rats. *Endocrinology* 117: 1914-1919.

Kato, Y., Y. Iwasaki, J. Iwasaki, H. Abe, N. Yanaihara, and H. Imura. 1978. Prolactin release by vasoactive intestinal polypeptide in rats. *Endocrinology* 103: 554-558.

Kato, Y., N. Matsushita, H. Ohta, K. Tojo, A. Shimatsu, and H. Imura. 1985. Regulation of prolactin secretion. In *The Pituitary Gland*, ed. H. Imura, 261-278. Raven Press, New York.

Larsson, L.-I. 1982. Localization of vasoactive intestinal polypeptide: A critical appraisal. In *Vasoactive Intestinal Peptide*, ed. S.I. Said, 51-63. Raven Press, New York.

Loren, I., P.C. Emson, J. Fahrenkrug, A. Bjorklund, J. Alumets, R. Hakanson, and F. Sundler. 1979. Distribution of vasoactive intestinal polypeptide in the rat and mouse brain. *Neuroscience* 4: 1953-1976.

Lumpkin, M.D., W.K. Samson, and S.M. McCann. 1983. Hypothalamic and pituitary sites of action of oxytocin to alter prolactin secretion in the rat. *Endocrinology* 112: 1711-1717.

Magistretti, P.J., M.M. Dietl, P.R. Hof, J.-L. Martin, J.M. Placios, N. Schaad, and M. Schorderet. 1988. Vasoactive intestinal peptide as a mediator of intercellular communication in the cerebral cortex. *Ann. N.Y. Acad. Sci.* 527: 110-129.

Marley, P., and P.C. Emson. 1982. VIP as a neurotransmitter in the central nervous system. In *Vasoactive Intestinal Peptide*, ed. S.I. Said, 341-360. Raven Press, New York.

Melin, P., J. Trojnar, B. Johansson, H. Vilhardt, and M. Akerlund. 1986. Synthetic antagonists of the myometrial response to vasopressin and oxytocin. *J. Endocrinol.* 111: 125-131.

Mezey, E., and J.Z. Kiss. 1985. Vasoactive intestinal peptide containing neurons in the paraventricular nucleus may participate in regulating prolactin secretion. *Proc. Natl. Acad. Sci.* 82: 247-254.

Mogg, R.J., and W.K. Samson. 1990. Interactions of dopaminergic and peptidergic factors in the control of prolactin release. *Endocrinology* 126: 728-735.

Morel, G., J.G. Chabot, and P.M. Dubois. 1988. Ultrastructural evidence for oxytocin in the rat anterior pituitary gland. *Acta Endocrinol.* 117: 307-314.

Murai, I., and N. Ben-Jonathan. 1987. Posterior pituitary lobectomy abolishes the suckling-induced rise in prolactin (PRL): Evidence for a PRL-releasing factor on the posterior pituitary. *Endocrinology* 121: 205-211.

Murai, I., S. Reichlin, and N. Ben-Jonathan. 1989. The peak phase of the proestrous prolactin surge is blocked by either posterior pituitary lobectomy or antisera to vasoactive intestinal peptide. *Endocrinology* 124: 1050-1055.

Nicosia, S., A. Spada, and G. Giannattasio. 1982. Effects of vasoactive intestinal polypeptide on the pituitary gland. In *Vasoactive Intestinal Peptide*, ed. S.I. Said, 263-275. Raven Press, New York.

Nishizawa, M., Y. Hayakawa, N. Yanaihara, and H. Okamoto. 1985. Nucleotide sequence divergence and functional constraint in VIP precursor mRNA evolution between human and rat. *FEBS Lett.* 183: 55-59.

Ohtsuka, S., A. Miyake, T. Nishizaki, K. Tasaka, and O. Tanizawa. 1988. Vasoactive intestinal peptide stimulates gonadotropin-releasing hormone release from rat hypothalamus *in vitro. Acta Endocrinol.* 117: 399-402.

Pandol, S.J., K. Dharmsatthaphorn, M.S. Schoeffield, W. Vale, and J. Rivier. 1986. Vasoactive intestinal peptide receptor antagonist (4Cl-d-Phe6, Leu17) VIP. *Am. J. Physiol.* 250: G553-557.

Phillis, J.W. 1982. Neuronal excitation by vasoactive intestinal polypeptide. In *Vasoactive Intestinal Peptide*, ed. S.I. Said, 299-305. Raven Press, New York.

Polak, J.M., and S.R. Bloom. 1982. Distribution and tissue localization of VIP in the central nervous system and in seven peripheral organs. In *Vasoactive Intestinal Peptide*, ed. S.I. Said, 107-120. Raven Press, New York.

Raiteri, M., M. Marchi, and P. Paudice. 1987. Vasoactive intestinal peptide (VIP) potentiates the muscarinic stimulation of phosphoinositide turnover in rat cerebral cortex. *Eur. J. Pharmacol.* 133: 127-128.

Reichlin, S. 1988. Neuroendocrine significance of vasoactive intestinal polypeptide. *Ann. N.Y. Acad. Sci.* 527: 431-449.

Rivier, C., and W. Vale. 1974. *In vivo* stimulation of prolactin secretion in the rat by thyrotropin releasing factor, related peptides and hypothalamic extracts. *Endocrinology* 95: 978-983.

Robberecht, P., P. Chatelain, M. Waelbroeck, and J. Christophe. 1982. Heterogeneity of VIP-recognizing binding sites in rat tissues. In *Vasoactive Intestinal Peptide*, ed. S.I. Said, 323-332. Raven Press, New York.

Roberts, G.W., P.L. Woodhams, M.G. Bryant, T.J. Crow, S.R. Bloom, and J.M. Polak. 1980. VIP in the rat brain: Evidence for a major pathway linking the amygdala and hypothalamus via the stria terminalis. *Histochemistry* 65: 103-119.

Rosselin, G., A. Anteunis, A. Astesano, C. Boissard, P. Gali, G. Hejblum, and J.C. Marie. 1988. Regulation of the vasoactive intestinal peptide receptor. *Ann. N.Y. Acad. Sci.* 527: 220-237.

Rotsztejn, W.H., L. Benoist, J. Besson, G. Beraud, M.T. Bluet-Pajot, C. Kordon, G. Rosselin, and J. Duval. 1980. Effect of vasoactive intestinal peptide (VIP) on the release of adenohypophyseal hormones from purified cells obtained by unit gravity sedimentation: Inhibition by deaxmethasone of VIP-induced prolactin release. *Neuroendocrinology* 31: 282-286.

Ruberg, M., W.H. Rotsztejn, S. Arancibia, J. Besson, and A. Enjalbert. 1978. Stimulation of prolactin release by vasoactive intestinal peptide (VIP). *Eur. J. Pharmacol.* 51: 319-321.

Said, S.I., and V. Mutt. 1970. Polypeptide wide broad biological activity: Isolation from small intestine. *Science* 169: 1217-1218.

Said, S.I., and V. Mutt. 1972. Isolation from porcine intestinal wall of a vasoactive octacosapeptide related to secretin and glucagon. *Eur. J. Biochem.* 28: 199-204.

Said, S.I., and V. Mutt. 1988. Vasoactive Intestinal Peptide and Related Peptides. *Ann. N.Y. Acad. Sci.* vol. 527, New York.

Said, S.I., and J.C. Porter. 1979. Vasoactive intestinal polypeptide: Release into the hypophyseal portal blood. *Life Sci.* 24: 227-230.

Samson, W.K. 1982. Radioimmunological localization of VIP in the mammalian brain. In *Vasoactive Intestinal Peptide*, ed. S.I. Said, 91-105. Raven Press, New York.

Samson, W.K., R. Bianchi, R.J. Mogg, J. Rivier, W. Vale, and P. Melin. 1989. Oxytocin mediates the hypothalamic action of vasoactive intestinal peptide to stimulate prolactin secretion. *Endocrinology* 124: 812-819.

Samson, W.K., K.P. Burton, J.P. Reeves, and S.M. McCann. 1981. Vasoactive intestinal peptide stimulates luteinizing hormone-releasing hormone release from median eminence synaptosomes. *Regul. Peptides* 2: 253-264.

Samson, W.K., M.D. Lumpkin, and S.M. McCann. 1984. Presence and possible site of action of secretin in the rat pituitary and hypothalamus. *Life Sci.* 34: 155-163.

Samson, W.K., M.D. Lumpkin, and S.M. McCann. 1986. Evidence for a physiological role for oxytocin in the control of prolactin secretion. *Endocrinology* 119: 554-560.

Samson, W.K., M.D. Lumpkin, J.K. McDonald, and S.M. McCann. 1983. Prolactin-releasing activity of porcine intestinal peptide (PHI-27). *Peptides* 4: 817-819.

Samson, W.K., L. Martin, R.J. Mogg, and R.J. Fulton. 1990. A non-oxytocinergic prolactin releasing factor and a non-dopaminergic prolactin inhibiting factor in bovine neurointermediate lobe extracts: *In vitro* and *vivo* studies. *Endocrinology* 126: 1610-1617.

Samson, W.K., J.K. McDonald, and M.D. Lumpkin. 1985. Naloxone-induced dissociation of oxytocin and prolactin releases. *Neuroendocrinology* 40: 68-70.

Samson, W.K., and R.J. Mogg. 1990. Peptidergic control of prolactin secretion. *Biomed. Res.* 10(S3): 107-116.

Samson, W.K., S.I. Said, and S.M. McCann. 1979. Radioimmunologic localization of vasoactive intestinal polypeptide in hypothalamic and extrahypothalamic sites in the rat brain. *Neurosci. Lett.* 12: 265-269.

Samson, W.K., G. Snyder, S.I. Said, and S.M. McCann. 1980. *In vitro* stimulation of prolactin release by vasoactive intestinal polypeptide. *Peptides* 1: 325-332.

Sarkar, D.P. 1989. Evidence for prolactin feedback actions on hypothalamic oxytocin, vasoactive intestinal peptide and dopamine secretion. *Neuroendocrinology* 49: 520-524.

Sarkar, D.P. and D.M. Gibbs. 1984. Cyclic variation of oxytocin in the blood of pituitary vessels of rats. *Neuroendocrinology* 39: 481-483.

Shimatsu, A., Y. Kato, T. Inoue, N.D. Christofides, S.R. Bloom, and H. Imura. 1983. Peptide histidine isoleucine and vasoactive intestinal polypeptide-like immunoreactivity coexist in rat hypophyseal portal blood. *Neurosci. Lett.* 43: 259-262.

Shimatsu, A., Y. Kato, N. Matsushita, H. Katakami, N. Yanaihara, and H. Imura. 1981. Immunoreactive vasoactive intestinal polypeptide in rat hypophyseal-portal blood. *Endocrinology* 108: 395-398.

Shimatsu, A., Y. Kato, N. Matsushita, H. Katakami, N. Yanaihara, and H. Imura. 1982. Stimulation by serotonin of vasoactive intestinal polypeptide release into rat hypophyseal portal blood. *Endocrinology* 111: 338-340.

Shimatsu, A., Y. Kato, N. Matsushita, H. Ota, Y. Kabayama, N. Yanaihara, and H. Imura. 1984. Prostaglandin D₂ stimulates vasoactive intestinal polypeptide release into rat hypophyseal portal blood. *Peptides* 5: 395-398.

Sims, K.B., D.L. Hoffman, S.I. Said, and E.A. Zimmerman. 1980. Vasoactive intestinal polypeptide (VIP) in mouse and rat brain: An immunocytochemical study. *Brain Res.* 186: 165-183.

Strauss, E., T.N. Keltz, and R.S. Yalow. 1982. Enzymatic degradation of VIP. In *Vasoactive Intestinal Peptide*, ed. S.I. Said, 333-339. Raven Press, New York.

Tashjian, A.H., N.J. Barowsky, and D.K. Jensen. 1971. Thyrotropin releasing hormone: Direct evidence for stimulation of prolactin production by pituitary cells in culture. *Biochem. Biophys. Res. Commun.* 43: 516-523.

Tatemoto, K., and V. Mutt. 1981. Isolation and characterization of the intestinal peptide porcine PHI (PHI-27), a new member of the glucagon-secretin family. *Proc. Natl. Acad. Sci. U.S.A.* 78: 6603-6607.

Vale, W., J. Vaughn, M. Smith, G. Yamamoto, J. Rivier, and C. Rivier. 1983. Effects of synthetic ovine corticotropin releasing factor, glucocorticoids, catecholamines, neurohypophyseal peptides and other substances on cultured corticotropic cells. *Endocrinology* 113: 1121-1131.

Vijayan, E., W.K. Samson, S.I. Said, and S.M. McCann. 1979. Vasoactive intestinal peptide: Evidence for a hypothalamic site of action to release growth hormone, luteinizing hormone, and prolactin in conscious ovariectomized rats. *Endocrinology* 104: 53-57.

Werner, S.A., A.L. Hulting, T. Hökfelt, P. Eneroth, K. Tatemoto, V. Mutt, L. Marode, and E. Wunsch. 1983. Effect of PHI-27 on prolactin release *in vitro*. *Neuroendocrinology* 37: 476-478.

Yamagami, T., K. Ohsawa, M. Nishizawa, C. Inoue, E. Gotoh, N. Yanaihara, H. Yamamoto, and H. Okamoto. 1988. Complete nucleotide sequence of human vasoactive intestinal peptide/PHM-27 gene and its inducible promoter. *Ann. N.Y. Acad. Sci.* 527: 87-102.

REPRODUCTIVE NEUROENDOCRINE EFFECTS OF NEUROPEPTIDE Y AND RELATED PEPTIDES

John K. McDonald, Ph.D.
Department of Anatomy and Cell Biology
Emory University School of Medicine
Atlanta, Georgia

I.   Introduction

II.   Background

    A.   Pancreatic Polypeptide

    B.   Neuropeptide Y

III. Distribution of Neuropeptide Y

    A.   Introduction

    B.   Reproductive Neuroendocrine Axis

IV.   Function of Neuropeptide Y in the Reproductive Neuroendocrine Axis

    A.   Hypothalamus

    B.   Anterior Pituitary Gland

V.   Summary

VI.   Directions for Future Research

Introduction

This chapter will summarize current research on the reproductive neuroendocrine effects of neuropeptide Y (NPY) and related peptides, especially pancreatic polypeptide (PP). A brief background section will present the history of research in this area and basic information concerning NPY biosynthesis, NPY receptors and the relationship of NPY to homologous peptides. Next, the distribution of NPY in the brain and reproductive neuroendocrine axis will be presented. References to NPY distribution in the gastrointestinal system will also be provided in this chapter. The distribution and function of NPY in other neuroendocrine systems and in the cardiovascular system is beyond the scope of this chapter and will not be described. Our primary focus is on the distribution and function of NPY in the reproductive neuroendocrine axis; the major points will be summarized and directions for future research proposed. Several reviews on NPY and related peptides have recently been published; the reader is referred to them for additional information (Agnati *et al.*, 1986; Allen and Bloom, 1986; Chance *et al.*, 1979; Edvinsson, 1988; Emson and de Quidt, 1984; Fuxe *et al.*, 1987; Gray and Morley, 1986; Hazelwood, 1981; Lin, 1980; Maccarrone and Jarrott, 1986a; McCann *et al.*, 1989; McDonald, 1988; Polak and Bloom, 1984; Potter, 1988; Solomon, 1985; Sundler *et al.*, 1986; Wahlestedt *et al.*, 1989).

Background

Pancreatic Polypeptide

NPY research actually began with studies of avian (a) and bovine (b) pancreatic polypeptide (PP) isolated from pancreatic extracts (Chance *et al.*, 1979; Kimmel *et al.*, 1968; Kimmel *et al.*, 1975). Immunohistochemical studies of aPP and bPP revealed a widespread distribution in the central and peripheral nervous system (Card *et al.*, 1983; Fujii *et al.*, 1982; Hökfelt *et al.*, 1981; Loren *et al.*, 1979; Lundberg *et al.*, 1982a; Olschowka *et al.*, 1981). Especially high concentrations of PP-like immunoreactivity were observed in the cerebral cortex, (McDonald *et al.*, 1982) striatum, (Vincent *et al.*, 1982a; 1982b; Vincent and Johansson, 1983; Vincent *et al.*, 1983) hypothalamus, (Card and Moore, 1982; Card and Moore, 1984; Card *et al.*, 1983) and also in sympathetic ganglia (Loren *et al.*, 1979). One notable feature was the localization of PP-like immunoreactivity in certain hypothalamic nuclei (arcuate, paraventricular, suprachiasmatic, periventricular, and medial preoptic) which are known to represent important areas of neuroendocrine regulation of hormone release from the anterior pituitary (Card *et al.*, 1983; Olschowka *et al.*, 1981).   PP-like immunoreactivity was localized not only in intrinsic neurons in the cerebral cortex, (McDonald *et al.*, 1982; Vincent *et al.*, 1982b) hypothalamic arcuate nucleus and striatum (Card *et al.*, 1983; Olschowka *et al.*, 1981), but also in catecholaminergic neurons in the brainstem and in noradrenergic neurons in sympathetic preaortic and paraspinal ganglia (Hunt *et al.*, 1981; Jacobowitz and Olschowka, 1982a; 1982b; Kobayashi *et al.*, 1983; Lundberg *et al.*, 1980a; Lundberg *et al.*, 1982a; Olschowka and Jacobowitz, 1983; Uddman *et al.*, 1982; Vincent *et al.*, 1982a).   Thus, PP-like immunoreactivity was distributed in a wide variety of target locations intimately involved in the central and peripheral control of several neuroendocrine and autonomic functions.

The earliest report of neuroendocrine activity of PP was provided in 1983 (McDonald and Lumpkin).   Injections of aPP and bPP into the third cerebral ventricle (3V) of ovariectomized (OVX) rats decreased the secretion of luteinizing hormone (LH) and growth hormone (GH) from the anterior pituitary (McDonald and Lumpkin, 1983; McDonald *et al.*, 1985b).   These inhibitory effects of aPP and bPP on LH release were potent and dose-related.   GH levels increased significantly following the initial decline in animals treated with the lower dose of bPP and rebounded to greater levels than those preceding injection. The minimal effective dose to suppress GH secretion was higher than that required for LH.   It is noteworthy that the PPs had no effect on thyroid stimulating hormone (TSH) or prolactin (PRL).   Injection of the C-terminal hexapeptide amide of human (h)PP, which is identical at four of six amino acid sites to aPP and bPP, into the 3V of OVX rats did not affect the release of LH, GH, TSH or PRL, suggesting that more than the C-terminal hexapeptide of the molecule is required for bioactivity.   These investigators also examined the potential effects of the PPs on hormone release from overnight cultures of dispersed anterior pituitary cells.   No effects on LH, GH, TSH, or PRL secretion were observed (McDonald *et al.*, 1985b).

The effects of aPP and bPP on LH and GH secretion were postulated to occur either directly by modulating the secretion of the hypophysiotropic peptides, luteinizing hormone-releasing hormone (LHRH), somatostatin (SRIF) and/or growth hormone-releasing hormone (GHRH), into the pituitary portal vessels, or indirectly by altering catecholaminergic influences on the secretion of these hypophysiotropic hormones (McDonald *et al.*, 1985b).   Others soon reported that 3V injection of hPP into OVX rats exerted similar suppressive effects on LH release; however, administration of estrogen and progesterone to OVX rats changed the direction of these effects so that hPP mildly stimulated LH release

(Kalra and Crowley, 1984a). Gonadal steroids also change the direction of noradrenergic effects on LH secretion.

In addition to these early reports of reproductive neuroendocrine activity of PP, Albers *et al.* (1984) demonstrated that injection of aPP into the suprachiasmatic nucleus of hamsters shifted the phase of circadian activity rhythms. These early studies of the hypothalamic activity of PP influencing hormone secretion and circadian rhythms constituted the first indication that PPs might play a central role in hypothalamic function.

During the course of these studies, Tatemoto and colleagues (1982) isolated a peptide of 36 residues from porcine brain. The peptide was named neuropeptide Y (NPY) to indicate the neural origin, and the high content of tyrosine by using the biochemical letter designation Y (Tatemoto, 1982b). NPY showed extensive sequence homology with the various forms of PP, (Chance *et al.*, 1979; Kimmel *et al.*, 1975; Kimmel *et al.*, 1984; Kimmel *et al.*, 1986) and also with peptide YY (PYY), isolated from extracts of porcine intestine (see Table 1) (Tatemoto, 1982a). Accordingly, NPY, PP, and PYY are considered as a family of peptides.

One troubling issue for the investigators working with PP was that PP-like immunoreactivity had been localized with immunohistochemical methods, yet it could not be isolated biochemically from the brain. The degree of sequence

Table 1

THE NPY - PP - PYY FAMILY

		AMINO ACID SEQUENCES OF NPY AND RELATED PEPTIDES	HOMOLOGY TO PORCINE NPY	
		1 6 12 18 24 30 36	# of AMINO ACIDS	% HOMOLOGY
NPY	Porcine	Y P S K P D N P G E D A P A E D L A R Y Y S A L R H Y I N L I T R Q R Y-NH$_2$	36	100
	Rat	Y - S K - D N - - E D - P A - D M A R Y Y S A L R H - I - L I - - Q - - -	35	97
	Human	Y - S K - D N - - E D - P A - D M A R Y Y S A L R H - I - L I - - Q - - -	35	97
PP	Avian (Chicken)	G - S Q - T Y - - D D - P V - D L I R F Y D N L Q Q - L - V V - - H - - -	20	56
	Bovine	A - L E - E Y - - D M - T P - Q M A Q Y A A E L R R - I - M L - - P - - -	17	47
	Rat	A - L E - M Y - - D Y - T H - Q R A Q Y E T Q L R R - I - T L - - P - - -	17	47
	Human	A - L E - V Y - - D M - T P - Q M A Q Y A A D L R R - I - M L - - P - - -	17	47
	Salmon	Y - P K - E N - - E D - P P - E L A K Y Y T A L R H - I - L I - - Q - - -	30	84
	Anglerfish (aPY, peptide YG)	Y - P K - E T - - S M - S P - D M A S Y Q A A V R H - V - L I - - Q - - G-COOH	23	64
PYY	Porcine	Y - A K - E A - - E D - S P - E L S R Y Y A S L R H - L - L V - - Q - - -	25	69

homology between NPY and PP prompted immunohistochemists to preabsorb aPP and bPP antisera with NPY before use (Lundberg *et al.*, 1984b; Moore *et al.*, 1984). The results showed that "PP-like" immunoreactivity was eliminated by this treatment. It was apparent that previous observations of "PP-like" staining were actually due to cross-reactivity of these antisera with the endogenous peptide NPY (Di Maggio *et al.*, 1985). Soon investigators began performing immunohistochemical studies with NPY antisera and physiological studies with NPY and found virtually identical results to experiments using PP and PP antisera (Albers and Ferris, 1984; Albers *et al.*, 1984; Card *et al.*, 1983; Chronwall *et al.*, 1985; Clark *et al.*, 1984; de Quidt and Emson, 1986a, 1986b; Di Maggio *et al.*, 1985; Kalra and Crowley, 1984a, 1984b, 1984c; Lundberg *et al.*, 1980b, 1984b; Lundberg and Tatemoto, 1982; McDonald and Lumpkin, 1983; McDonald *et al.*, 1984, 1985a, 1985b; Moore *et al.*, 1984; Ohhashi and Jacobowitz, 1983).

Neuropeptide Y

NPY is a 36 amino-acid peptide that belongs to the NPY-PP-PYY family of peptides (Andrews *et al.*, 1985; Kimmel *et al.*, 1984; Tatemoto, 1982a, 1982b; Tatemoto *et al.*, 1982). The amino acid sequences of these extensively homologous peptides are presented in Table 1. Porcine NPY differs from human and rat NPY, which are identical, only at position 17 where a leucine replaces the methionine. NPY is very similar to PYY (69% homology), which is found predominantly in the gastrointestinal system, although PYY-like immunoreactivity has been observed in the brainstem and spinal cord (Broome *et al.*, 1987; Lundberg *et al.*, 1982b; Taylor 1985). The mammalian forms of PP exhibit 47% homology with pNPY, while avian PP is 56% homologous. Piscine forms of PP are more similar to mammalian NPY than the mammalian PPs. Anglerfish PP (aPY) is 64% homologous to pNPY and may also be found in anglerfish brains (Milgram *et al.*, 1989). In contrast, the carboxyl terminal fragments of human prepro-NPY and anglerfish prepro-aPY exhibit only 33% identity (Andrews and Dixon, 1986). Salmon PP is striking in its similarity to pNPY at 23 of 36 sites (84%) (Kimmel *et al.*, 1986). The amino acid sequence of salmon PP is more similar to mammalian NPY than any other peptide. The conservation of the amino acid structure of NPY throughout evaluation allows NPY-like immunoreactivity to be found in species ranging from insects to humans (Andrews *et al.*, 1985; Corder *et al.*, 1988; Falkmer *et al.*, 1985; Kimmel *et al.*, 1986; Noe *et al.*, 1986, 1989b; Yui *et al.*, 1985).

Studies by Minth and colleagues (1984) demonstrate that NPY is derived from a precursor (prepro-NPY) that is 97 amino acids long and contains two sites of proteolytic processing. Prepro-NPY is processed to generate three peptides, an N-terminal signal peptide of 28 residues, NPY itself composed of 36 amino acids, and a C-terminal flanking peptide of NPY called C-terminal flanking peptide of NPY (CPON). These investigators have also characterized, sequenced and expressed the cloned hNPY gene (Minth *et al.*, 1986). The genes encoding PP and NPY are located on human chromosomes 17 and 7, respectively (Takeuchi *et al.*, 1986). Of particular relevance to this chapter is the localization, using *in situ* hybridization techniques, of high levels of NPY mRNA in the arcuate nucleus of the hypothalamus, a nucleus rich in NPY perikarya (Gehlert *et al.*, 1987). These new insights into the molecular biology of NPY biosynthesis will undoubtedly facilitate our understanding of the nature of gonadal steroid effects on NPY levels in several hypothalamic nuclei (Crowley *et al.*, 1985). CPON is colocalized with NPY in the nervous system and in other tissues although the function of CPON, if any, remains unknown (Allen *et al.*, 1985b; Gulbenkian *et al.*, 1985). Virtually nothing has been reported about the enzymes involved in post-translational processing of prepro-NPY or their potential regulation by gonadal steroids.

A serious limitation in NPY research at this time is the lack of specific NPY receptor agonists and antagonists to study the physiology and pharmacology of NPY receptors. The biochemical and anatomical studies of NPY receptors have relied on a molar excess of NPY to displace binding of radiolabeled NPY (Busch-Sorensen *et al.*, 1989; Chang *et al.*, 1985, 1986; Härfstrand *et al.*, 1986a; Martel *et al.*, 1986; Nakajima *et al.*, 1986; Saria *et al.*, 1985; Unden and Bartfai, 1984; Unden *et al.*, 1984). Investigations of NPY binding to brain membranes show high levels in the hippocampus and moderate levels in the hypothalamus, cortex, midbrain, and striatum (Busch-Sorensen *et al.*, 1989; Chang *et al.*, 1985). Others have observed a high binding density on hypothalamic membranes (Unden *et al.*, 1984). In contrast, the results obtained using autoradiographic techniques show moderate to low levels of binding in the hypothalamus (Martel *et al.*, 1986). Agreement between the two techniques was found in the hippocampus and cerebral cortex.

It is interesting to note that the anterior pituitary gland displayed no binding of labeled NPY while high levels were observed in the posterior pituitary (Busch-Sorensen *et al.*, 1989; Saavedra and Cruciani, 1988). The effects of NPY on hormone release from the anterior pituitary are discussed below. NPY levels in the neurohypophysis are responsive to osmotic stimulation (Hooi *et al.*, 1989).

Development of a potent NPY agonist or antagonist would facilitate the analysis of NPY receptors and perhaps resolve apparent differences in results obtained by various investigators. Recently, Wahlestedt and co-workers (1986) have defined subclasses of NPY receptors (Y_1 and Y_2) based on the use of NPY and PYY fragments of different lengths (Heilig *et al.*, 1988). These fragments may help elucidate the sites of NPY activity and distinguish between pre- and postsynaptic effects of the peptide (Heilig *et al.*, 1988; Wahlestedt *et al.*, 1986).

NPY has been implicated in the control of a wide variety of physiological functions. One of its major roles is in the cardiovascular system where it affects vascular tone and cardiac function (Abel *et al.*, 1988; Agnati *et al.*, 1986; Edvinsson, 1988; Lundberg and Tatemoto, 1982; McDonald, 1988; Zukowska-Grojec and Vaz, 1988). NPY fibers innervate pancreatic islets and NPY exerts direct and indirect effects on insulin secretion (Carlei *et al.*, 1985; Moltz and McDonald, 1985; Noe *et al.*, 1986; Pettersson *et al.*, 1987a, 1987b). It affects gastric acid release through central mechanisms and is one of the most potent stimulators of feeding behavior (Clark *et al.*, 1985, 1988; Humphreys *et al.*, 1988; Kalra *et al.*, 1988; Levine and Morley, 1984; Morley *et al.*, 1985; Sahu *et al.*, 1988; Stanley and Leibowitz, 1984, 1985; Stanley *et al.*, 1985a, 1985b, 1986). NPY fibers densely innervate the paraventricular nucleus of the hypothalamus and stimulate secretion of adrenocorticotropic hormone and corticosterone, perhaps through activation of corticotropin releasing factor (CRF)-neurons (Härfstrand *et al.*, 1986b; Liposits *et al.*, 1988; Sawchenko *et al.*, 1985; Sawchenko and Pfeiffer, 1988; Wahlestedt *et al.*, 1987). A detailed analysis of these various NPY functions is beyond the objectives of this chapter which focuses specifically on the distribution and function of NPY in the reproductive neuroendocrine axis.

Distribution of Neuropeptide Y

Introduction

NPY is widely distributed throughout the central, peripheral and enteric nervous system, and may be more abundant than somatostatin. A detailed description of the distribution of NPY in these areas is beyond the scope of this chapter, but has been recently reviewed (McDonald, 1988). Two tables have been included for the convenience of the reader: Table 2 contains references on the distribution of NPY in the central nervous system and also in the gastrointestinal system; Table 3 provides an index to NPY studies in the reproductive neuroendocrine axis.

Table 2. Distribution of NPY in the central nervous system and gastrointestinal system: Index to references. *Review articles.

Central Nervous System

General Distribution
Adrian *et al.*, 1983; Allen *et al.*, 1983; Beal *et al.*, 1986a, 1987; Chronwall *et al.*, 1985; Dawbarn *et al.*, 1984; de Quidt and Emson, 1986a,b; Everitt *et al.*, 1984; Gray and Morley, 1986; Lundberg *et al.*, 1984b; *McDonald, 1988; McDonald *et al.*, 1988a; Nakagawa *et al.*, 1985; Noe *et al.*, 1989a,b; *O'Donohue *et al.*, 1985; Polak and Bloom, 1984; Smith *et al.*, 1985; Sundler *et al.*, 1986; *Wahlestedt *et al.*, 1989

Eye
Allen *et al.*, 1983; Björklund *et al.*, 1985; Bruun *et al.*, 1984, 1985; Kumoi *et al.*, 1985; Muske *et al.*, 1987; Osborne *et al.*, 1985; Stone, 1986; Stone *et al.*, 1986; Terenghi *et al.*, 1983; Zhang *et al.*, 1984

Olfactory Bulb
Gall *et al.*, 1986; Scott *et al.*, 1987

Spinal Cord
Gibson *et al.*, 1984; Katagiri *et al.*, 1986; Maccarrone and Jarrott, 1985, 1986b; Sasek and Elde, 1985; VanDongen *et al.*, 1985

Brain Stem
Agnati *et al.*, 1988; Bergstrom *et al.*, 1984; Blessing *et al.*, 1986; Dai *et al.*, 1986; Everitt *et al.*, 1984; Foster *et al.*, 1984; Fuxe *et al.*, 1986; Gray *et al.*, 1986; Härfstrand *et al.*, 1987a; Hökfelt *et al.*, 1983a,b; Holets *et al.*, 1988; Maccarrone and Jarrott, 1985; Maccarrone *et al.*, 1986; Sawchenko *et al.*, 1985; Senba *et al.*, 1985; Smialowska, 1988; Smith *et al.*, 1985; Sutin and Jacobowitz, 1988; Yamazoe *et al.*, 1985

Substantia Innominata
Allen *et al.*, 1984b

Nucleus Accumbens
Massari *et al.*, 1988

Amygdala
Allen *et al.*, 1984d; Gray *et al.*, 1986; Gustafson *et al.*, 1986

Striatum
Aoki and Pickel, 1988; Beal *et al.*, 1986a,c; Chattha and Beal, 1987; Christie *et al.*, 1986; Dawbarn *et al.*, 1985; Kerkerian *et al.*, 1986; Smith and Parent, 1986

Table 2 (cont'd)

Hypothalamus	Bai *et al.*, 1985; Calka and McDonald, 1989; Card and Moore, 1984, 1988; Chronwall *et al.*, 1984; Ciofi *et al.*, 1987; Gray *et al.*, 1986; Guy and Pelletier, 1988; Härfstrand *et al.*, 1986b; Harrington *et al.*, 1985; Hisano *et al.*, 1988; Holets *et al.*, 1988; Kerkerian and Pelletier, 1986; Khorram *et al.*, 1987b; Li and Pelletier, 1986; Liposits *et al.*, 1988; McDonald *et al.*, 1988a; Moore *et al.*, 1984; Nakagawa *et al.*, 1985; Pelletier *et al.*, 1984; Sabatino *et al.*, 1987; Sahu *et al.*, 1987, 1988; Sawchenko *et al.*, 1985; Sawchenko and Pfeiffer, 1988; Simerly and Swanson, 1987; Tigges *et al.*, 1987; Ueda *et al.*, 1986
Thalamus	Harrington *et al.*, 1985; Mantyh and Kemp, 1983; Moore *et al.*, 1984; Ueda *et al.*, 1986
Hippocampus	Allen *et al.*, 1985c; Chan-Palay *et al.*, 1986a,b; Kohler *et al.*, 1986
Cerebral Cortex	Allen *et al.*, 1984b, 1985c; Beal and Martin, 1986; Beal *et al.*, 1986b; Chan-Palay *et al.*, 1985a,b; Chattha and Beal, 1987; Chronwall *et al.*, 1984; Dawbarn *et al.*, 1986; de Quidt *et al.*, 1985; Foster *et al.*, 1986; Hendry *et al.*, 1984a,b; Maccarrone and Jarrott, 1985, 1986b; McDonald *et al.*, 1982; Nakamura and Vincent, 1986; Rosser *et al.*, 1986; Tigges *et al.*, 1989; Vincent *et al.*, 1982b; Wahle *et al.*, 1986
Development	Allen *et al.*, 1984a; Foster and Schultzberg, 1984; Foster *et al.*, 1984; McDonald *et al.*, 1982; Senba *et al.*, 1985
Gastrointestinal System	Allen *et al.*, 1984c; Carlei *et al.*, 1985; Daniel *et al.*, 1985; *Ekblad *et al.*, 1984a,b, 1985, 1987; Falkmer *et al.*, 1985; Feher and Burnstock, 1986a,b; Furness *et al.*, 1983, 1984, 1985; Keast *et al.*, 1985; Lee *et al.*, 1985; Lindh *et al.*, 1986; Noe *et al.*, 1986, 1989a,b; Pettersson *et al.*, 1987a,b; Sundler *et al.*, 1983, *1986; Wang *et al.*, 1987

Table 3. Studies of NPY and related peptides in the reproductive neuroendocrine axis: Index to references. *Review articles.

Hypothalamus	Allen *et al.*, 1985d; Allen *et al.*, 1987; Bai *et al.*, 1985; Calka and McDonald, 1989; Card and Moore, 1982, 1984, 1988; Card *et al.*, 1983; Chronwell *et al.*, 1984, 1985; Ciofi *et al.*, 1987, 1988; Clark *et al.*, 1985; Crowley *et al.*, 1985; Crowley and Kalra, 1987; Fuxe *et al.*, 1987; Gehlert *et al.*, 1987; Gray *et al.*, 1986; Guy and Pelletier, 1988; Guy *et al.*, 1988; Härfstrand *et al.*, 1986b, 1987b; Harrington *et al.*, 1985; Hisano *et al.*, 1988; Holets *et al.*, 1988; Kalra and Crowley, 1984 a,b,c; Kalra *et al.*, 1987, 1988; Kerkerian *et al.*, 1985; Kerkerian and Pelletier, 1986; Khorram *et al.*, 1987a,b, 1988; Li and Pelletier, 1986; *McCann *et al.*, 1989; *McDonald, 1988; McDonald and Lumpkin, 1983; McDonald *et al.*, 1984, 1985a,b, 1986, 1988a,b,c, 1989; Moore *et al.*, 1984; Nakagawa *et al.*, 1985; Pau *et al.*, 1989; Pelletier *et al.*, 1984; Rodriguez-Sierra *et al.*, 1987; Sabatino and McDonald, 1987, 1988; Sabatino *et al.*, 1987, 1989a,b; Sahu *et al.*, 1987, 1988, 1989; Simerly and Swanson, 1987; Tigges *et al.*, 1987; Wehrenberg *et al.*, 1988; Woller *et al.*, 1989
Pituitary Portal System	McDonald *et al.*, 1987a,b, 1988c; Sutton *et al.*, 1987, 1988a,b
Pituitary	Crowley *et al.*, 1987; Fujii *et al.*, 1982; Jones *et al.*, 1989; Kerkerian *et al.*, 1985; Khorram *et al.*, 1988; McDonald *et al.*, 1984, 1985a, 1986, 1988c; Rodriguez-Sierra *et al.*, 1987; Saavedra and Cruciani, 1988; Sutton *et al.*, 1987, 1988a,b
Gonads & Genital Tract	
Female	Allen *et al.*, 1985a; Fried *et al.*, 1985; Heinrich *et al.*, 1986; Huang *et al.*, 1984; Kannisto *et al.*, 1986; McDonald *et al.*, 1987c; McNeill and Burden, 1986; Morris *et al.*, 1985; Papka *et al.*, 1985; Papka and Traurig, 1988; Samuelson and Dalsgaard, 1985; Stjernquist *et al.*, 1983; Tenmoku *et al.*, 1988
Male	Adrian *et al.*, 1984; Chang *et al.*, 1985; Gu *et al.*, 1983; Hedlund and Andersson, 1985; Huidoboro-Toro, 1985; Lundberg and Stjarne, 1984; Lundberg *et al.*, 1984a; Ohhashi and Jacobowitz, 1983; Stjarne and Lundberg, 1986; Stjarne *et al.*, 1986; Wahlestedt *et al.*, 1986

One general feature of NPY in the nervous system is its presence in local circuit neurons in several regions and also in brainstem noradrenergic and adrenergic neurons which project fibers throughout the brain, especially to hypothalamic regulatory nuclei (Blessing *et al.*, 1986; Dai *et al.*, 1986; Everitt *et al.*, 1984; Foster *et al.*, 1984; Hökfelt *et al.*, 1983a, 1983b; Sawchenco *et al.*, 1985; Smialowska, 1988; Sutin and Jacobowitz, 1988). In the periphery, NPY is found in local neurons and also in noradrenergic paraspinal and preaortic sympathetic ganglia. Thus, NPY is often colocalized with catecholamines, both centrally and peripherally. Therefore, NPY occupies a key position as a regulator of endocrine, cardiovascular and autonomic function through modulation of catecholaminergic efficacy. The discussion which follows is limited to the reproductive neuroendocrine axis.

Reproductive Neuroendocrine Axis

Hypothalamus. High concentrations of NPY immunoreactivity have been observed in several hypothalamic nuclei and in the preoptic area of many species. NPY-immunoreactive perikarya are concentrated in the ventromedial aspect of the arcuate nucleus (Fig. 1) (Chronwall *et al.*, 1984, 1985; de Quidt *et al.*, 1985; de Quidt and Emson, 1986a; Sabatino *et al.*, 1987), yet are distinct from the tuberoinfundibular dopaminergic neurons located in the dorsolateral part of the arcuate nucleus. However, Guy and Pelletier (1988) have observed contacts between NPY and tyrosine hydroxylase immunoreactive structures in the arcuate nucleus. A subset of these arcuate NPY neurons also contains growth hormone-releasing hormone (GHRH) which may have some physiological importance in view of the central effects of NPY on growth hormone secretion (Ciofi *et al.*, 1987, 1988; McDonald *et al.*, 1984, 1985a).

Dense fields of NPY-immunoreactive fibers are present in many areas including the suprachiasmatic, paraventricular, periventricular, arcuate, dorsomedial, anterior and posterior hypothalamic nuclei. The bed nucleus of the stria terminalis and the medial preoptic nucleus are also densely innervated. The

Fig. 1A-C. [Opposite page] Distribution of NPY-immunoreactivity in the infundibular nucleus (IN) and upper infundibular stem (UIS) of the rhesus monkey. A. The dorsal UIS contains many vessels (v) ventral to the IN. Bar=100 μm x 58. B. High-magnification view of A shows the relationship of NPY-labeled perikarya in the IN and varicose fibers streaming ventrally to surround portal vessels (v). Bar=100 μm x 92. C. NPY-immunoreactive axons surround portal capillaries in the middle IS (MIS) adjacent to the dorsal aspect of the anterior pituitary gland (AP). Bar=100 μm x 92. From *Cell Tissue Res.* 254: 499-509, 1988 with permission.

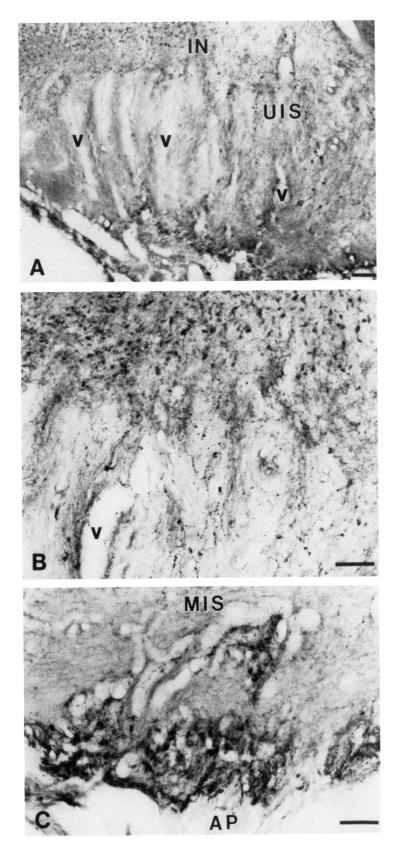

median eminence (ME) contains NPY-labeled fibers situated primarily in the subependymal and internal zones although other fibers surround capillary loops of the hypothalamo-hypophysial portal vessels (McDonald *et al.*, 1987a, 1988c). Species differences exist in the density of NPY-containing fibers around portal vessels; the rhesus monkey displays a high concentration while the rat and hamster show a lower concentration of these fibers (Figs. 1, 2) (McDonald *et al.*, 1987a, 1987b; Sabatino *et al.*, 1987). Some of the fibers in the ME probably release NPY into the portal vessels while other fibers are destined for the neural lobe. High levels of NPY have been measured in blood collected from the hypophysial stalk (McDonald *et al.*, 1987a, 1987b; Sutton *et al.*, 1987, 1988a, 1988b). The precise cellular origin of the NPY-labeled fibers in the hypothalamic nuclei and the ME is not completely understood at this time although perikarya in brainstem catecholaminergic nuclei and the arcuate nucleus probably account for the majority of the fibers.

Pituitary. NPY-immunoreactive fibers are located throughout the neurohypophysis of the rhesus monkey (Figs. 1, 2) (McDonald *et al.*, 1988c). These fibers course down the infundibular stalk and into the neural lobe where many are observed in the vicinity of vessels. Accordingly the neural lobe has been postulated to be a site of NPY secretion. NPY released here could enter the systemic circulation or travel via other vascular connections to reach the anterior lobe. NPY immunoreactive perikarya have also been described in the intermediate lobe of the rhesus monkey (McDonald *et al.*, 1988c).

Recently in the rat NPY mRNA has been detected in the anterior pituitary and NPY immunoreactivity localized in thyrotropes (Jones *et al.*, 1989). In addition to the local NPY production by these cells, NPY is also delivered in high concentrations to the anterior pituitary via the portal vessels. The effects of NPY on hormone secretion by the pituitary are discussed below in Section IV B.

Gonad. The ovary receives a dense innervation by NPY-containing fibers which are primarily associated with the vasculature, although additional fibers

Fig. 2A-C. [Opposite Page] Relationship of NPY-immunoreactive Fibers to Portal Vessels Within the MIS and Adjacent to the AP of the Rhesus Monkey. A. Dense accumulation of NPY-positive varicosities surrounding several small and large vessels in the MIS. The labeled vessel (v) distributes within the AP. Bar=50 μm x 169. **B.** NPY-immunoreactive varicosities (arrow) in relationship to a vessel (v) leading into the AP. Bar=20 μm x 264. **C.** NPY-labeled axons encircle this long portal vessel (v) in the MIS adjacent to the AP. Note the fibers that continue inferiorly in the IS. Bar=20 μm x 264. Abbreviations as in legend to Fig. 1. From *Cell Tissue Res.* 254: 499-509, 1988 with permission.

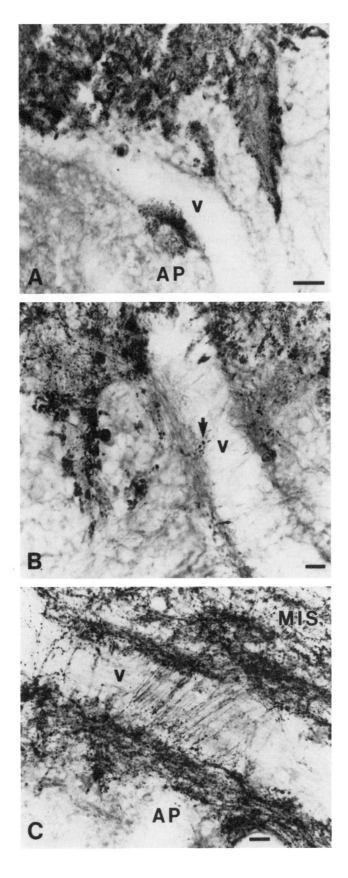

course in the interstitial tissue and in the vicinity of follicles (Allen *et al.*, 1985a; Huang *et al.*, 1984; Kannisto *et al.*, 1986; McDonald *et al.*, 1987c; Papka *et al.*, 1985; Stjernquist *et al.*, 1983). These axons are carried to the rat ovary via the plexus nerve and are probably involved in regulation of vasomotor tone (McDonald *et al.*, 1987c; Papka *et al.*, 1985; Stjernquist *et al.*, 1983). NPY does not appear to affect basal or FSH-stimulated estrogen and progesterone production from cultured granulosa cells *in vitro* (McDonald *et al.*, 1987c). However, in view of the location of NPY fibers around follicles and in the interstitial tissue, a potential endocrine role cannot be ruled out.

The testis receives a sparse innervation by NPY fibers which are associated with a few blood vessels and the tunica albuginea. NPY-containing fibers do not appear to be distributed within the testis itself.

Function of Neuropeptide Y in the Reproductive Neuroendocrine Axis

In this section the effects of NPY on reproductive hormone secretion from the hypothalamus and pituitary gland are presented.

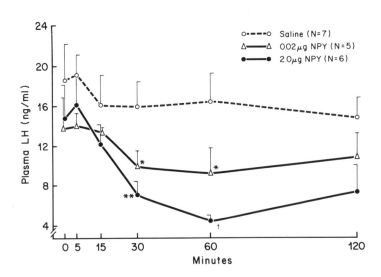

Fig. 3. Effect of the 3V injection of 0.9% saline and NPY on plasma levels of LH in conscious, unrestrained, ovariectomized rats. In this figure, points and vertical bars represent mean values ± SEM. Symbols adjacent to points represent the level of significance when compared to saline-injected controls: *, $p < 0.05$; **, $p < 0.025$; t, $p < 0.005$. NPY doses were 2.0 and 0.02 µg. From *Proc. Natl. Acad. Sci.* 82: 561-564, 1985 with permission.

Hypothalamus

Effects of Neuropeptide Y on hormone release in vivo. As described above, NPY was isolated from porcine brain by Tatemoto and colleagues while functional studies of reproductive neuroendocrine effects of PP were in progress. When NPY became commercially available, its potential neuroendocrine activity was evaluated in similar test systems; the first reports appeared in 1984 (Kalra and Crowley, 1984b, 1984c; McDonald *et al.*, 1984). Injection of pNPY into the 3V of OVX rats produced a prolonged and dose-related decrease in plasma levels of LH (McDonald and Lumpkin, 1983; McDonald *et al.*, 1985b). These inhibitory effects of high doses of NPY (5.0 μg) persisted for up to 3 h. Lower doses were also effective; suppression of LH levels was obtained after a 3V injection of 20 ng (Fig. 3) (McDonald *et al.*, 1984, 1985a). Plasma concentrations of follicle stimulating hormone (FSH) in these animals were not altered, lending further evidence to support the existence of a separate FSH-releasing factor. Growth hormone (GH) levels also declined in these animals; however, the minimal effective dose of NPY was higher than that required to decrease LH secretion (McDonald *et al.*, 1984, 1985a). These effects of NPY were postulated to occur by direct inhibition of LHRH secretion or through modulatory effects on catecholaminergic fibers which, in turn, influence LHRH secretion (Kalra and Crowley, 1984c; McDonald *et al.*, 1984, 1985a). The suppression of GH release was suggested to occur through stimulation of SRIF release, inhibition of GHRH release, through some modulation of catecholaminergic tone on SRIF/GHRH systems, or by a combination of these (McDonald *et al.*, 1984, 1985a).

Other investigators observed similar inhibitory effects of NPY on LH secretion in OVX rats but performed another experiment in which estrogen and progesterone were administered to OVX rats before injection of NPY (Kalra and Crowley, 1984c; Kalra *et al.*, 1987). The presence of these gonadal steroids changed the direction of the effect of NPY to a mild and brief stimulation of LH secretion. The effects of norepinephrine in this system are modulated by gonadal steroids in a similar manner. These data are particularly interesting in view both of the partial colocalization of NPY in noradrenergic and adrenergic fibers which innervate the hypothalamus and the numerous examples of NPY modulation of noradrenergic efficacy in other systems (Abel *et al.*, 1988; Huidoboro-Toro, 1985; Lundberg and Stjarne, 1984; Lundberg *et al.*, 1984a; Stjarne and Lundberg, 1986; Stjarne *et al.*, 1986). A recent study demonstrates that blockade of α_2-adrenergic receptors with the drug yohimbine, before cerebroventricular injection of NPY in OVX steroid-primed rats, suppressed the stimulatory effectiveness of NPY on

LH secretion (Allen *et al.*, 1987). Administration of the α_1 antagonist, prazosin, the β receptor antagonist, propranolol, the dopamine antagonist, pimozide, or the opiate receptor agonist morphine, did not attenuate the NPY-induced LH secretion (Allen *et al.*, 1987). Perhaps NPY modulates noradrenergic neurotransmission through direct or indirect effects on presynaptic α_2 receptors.

Guy *et al.* (1988) treated OVX rats with α-methylparatyrosine at 18 and 4 h before cerebroventricular injection of NPY (2 μg). Blockade of catecholamine synthesis had no effect on NPY-induced suppression of LH release, suggesting that catecholamines are not required for the inhibitory effects of NPY on LH secretion in OVX rats. However these investigators found that treatment of OVX rats with the serotonin neurotoxin, 5,7-dihydroxytryptamine, not only prevented NPY suppression of LH secretion but increased plasma levels of LH (Guy *et al.*, 1988). The authors suggested that NPY modulation of LHRH release may be direct or mediated by stimulation of serotonin release which may then inhibit LHRH secretion (Guy *et al.*, 1988). Additional studies are necessary to determine if catecholamines mediate the effects of NPY only in the presence of gonadal steroids and if NPY inhibition of LH release in OVX animals is direct or mediated through another transmitter or peptide system.

The neuroendocrine function of NPY has also been investigated in male rats. Injection of a large dose of NPY (15 μg) into the lateral ventricle of anesthetized and castrated rats reduced plasma levels of LH while exerting no effect on FSH (Kerkerian *et al.*, 1985). Administration of a 10 μg dose of NPY into the third ventricle of intact male rats significantly increased plasma levels of LH. A dose of 25 μg of NPY caused a slight stimulation of LH release, whereas both 2 and 0.5 μg had no effect (Allen *et al.*, 1985). In contrast, Härfstrand *et al.* (1987b) observed no effect of lateral ventricular injection of NPY on plasma levels of LH in intact male rats or rats pre-treated with α-methylparatyrosine, an inhibitor of tyrosine hydroxylase, the biosynthetic enzyme for catecholamines.

Methodological differences such as the location of the ventricular cannula, the method of blood collection (trunk vs. intra-atrial cannula), and the recovery period after ventricular cannulation may account for some of these apparently conflicting results. In a separate experiment, Härfstrand *et al.* (1986b) injected 1.25 nmol of NPY into the lateral ventricle of intact males rats. Plasma levels of LH significantly decreased; however, inhibition of tyrosine hydroxylase before NPY injection prevented this decline in LH. Although FSH levels were not altered by this high dose of NPY, levels of vasopressin, aldosterone, angiotensin II, adrenocorticotropic hormone, and corticosterone increased, while prolactin and TSH declined. These authors suggested that ventricular injection of NPY increases activity of tubero-infundibular dopaminergic neurons. In contrast, Allen *et al.* (1987) found that pimozide had no effect on the NPY-induced increase in LH levels in steroid-primed OVX rats. These authors injected 0.47 nmol, approximately one-third of the dose given by Härfstrand *et al.* (1986b).

Li and Pelletier (1986) observed that the number of NPY immunoreactive neurons in the arcuate nucleus of intact female rats increased following treatment with haloperidol, a dopaminergic receptor antagonist, or α-methylparatyrosine. They suggested that dopaminergic neurons may exert a tonic inhibitory influence on NPY-containing neurons in the arcuate nucleus. The potential interactions of NPY and dopaminergic neurons in the arcuate nucleus deserve additional investigation.

Recent findings support the hypothesis advanced in these studies, namely that NPY affects LHRH secretion from the hypothalamus. It is well established that the pulsatile secretion of LH from the pituitary is directly influenced by episodic LHRH release. Injection of pNPY (5 µg) into the 3V of OVX rats

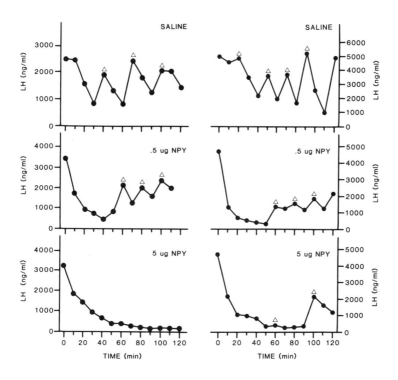

Fig. 4. Individual profiles of LH secretion (nanograms per ml) in OVX rats receiving a 3V injection of NPY (5.0 or 0.5 µg/2 µl) or saline (2 µl) at 0 min. Triangle symbol indicates a pulse of LH secretion. From *Endocrinology* 125: 186-191, 1989 with permission.

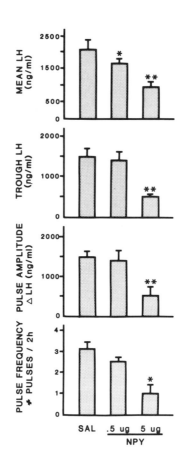

Fig. 5. This four-panel histogram summarizes the effects of NPY (5.0 or 0.5 μg) on the mean level, trough level, pulse amplitude, and pulse frequency of LH secretion over the 2 h sampling period (mean ± SEM). *, p < 0.05, **, p < 0.01 (vs. saline (SAL)-injected controls). From *Endocrinology* 125: 186-191, 1989 with permission.

suppressed the pulse frequency, pulse amplitude, and trough levels of LH secretion when compared to vehicle-injected controls (Figs. 4, 5) (McDonald *et al.*, 1989). In an additional experiment, the sensitivity of the anterior pituitary to an intravenous injection of LHRH (10 ng/100 g BW) was evaluated 60 min after 3V injection of NPY, when LH secretion is maximally suppressed, and compared to the response in controls receiving 3V saline (McDonald *et al.*, 1989). Control animals showed a 124% increase in LH release. In contrast, NPY-treated animals increased LH secretion by 1239% (Fig. 6). This hyper-responsiveness was probably caused by decreased delivery of endogenous LHRH to gonadotrophs through NPY inhibition of pulsatile LHRH release into the portal vessels. Perhaps the gonadotrophs increased intracellular pools of LH, and/or their sensitivity to LHRH, through changes in LHRH receptors or post-receptor transduction mechanisms (McDonald *et al.*, 1989).

Khorram *et al.* (1987a) employed push-pull perfusion of the medial basal hypothalamus to examine the effects of NPY on LHRH release in intact and OVX rabbits. NPY decreased the pulse frequency and amplitude of LHRH release in OVX rabbits and stimulated the mean level and pulse amplitude of LHRH release in intact animals (Khorram *et al.*, 1987a). The effects of NPY on

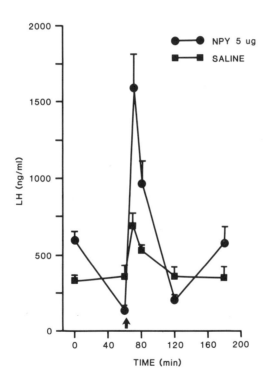

Fig. 6. Effect of NPY on the pituitary LH response to iv LHRH. Plasma levels of LH (mean ± SEM) in OVX rats receiving a 3V injection of NPY (5 μg; circle) or saline (square). At 60 min, when LH levels were significantly reduced in NPY-treated rats, LHRH (10 ng/100 g BW) was injected iv (arrow). Saline-injected animals showed a 124% increase in plasma LH 10 min after LHRH administration. In contrast, NPY-treated animals responded with a 1239% increase in LH release (p < 0.01 vs. controls). From *Endocrinology* 125: 186-191, 1989 with permission.

LHRH secretion may be direct or mediated through noradrenergic input to the pulse generator. Taken together, these data from *in vivo* studies suggest that NPY is a component in the LHRH pulse generating system.

Additional support for a central role of NPY in the control of LHRH secretion has been provided by Wehrenberg *et al.* (1988) who injected NPY antiserum into the cerebral ventricle of estrogen- and progesterone-treated OVX rats and blocked the surge of LH secretion. Presumably the NPY antiserum prevented endogenously released NPY from stimulating LHRH secretion before the LH surge. Guy *et al.* (1988) injected 10 or 30 μl of NPY antiserum into the lateral ventricle of OVX rats and observed no effect on plasma levels of LH within 1 h after the injection. Perhaps NPY immunoneutralization is effective only in steroid-treated rats. The site of action of NPY or NPY antiserum to affect LHRH release *in vivo* is not known. There are several possibilities, including: 1) the medial preoptic nucleus/diagonal band of BROCA region where NPY fibers overlap with LHRH perikarya (Calka and McDonald, 1989; McDonald *et al.*, 1988b); 2) the medial basal hypothalamus where NPY perikarya are concentrated and where several components of the LHRH pulse generator

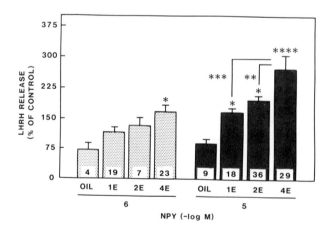

Fig. 7. Effects of estrogen (1E = 1 10mm implant; 2E = 2 10mm implants; 4E = 2 20mm implants of estradiol benzoate 235 μg/ml) or vehicle treatment on NPY (10^{-6} and 10^{-5} M)-stimulated LHRH release from ME fragments of OVX rats. ****, $p < 0.001$, 4E vs. OIL for 10^{-5} M NPY. ***, $p < 0.005$, 4E vs. 1E for 10^{-5} M NPY. **, $p < 0.01$, 4E vs. 2E for 10^{-5} M NPY. *, $p < 0.05$, 2E and 1E vs. OIL for 10^{-5} M NPY, and 4E vs. OIL for 10^{-6} M NPY. From *Endocrinology* 124: 2089-2098, 1989 with permission.

reside (Chronwall *et al.*, 1984, 1985; McDonald *et al.*, 1989); and 3) the ME where NPY and LHRH fibers overlap in the lateral region and internal zone (Calka and McDonald, 1989; McDonald *et al.*, 1988b). Insight into this issue has been provided from studies performed *in vitro* using medial basal hypothalamic (MBH) and ME fragments from rats and rabbits.

  Effects of neuropeptide Y on hormone release in vitro. Short-term incubations of MBH and ME tissues obtained from OVX and OVX-steroid treated animals have been used to assess the effects of NPY on LHRH secretion (Crowley and Kalra, 1987; Sabatino and McDonald, 1987, 1988; Sabatino *et al.*, 1989a, 1989b). NPY has no inhibitory effect on basal or potassium-stimulated LHRH release from MBH or ME tissues from OVX animals (Fig. 7). These studies suggest that additional circuits which are not contained in these tissue fragments may be needed to demonstrate the inhibitory effects of NPY in OVX rats. Perhaps 3V injections of NPY activate MBH neurons which project rostrally to contact LHRH perikarya. These central injections of NPY may activate NPY perikarya in the arcuate nucleus which could project to the preoptic region. NPY may also affect its own secretion in some form of ultrashort loop feedback

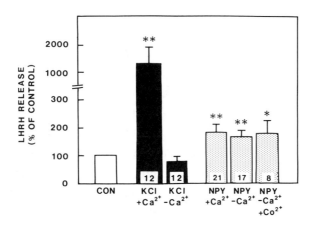

Fig. 8.   Effect of Ca^{2+}-free medium on NPY (10^{-5} M) and KCl (56 mM)-stimulated LHRH release form ME fragments obtained from 2E-treated OVX rats. Ca^{2+}-free medium is designated $-Ca^{2+}$, whereas normal medium containing 2.5mM Ca^{2+} is designated $+Ca^{2+}$. **, $p < 0.001$, KCl ($+Ca^{2+}$) vs. KCl ($-Ca^{2+}$) or control (CON) period and NPY ($+Ca^{2+}$ or $-Ca^{2+}$) vs. control period, *, $p < 0.05$, NPY [$-Ca^{2+}$ with Co^{2+} (2.5 mM)] vs. control period. There were no significant differences between NPY groups. From *Endocrinology* 124: 2089-2098, 1989 with permission.

as originally proposed (McDonald *et al.*, 1985a). Alternatively, serotonergic and/or noradrenergic neurons may be involved in mediating this inhibitory effect of NPY on LHRH release and the absence of such cell bodies in these fragments may account for the lack of NPY inhibition (Allen *et al.*, 1987; Guy *et al.*, 1988; Kalra and Crowley, 1984c; McDonald *et al.*, 1985a).

Gonadal steroids dramatically change the response of both MBH and ME fragments to NPY. Treatment of OVX rats with estrogen enhanced NPY-stimulated LHRH release from MBH fragments (Crowley and Kalra, 1987). Higher doses of estrogen increased the stimulatory effect of NPY (Crowley and Kalra, 1987). Khorram *et al.* (1988) tested MBH fragments from intact and OVX rabbits in a superfusion system. NPY had no effect on LHRH secretion from MBHs obtained from OVX rabbits; however, NPY stimulated LHRH release from MBH fragments of intact animals.

Sabatino and colleagues utilized the isolated ME preparation to investigate the effects of NPY on LHRH release (Sabatino and McDonald, 1987, 1988; Sabatino *et al.*, 1989a, 1989b). One significant advantage of this system, which has been employed extensively to study LHRH secretory dynamics, is that potential influences by opioid, dopaminergic, LHRH, NPY and other perikarya are eliminated. Administration of silastic capsules, containing estradiol benzoate in oil (235 µg/ml) to OVX rats produced physiological plasma levels of estrogen observed during the estrous cycle. NPY stimulated LHRH release in a dose-related manner from individual ME tissues and the stimulatory effectiveness of NPY was enhanced by increasing plasma estrogen levels within this physiological range (Figs. 7, 8) (Sabatino and McDonald, 1988; Sabatino *et al.*, 1989b). Thus, the LHRH fibers in the ME are sufficient to demonstrate NPY-stimulated LHRH release in OVX estrogen-treated rats. In contrast, additional circuitry beyond that contained within the ME, or within the MBH, seems to be involved in the NPY-induced suppression of LHRH secretion in OVX rats receiving vehicle.

Sabatino and colleagues also demonstrated that removing calcium from the culture medium had no effect on NPY-stimulated LHRH release, even in the presence of 2.5 mM cobalt, implying that influx of extracellular calcium is not an obligatory step in NPY-stimulated LHRH secretion (Fig. 8) (Sabatino and McDonald, 1988; Sabatino *et al.*, 1989b). Influx of extracellular calcium is also not required for a portion of the prostaglandin E_2-induced LHRH release from the ME. Translocation of calcium from intracellular storage sites is likely to be involved in NPY-stimulated LHRH secretion from the ME. Virtually nothing is known, at this time, about the intracellular pathways which mediate NPY effects on LHRH release. Also unknown at this time is whether the stimulatory effects of NPY on LHRH release from the ME are direct or mediated by effects on other transmitters such as norepinephrine. These studies strongly suggest that estrogen is necessary for NPY to stimulate LHRH release, and that increasing plasma estrogen levels within a physiological range significantly increases the efficacy of NPY.

We have recently investigated the hypothesis that progesterone enhances the stimulatory effectiveness of NPY on LHRH secretion from ME fragments of OVX rats treated with estrogen (Sabatino *et al.*, 1989a). Animals were treated with silastic capsules containing estrogen as described above. Three days later at 8:30 a.m., rats received a subcutaneous injection of 1, 2, or 19 mg of progesterone or vehicle. Three hours later the animals were sacrificed, the ME dissected and placed in short-term culture. The 1 and 2 mg doses of progesterone produced high physiological plasma levels of progesterone while pharmacological levels were obtained after the 19 mg injection. The basal release and ME content of LHRH were significantly increased by some of these progesterone treatments. Physiological doses of progesterone were without effect on NPY-induced LHRH release in rats with diestrous or early proestrous plasma levels of estrogen. Administration of physiological doses of progesterone to rats with high proestrous levels of estrogen slightly, although not significantly, decreased the stimulatory effectiveness of NPY on LHRH release. Treatment with the pharmacological dose of progesterone significantly decreased NPY-stimulated LHRH secretion in animals with high proestrous levels of estrogen, in spite of significantly elevated content and basal release of LHRH from the ME (Sabatino *et al.*, 1989a). These results suggest that progesterone does not enhance the effectiveness of NPY and that under certain conditions progesterone may actually decrease NPY-stimulated LHRH release. In contrast, Crowley and Kalra (1987) found a slight but statistically nonsignificant increase in NPY-stimulated LHRH release from MBH fragments of rats treated with estrogen and progesterone when compared to rats receiving estrogen alone. These animals received estrogen capsules and a subcutaneous injection of progesterone (15 mg) two days before sacrifice. Therefore, the time of sacrifice following progesterone administration may be an important factor in the assessment of progesterone effects on the modulation of NPY activity. Additional research is needed to determine the physiological importance, if any, of progesterone in altering the nature of NPY effects on LHRH release throughout the day of proestrous.

The data presented above show that NPY modulates LHRH secretion from the hypothalamus both *in vivo* and *in vitro* and that the direction and magnitude of these effects in females are related to the presence and level of circulating estrogen.

Gonadal steroids affect hypothalamic levels of neuropeptide Y. Alterations in plasma levels of gonadal steroids also affect the concentration of NPY in the hypothalamus. Crowley *et al.* (1985) examined the effects of bilateral OVX followed by treatment with estrogen alone or the combination of estrogen and progesterone on NPY levels in several brain regions. Injection of 50 μg of estradiol benzoate significantly decreased NPY levels in the ME 48 h later. Injection of a single dose of progesterone (2.5 mg) at 10:00 AM into OVX and estrogen-treated animals induced a significant increase in NPY levels in the ME 1 h later followed by a gradual decline. The pattern of NPY alterations in the ME after progesterone administration was similar to that for LHRH. These changes in the levels of both peptides preceded the afternoon rise in plasma LH. A significant decline in NPY concentration was detected in the interstitial nucleus of the stria terminalis and in the arcuate nucleus following estrogen treatment, while no changes were observed in the preoptic, periventricular and ventromedial nuclei. Progesterone treatment further decreased NPY levels in the interstitial nucleus of the stria terminalis and in the ventromedial nucleus while transient declines were seen in the preoptic and arcuate nuclei (Crowley *et al.*, 1985).

Sabatino *et al.* (1989b) measured NPY and LHRH concentrations in the ME of OVX rats implanted for three days before sacrifice with silastic tubes containing estradiol benzoate in oil. NPY levels in the ME increased slightly, although not significantly, in animals with diestrous or early proestrous levels of estrogen (17 and 30 pg/ml), whereas a significant increase was seen in rats with plasma estrogen levels of 73 pg/ml. LHRH levels in the ME were more responsive to these estrogen treatments and were elevated significantly when plasma estrogen concentrations were 30 and 73 pg/ml. These results contrast with those described above which showed a significant decrease in NPY levels in the ME following estrogen injection 48 h before sacrifice (Crowley *et al.*, 1985). It is probable that the methods of estrogen administration in these two studies produced very different patterns and concentrations of circulating estrogen which may have had dissimilar effects on NPY levels in the ME. A single subcutaneous injection of 1 or 2 mg of progesterone at 8:00 AM to OVX rats implanted with silastic capsules containing estrogen, which produced plasma levels of 73 pg/ml, had no effect on the ME content of NPY and slightly increased LHRH content at 11:00 AM when compared to vehicle-injected controls (Sabatino *et al.*, 1989a). These doses of progesterone produced plasma levels of 30 and 64 ng/ml which are similar to those observed on proestrous. Injection of a pharmacological dose (19 mg) of progesterone, which resulted in plasma levels of 534 ng/ml, significantly increased the ME content of NPY and LHRH. These observations on the effects of progesterone also differ somewhat from those of Crowley *et al.* (1985) although the method of estrogen administration and the time of sampling after progesterone injection were not the same. These results emphasize both the need to carefully evaluate the method of steroid administration when comparing studies about the effectiveness of NPY on LHRH release and to consider frequent testing throughout the day of the experiment.

Castration of male rats two weeks prior to sacrifice significantly decreased the concentration of NPY in the ME, arcuate nucleus, and ventromedial nucleus and had no effect on NPY levels in the medial preoptic, paraventricular and dorsomedial nuclei (Sahu *et al.*, 1987, 1989). Restoration of plasma testosterone levels with two 15 mm subcutaneous testosterone implants for ten days increased NPY levels in the ME, arcuate and ventromedial nuclei and had no effect on levels in the medial preoptic, paraventricular and dorsomedial nuclei.

The site(s) of action where testosterone or estrogen induce these regionally specific changes in NPY levels is currently unknown. As described in a previous section, NPY perikarya are concentrated in the arcuate nucleus, and in brainstem catecholaminergic neurons which project to the hypothalamus as well as other locations. One of the challenges in this area of research is to determine the means by which gonadal steroids affect hypothalamic NPY levels in a site-specific manner. Do neurons which synthesize NPY express receptors for estrogen and/or progesterone or are the effects of these steroids mediated by other neurons? In contrast to the results obtained in rats, Khorram *et al.* (1987b) reported that chronic treatment (6 weeks) of castrated rhesus monkeys with testosterone administered in subcutaneous silastic capsules had no effect on NPY levels in several hypothalamic and limbic regions. These discrepancies between studies performed in rats and monkeys may represent true species differences in steroid sensitivity or may reflect differences in rates of biosynthesis, transport and secretion.

Hypothalamic secretion of neuropeptide Y. The factors which control NPY secretion from the hypothalamus are poorly understood at this writing. Recently, NPY release has been examined *in vivo* using push-pull perfusion of the basal hypothalamus. Pau *et al.* (1989) measured low NPY levels in push-pull perfusates of the MBH and 3V of unanesthetized, intact female rabbits. Intravenous and intraventricular administration of cupric acetate increased levels of NPY and LHRH released into the MBH and 3V. Plasma LH and FSH levels rose gradually following these initial waves of NPY and LHRH secretion. It remains to be determined, as suggested by the authors, if the release of NPY induced by cupric acetate caused the increase in LHRH secretion or if these two secretory events were produced by related, but independent, mechanisms (Pau *et al.*, 1989).

The *in vivo* release of NPY in the stalk-ME of castrated rhesus monkeys has recently been characterized using the push-pull perfusion method (Woller *et al.*, 1989). NPY release was pulsatile in both male and female monkeys and occurred with a similar frequency to that reported for LHRH release. A careful examination of the potential relationship between pulsatile NPY and LHRH release in castrated monkeys and rats is essential in view of the suppressive effects of 3V injection of NPY on pulsatile LH release in OVX rats (McDonald *et al.*, 1989).

Sahu *et al.* (1989) examined the effects of castration and testosterone replacement on the basal and potassium-stimulated release of NPY *in vitro* from short-term cultures of rat MBH fragments. Castration decreased potassium-stimulated NPY release when compared to that in intact rats. Testosterone treatment of castrated rats significantly increased NPY release to levels slightly greater than those observed in intact animals. Application of prostaglandin E_2 (PGE_2) did not significantly modify NPY release from MBH fragments of intact or castrated rats, but increased LHRH secretion from MBHs obtained from intact animals. Perhaps different intracellular pathways are responsible for the secretion of NPY and LHRH.

The extracellular factors (neurotransmitters, neuromodulators and other neuropeptides) which affect NPY release from the hypothalamus and the intracellular pathways which mediate NPY secretion are currently unknown and offer a new challenge for investigators in this field. Unraveling these mysteries will enhance our understanding of the role of NPY in the neuroendocrine regulation of LHRH secretion from the hypothalamus and in the regulation of gonadotropin release from the anterior pituitary gland.

Anterior Pituitary

In addition to this central modulation of LHRH secretion, NPY is also secreted into the pituitary portal blood and directly affects hormone secretion from the anterior pituitary gland (Crowley *et al.*, 1987; Khorram *et al.*, 1988; McDonald *et al.*, 1985a, 1987a, 1987b; Sutton *et al.*, 1987, 1988a, 1988b).

Several investigators have examined the potential activity of NPY on pituitary hormone secretion using different test systems. McDonald *et al.* (1984;

1985a) demonstrated that NPY stimulated the secretion of LH, FSH and GH from anterior pituitary cells which had been dispersed and cultured overnight before being loaded into syringe columns containing Biogel P-2 (Fig. 9). However, no effects of NPY were detected when the cells were floating freely in test tubes. Kerkerian *et al.* (1985) applied NPY (1 μM-10 pM) to 4-day primary cultures of anterior pituitary cells from adult female rats and observed no effect on the release of LH, FSH, GH, and ACTH in the absence and presence of LHRH (0.3 nM) during a 4 h exposure period. Rodriguez-Sierra *et al.* (1987) incubated hemipituitary fragments for 1 h and found no effect of NPY (10 nM-10 pM) on

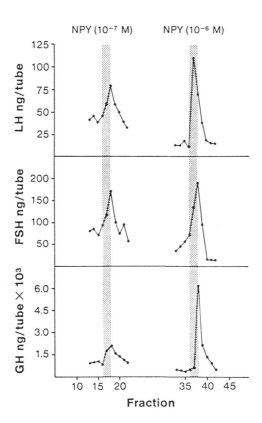

Fig. 9. LH, FSH, and GH released from a perifused cell column (Bio-Gel P-2, 0.4 x 1.5 cm) loaded with 7 x 10^6 dispersed anterior pituitary cells (ovariectomized female rat donors). Five-minute fractions were collected and NPY (10^{-6} and 10^{-7} M) was included in the medium for the 10-min exposure periods indicated by the stippled bars. From *Proc. Natl. Acad. Sci.* 82: 561-564, 1988 with permission.

LH, FSH or TSH secretion. In view of these apparently conflicting results it seems that the effects of NPY on anterior pituitary hormone release may be transient and detected more efficiently using a continuous flow system.

Khorram *et al.* (1988) quartered anterior pituitaries from OVX and intact rabbits and placed them in a perfusion system. NPY, at a dose of 80 nM, caused a transient stimulation of LH and FSH release from pituitaries of OVX rabbits and a sustained stimulation of LH and FSH secretion from glands taken from intact rabbits. This study provided the first indication that ovarian steroids modulate the response of the anterior pituitary gland to NPY. Thus, NPY effects on reproductive hormone secretion from both the hypothalamus and anterior pituitary gland are modulated by the presence of circulating ovarian steroids.

Crowley *et al.* (1987) employed two different test systems to analyze the effects of NPY on LH secretion from pituitaries of OVX rats. NPY was applied to hemipituitary fragments at concentrations of 0.1 and 1 μM and caused a slight but statistically significant increase in the proportion of LH released when compared to the release in tubes containing untreated control fragments. NPY also augmented the effect of LHRH on LH release from these fragments. In a separate experiment, dispersed pituitary cells were cultured for three days before testing. NPY was applied to the cells which were attached to culture plates. In this experiment, NPY did not alter basal release of LH but potentiated the response of these cells to 1 and 10 nM LHRH. This effect of NPY was dose related (10 and 100 nM) in the presence of 10 nM LHRH but not 1 nM LHRH. The authors suggested that these effects of NPY are consistent with the view that NPY acts primarily as a modulator of the current pituitary response to LHRH (Crowley *et al.*, 1987). Since high concentrations of NPY are present in the pituitary portal circulation (McDonald *et al.*, 1987a; Sutton *et al.*, 1988a, 1988b), it is conceivable that NPY may play a physiological role to modulate pituitary sensitivity to LHRH on the afternoon of proestrous in intact cycling animals. NPY may also act at the level of the pituitary to affect the magnitude of LH secretion caused by each secretory pulse of LHRH.

Additional support for the concept of NPY modulation of the pituitary response to LHRH was provided by Sutton *et al.* (1988b) who reported that peripheral injection of NPY antiserum blocked the LH surge in OVX estrogen- and progesterone-treated rats. Perhaps NPY provides a modulatory tone on pituitary gonadotrophs to change their sensitivity to LHRH. Of course this situation is further complicated by the fact that NPY secretion into the portal blood also changes, in part due to the influence of gonadal steroids (Sutton *et al.*, 1987, 1988b). Crowley *et al.* (1985) observed that the ME concentration of NPY and LHRH changed in a parallel fashion before the LH surge and suggested that these peptides may be coreleased into the portal circulation. The hypophysial-portal plasma concentrations of NPY and LHRH change in a parallel manner during the estrous cycle. Sutton *et al.* (1988b) have suggested that NPY serves to prime gonadotrophs to LHRH. Additional evidence indicates a temporal relationship in the postnatal changes in the concentration of NPY and LHRH in the preoptic area, and also in the hypophysial portal levels of NPY and LHRH around the time of vaginal opening (Sutton *et al.*, 1988a). NPY levels in the portal blood increase significantly in the afternoon preceding the day of vaginal opening and these changes occur at approximately the time of the LHRH and LH surges.

Some investigators have injected NPY subcutaneously or intravenously to examine its activity on pituitary hormone release although these routes of NPY administration may produce undesirable cardiovascular effects. Kerkerian *et al.* (1985) injected 300 μg of NPY into the jugular vein of intact male rats and produced a sustained reduction in circulating LH levels without any effect on FSH. However, plasma levels of prolactin were elevated 4-fold following the injection of this large dose of NPY, perhaps indicating a stress effect (Zukowska-Grojec and Vaz, 1988). In another study, subcutaneous injection of 10 μg of NPY into castrated rats caused a doubling of plasma LH levels without affecting FSH or TSH (Rodriguez-Sierra *et al.*, 1987).

Taken together, these studies implicate NPY in the control of anterior pituitary function, yet it seems surprising that NPY receptors have not been observed in the anterior lobe of the pituitary (Busch-Sorensen *et al.*, 1989; Saavedra and Cruciani, 1988).

Although additional studies are needed to clarify the role of NPY in the secretion of gonadotropins and other hormones directly from the anterior pituitary gland, the majority of data suggest that NPY probably modulates hormone release at this site and may affect the cellular response to other hypophysiotropic factors such as LHRH. In addition, gonadal steroids appear to exert an additional level of modulation on the effectiveness of NPY on the anterior pituitary gland.

Summary

NPY, distributed extensively throughout the central, peripheral, and enteric nervous systems, is localized in local circuit neurons, in noradrenergic and adrenergic neurons in the brainstem, and in paraspinal and preaortic ganglia.

NPY is a multifunctional peptide which often modulates catecholaminergic activity in neuroendocrine, endocrine, cardiovascular and gastrointestinal systems.

NPY plays an influential modulatory role in the reproductive neuroendocrine axis, especially in the hypothalamus and anterior pituitary gland. NPY probably affects vasomotor function in the ovary although the proximity of NPY-containing fibers to ovarian follicles and interstitial tissue suggests that some unknown endocrine function remains to be discovered. High concentrations of NPY are found in several hypothalamic nuclei, the ME and neurohypophysis. The arcuate nucleus, which contains many NPY immunoreactive neurons, and brainstem neurons which contain NPY and catecholamines, contribute to the innervation of the hypothalamus although additional cell bodies are probably involved. High concentrations of NPY are present in the pituitary portal blood and appear to change in a similar fashion to LHRH in response to gonadal steroids. Gonadal steroids affect the concentration of NPY in hypothalamic nuclei as well as in the portal blood. Recently, the NPY prohormone messenger RNA has been found in pituitary thyrotrophs. The posterior pituitary receives a robust NPY innervation and may represent another release site by which NPY can gain access to the anterior pituitary as well as the systemic circulation.

NPY affects hypothalamic secretion of LHRH and qualitative as well as quantitative charactertistics of this influence are related to circulating gonadal steroid levels. NPY stimulates LHRH secretion when gonadal steroids are present and inhibits LHRH release in gonadectomized animals. Some of these effects may involve activation of noradrenergic systems. NPY suppresses pulsatile LHRH release in OVX rats and probably represents an important component of the LHRH pulse generator. Physiological levels of circulating estrogen are sufficient to demonstrate NPY stimulation of LHRH secretion from the MBH and ME *in vitro*. Physiological levels of progesterone in the presence of physiological levels of circulating estrogen do not appear to affect NPY-stimulated LHRH release from the ME of OVX, estrogen-primed rats.

NPY directly affects LH and FSH release from the anterior pituitary gland *in vitro* and enhances the gonadotropin secretory response to LHRH under certain conditions. These hypothalamic and pituitary actions of NPY support an important role for this neuropeptide in the proper coordination of neuroendocrine events which culminate in ovulation.

Directions for Future Research

Our understanding of the role of NPY in the control of reproductive hormone secretion from the hypothalamus and anterior pituitary gland has increased considerably in recent years although some of the most challenging questions remain to be answered:

1. How does NPY modulate LHRH secretion from the hypothalamus? Is this a direct effect of NPY on LHRH axons and/or cell bodies, an indirect effect occurring through catecholaminergic systems, or a combination of these two possibilities? Are other neurotransmitter (dopaminergic, serotonergic, noradrenergic) or neuropeptide systems (opioid, galanin) involved in the mediation of NPY activity? What is the role of NPY in the LHRH pulse generator? Through which mechanism(s) do gonadal steroids, especially estrogen and testosterone, affect the NPY modulation of LHRH secretion? Which intracellular pathways in LHRH neurons transduce the NPY signal?

2. Which neurotransmitter and neuropeptide systems are involved in the control of NPY biosynthesis and secretion from the hypothalamus? How do gonadal steroids affect NPY biosynthesis and secretion? Does this regulation occur through brainstem NPY-catecholaminergic neurons, arcuate NPY-neurons or interneurons? Is NPY release into the pituitary portal vessels coordinated with LHRH secretion?

3. Through which mechanisms does NPY affect LH and FSH release from gonadotrophs or enhance the response to LHRH? Is the sensitivity of gonadotrophs to NPY modified during the estrous cycle and especially before the LH surge? Are these effects of NPY receptor-mediated? Anatomical and biochemical studies indicate that NPY receptors are not present in the anterior pituitary. How then do these effects of NPY occur?

4. Finally, perhaps the most important question is whether the observed effects of NPY on the hypothalamus and anterior pituitary gland are physiologically significant in the control of ovulation.

ACKNOWLEDGMENTS

The author expresses his sincere thanks to Shirley Cox and Karen Cooper for their excellent secretarial assistance. The author gratefully acknowledges the support of a Basil O'Connor Scholar Research Grant (5-524) and a Basic Research Grant (1-1118) from the March of Dimes Birth Defects Foundation, a Research Career Development Award (HD-00727) and research grants (HD-19731 and BRSG S07-RR-05364) from the National Institutes of Health.

REFERENCES

Abel, P.W., C. Han, B.D. Noe and J.K. McDonald. 1988. Neuropeptide Y: Vasoconstrictor effects and role in cerebral vasospasm after experimental subarachnoid hemorrhage. *Brain Res.* 463: 250-258.

Adrian, T.E., J.M. Allen, S.R. Bloom, M.A. Ghatei, M.N. Rossor, G.W. Roberts, T.J. Crow, K. Tatemoto and J.M. Polak. 1983. Neuropeptide Y distribution in human brain. *Nature* 306: 584-586.

Adrian, T.E., J. Gu, J.M. Allen, K. Tatemoto, J.M. Polak and S.R. Bloom. 1984. Neuropeptide Y in the human male genital tract. *Life Sci.* 35: 2643-2648.

Agnati, L.F., K. Fuxe, R. Grimaldi, A. Härfstrand, M. Zoli, I. Zini, D. Ganten and P. Bernardi. 1986. Aspects on the role of neuropeptide Y and atrial peptides in control of vascular resistance. In *Central and Peripheral Mechanisms of Cardiovascular Regulation*, eds. A. Magro, W. Osswald, D. Reis and P. Vanhoutte, 503-526. Plenum Publishing, New York.

Agnati, L.F., K. Fuxe, M. Zoli, I. Zini, A. Härfstrand, G. Toffano and M. Goldstein. 1988. Morphometrical and microdensitometrical studies on phenylethanolamine-N-methyltransferase and neuropeptide Y immunoreactive neurons in the rostral medulla oblongata of the adult and old male rat. *Neuroscience* 26: 461-478.

Albers, H.E. and C.F. Ferris. 1984. Neuropeptide Y: Role in light-dark cycle entrainment of hamster circadian rhythms. *Neurosci. Lett.* 50: 163-168.

Albers, H.E., C.F. Ferris, S.E. Leeman and B.D. Goldman. 1984. Avian pancreatic polypeptide phase shifts hamster circadian rhythms when microinjected into the suprachiasmatic region. *Science* 223: 833-835.

Allen, J.M., G.P. McGregor, T.E. Adrian, S.R. Bloom, S.Q. Zhang, K.W. Ennis and W.G. Unger. 1983. Reduction of neuropeptide Y (NPY) in the rabbit iris-ciliary body after chronic sympathectomy. *Exp. Eye Res.* 37: 213-215.

Allen, J.M., G.P. McGregor, P.L. Woodhams, J.M. Polak and S.R. Bloom. 1984a. Ontogeny of a novel peptide, neuropeptide Y (NPY) in rat brain. *Brain Res.* 303: 197-200.

Allen, J.M., I.N. Ferrier, G.W. Roberts, A.J. Cross, T.E. Adrian, T.J. Crow and S.R. Bloom. 1984b. Elevation of neuropeptide Y (NPY) in substantia innominata in Alzheimer's type dementia. *J. Neurological Sci.* 64: 325-331.

Allen, J.M., J. Gu, T.E. Adrian, J.M. Polak and S.R. Bloom. 1984c. Neuropeptide Y in the guinea-pig biliary tract. *Experientia* 40: 765-767.

Allen, J.M., J.C. Yeats, M.A. Blank, G.P. McGregor, J. Gu, J.M. Polak and S.R. Bloom. 1985a. Effect of 6-hydroxydopamine on neuropeptides in the rat female genitourinary tract. *Peptides* 6: 1213-1217.

Allen, J.M., J.M. Polak and S.R. Bloom. 1985b. Presence of the predicted C-flanking peptide of neuropeptide Y (CPON) in tissue extracts. *Neuropeptides* 6: 95-100.

Allen, J.M., A.J. Cross, T.J. Crow, F. Javoy-Agid, Y. Agid and S.R. Bloom. 1985c. Dissociation of neuropeptide Y and somatostatin in Parkinson's disease. *Brain Res.* 337: 197-200.

Allen, J.M. and S.R. Bloom. 1986. Neuropeptide Y: A putative neurotransmitter. *Neurochem. Int.* 8: 1-8.

Allen, L.G., P.S. Kalra, W.R. Crowley and S.P. Kalra. 1985d. Comparison of the effects of neuropeptide Y and adrenergic transmitters on LH release and food intake in male rats. *Life Sci.* 37: 617-623.

Allen L.G., W.R. Crowley and S.P. Kalra. 1987. Interactions between neuropeptide Y and adrenergic systems in the stimulation of luteinizing hormone release in steroid-primed ovariectomized rats. *Endocrinology* 121: 1953-1959.

Allen, Y.S., T.E. Adrian, J.M. Allen, K. Tatemoto, T.J. Crow, S.R. Bloom and J.M. Polak. 1983. Neuropeptide Y distribution in the rat brain. *Science* 221: 877-879.

Allen, Y.S., G.W. Roberts, S.R. Bloom, T.J. Crow and J.M. Polak. 1984d. Neuropeptide Y in the stria terminalis: Evidence for an amygdalofugal projection. *Brain Res.* 321: 357-362.

Andrews, P.C., D. Hawke, J.E. Shively and J.E. Dixon. 1985. A nonamidated peptide homologous to porcine peptide YY and neuropeptide YY. *Endocrinology* 116: 2677-2681.

Andrews, P.C. and J.E. Dixon. 1986. Isolation and structure of the second of two major peptide products from the precursor to an anglerfish peptide homologous to neuropeptide Y. *J. Biol. Chem.* 261: 8674-8677.

Aoki, C. and V.M. Pickel. 1988. Neuropeptide Y-containing neurons in the rat striatum: Ultrastructure and cellular relations with tyrosine hydroxylase-containing terminals and with astrocytes. *Brain Res.* 459: 205-225.

Bai, F.L., M. Yamano, Y. Shiotani, P.C. Emson, A.D. Smith, J.F. Powell and M. Tohyama. 1985. An arcuato-paraventricular and -dorsomedial hypothalamic neuropeptide Y-containing system which lacks noradrenaline in the rat. *Brain Res.* 331: 172-175.

Beal, M.F. and J.B. Martin. 1986. Neuropeptides in neurological disease. *Ann. Neurol.* 20: 547-565.

Beal, M.F., G.K. Chattha and J.B. Martin. 1986a. A comparison of regional somatostatin and neuropeptide Y distribution in rat striatum and brain. *Brain Res.* 377: 240-245.

Beal, M.F., M.F. Mazurek, G.K. Chattha, C.N. Svendsen, E.D. Bird and J.B. Martin. 1986b. Neuropeptide Y immunoreactivity is reduced in cerebral cortex in Alzheimer's disease. *Ann. Neurol.* 20: 282-288.

Beal, M.F., R.C. Frank, E.W. Ellison and J.B. Martin. 1986c. The effect of neuropeptide Y on striatal catecholamines. *Neurosci. Lett.* 71: 118-123.

Beal, M.F., M.F. Mazurek and J.B. Martin. 1987. A comparison of somatostatin and neuropeptide Y distribution in monkey brain. *Brain Res.* 405: 213-219.

Bergstrom, L., H. Lagercrantz and L. Terenius. 1984. Post-mortem analyses of neuropeptides in brains from sudden infant death victims. *Brain Res.* 323: 279-285.

Björklund, H., T. Hökfelt, M. Goldstein, L. Terenius and L. Olson. 1985. Appearance of the noradrenergic markers tyrosine hydroxylase and neuropeptide Y in cholinergic nerves of the iris following sympathectomy. *J. Neurosci.* 5: 1633-1643.

Blessing, W.W., P.R.C. Howe, T.H. Joh, J.R. Oliver and J.O. Willoughby. 1986. Distribution of tyrosine hydroxylase and neuropeptide Y-like immunoreactive neurons in rabbit medulla oblongata, with attention to colocalization studies, presumptive adrenalin-synthesizing perikarya, and vagal preganglionic cells. *J. Comp. Neurol.* 248: 285-300.

Broome, M., T. Hökfelt and L. Terenius. 1987. Peptide YY (PYY)-immunoreactive neurons in the lower brain stem and spinal cord of rat. *Acta Physiol. Scand.* 125: 349-352.

Bruun, A., B. Ehinger, F. Sundler, K. Tornqvist and R. Uddman. 1984. Neuropeptide Y immunoreactive neurons in the guinea-pig uvea and retina. *Inves. Ophth. Vis. Science* 25: 1113-1123.

Bruun, A., B. Ehinger, V. Sytsma and K. Tornqvist. 1985. Retinal
    neuropeptides in the skates, Raja clavata, R. radiata, R. oscellata
    (Elasmobranchii). *Cell Tissue Res.* 241: 17-24.
Busch-Sorensen, M., S.P. Sheikh, M. O'Hare, O. Tortora, T.W. Schwartz and S.
    Gammeltoft. 1989. Regional distribution of neuropeptide Y and its receptor
    in the porcine central nervous system. *J. Neurochem.* 52: 1545-1552.
Calka, J. and J.K. McDonald. 1989. Neuropeptide-Y and LHRH in the rat
    hypothalamus: Anatomical relationships. *Soc. Neurosci. Abstr.* 15: 187.
Card, J.P. and R.Y. Moore. 1982. Ventral lateral geniculate nucleus efferents to
    the rat suprachiasmatic nucleus exhibit avian pancreatic polypeptide-like
    immunoreactivity. *J. Comp. Neurol.* 206: 390-396.
Card, J.P. and R.Y. Moore. 1984. The suprachiasmatic nucleus of the golden
    hamster: Immunohistochemical analysis of cell and fiber distribution.
    Neuroscience 13: 415-431.
Card, J.P. and R.Y. Moore. 1988. Neuropeptide Y localization in the rat
    suprachiasmatic nucleus and periventricular hypothalamus. *Neurosci. Lett.*
    88: 241-246.
Card, J.P., N. Brecha and R.Y. Moore. 1983. Immunohistochemical localization
    of avian pancreatic polypeptide-like immunoreactivity in the rat
    hypothalamus. *J. Comp. Neurol.* 217: 123-136.
Carlei, F., J.M. Allen, A.E. Bishop, S.R. Bloom and J.M. Polak. 1985.
    Occurrence, distribution and nature of neuropeptide Y in the rat pancreas.
    Experientia 41: 1554-1557.
Chan-Palay, V., Y.S. Allen, W. Lang, U. Haesler and J.M. Polak. 1985a. I.
    Cytology and distribution in normal human cerebral cortex of neurons
    immunoreactive with antisera against neuropeptide Y. *J. Comp. Neurol.* 238:
    382-389.
Chan-Palay, V., W. Lang, Y.S. Allen, U. Haesler and J.M. Polak. 1985b. II.
    Cortical neurons immunoreactive with antisera against neuropeptide Y are
    altered in Alzheimer's-type dementia. *J. Comp. Neurol.* 238: 390-400.
Chan-Palay, V., C. Kohler, U. Haesler, W. Lang and G. Yasargil. 1986a.
    Distribution of neurons and axons immunoreactive with antisera against
    neuropeptide Y in the normal human hippocampus. *J. Comp. Neurol.* 248:
    360-375.
Chan-Palay, V., W. Lang, U. Haesler, C. Kohler and G. Yasargil. 1986b.
    Distribution of altered hippocampal neurons and axons immunoreactive with
    antisera against neuropeptide Y in Alzheimer's-type dementia. *J. Comp.
    Neurol.* 248: 376-394.
Chance, R.E., M.G. Johnson, Hoffmann and T.M. Lin. 1979. Pancreatic
    polypeptide: A newly recognized hormone. In *Proinsulin, Insulin, C-Peptide,*
    eds. S. Baba, T. Kaneko and N. Yanihara, 419-425. Excerpta Medica,
    Amsterdam.
Chang, R.S., V.J. Lotti, T-B. Chen, D.J. Cerino and P.J. Kling. 1985.
    Neuropeptide Y (NPY) binding sites in rat brain labeled with ^{125}I-Bolton-
    Hunter NPY: Comparative potencies of various polypeptides on brain NPY
    binding and biological responses in the rat vas deferens. *Life Sci.* 37: 2111-
    2122.
Chang, R.S., V.J. Lotti and T-B. Chen. 1986. Increased neuropeptide Y (NPY)
    receptor binding in hippocampus and cortex of spontaneous hypertensive
    (SH) rats compared to normotensive (WKY) rats. *Neurosci. Lett.* 67: 275-278.
Chattha, G.K. and M.F. Beal. 1987. Effect of cysteamine on somatostatin and
    neuropeptide Y in rat striatum and cortical synaptosomes. *Brain Res.* 401:
    359-364.

Christie, M.J., P.M. Beart, J. Jarrott and C. Maccarrone. 1986. Distribution of neuropeptide Y immunoreactivity in the rat basal ganglia: Effects of excitotoxin lesions to caudate-putamen. *Neurosci. Lett.* 63: 305-309.

Chronwall, B.M., T.N. Chase and T.L. O'Donohue. 1984. Coexistence of neuropeptide Y and somatostatin in rat and human cortical and rat hypothalamic neurons. *Neurosci. Lett.* 52: 213-217.

Chronwall, B.M., D.A. DiMaggio, V.J. Massari, V.M. Pickel, Ruggiero and T.L. O'Donohue. 1985. The anatomy of neuropeptide-Y-containing neurons in rat brain. *Neuroscience* 15: 1159-1181.

Ciofi, P., D. Croix and G. Tramu. 1987. Coexistence of hGHRF and NPY immunoreactivities in neurons of the arcuate nucleus of the rat. *Neuroendocrinology* 45: 425-428.

Ciofi, P., D. Croix and G. Tramu. 1988. Colocalization of GHRF and NPY immunoreactivities in neurons of the infundibular area of the human brain. *Neuroendocrinology* 47: 469-472.

Clark, J.T., R.S. Gist, S.P. Kalra and P.S. Kalra. 1988. Alpha$_2$-adrenoceptor blockade attenuates feeding behavior induced by neuropeptide Y and epinephrine. *Physiol. Behav.* 43: 417-422

Clark, J.T., P.S. Kalra, W.R. Crowley and S.P. Kalra. 1984. Neuropeptide Y and human pancreatic polypeptide stimulate feeding behavior in rats. *Endocrinology* 115: 427-429.

Clark, J.T., P.S. Kalra and S.P. Kalra. 1985. Neuropeptide Y stimulates feeding but inhibits sexual behavior in rats. *Endocrinology* 117: 2435-2441.

Corder, R., R.C. Gaillard and P. Bohlen. 1988. Isolation and sequence of rat peptide YY and neuropeptide Y. *Regul. Pept.* 21: 253-261.

Crowley, W.R., R.E. Tessel, T.L. O'Donohue, B.A. Adler and S.P. Kalra. 1985. Effects of ovarian hormones on the concentrations of immunoreactive neuropeptide Y in discrete brain regions of the female rat: Correlation with serum luteinizing hormone (LH) and median eminence LH-releasing hormone. *Endocrinology* 117: 1151-1155.

Crowley, W.R. and S.P. Kalra. 1987. Neuropeptide Y stimulates the release of luteinizing hormone-releasing hormone from medial basal hypothalamus *in vitro*: Modulation by ovarian hormones. *Neuroendocrinology* 46: 97-103.

Crowley, W.R., A. Hassid and S.P. Kalra. 1987. Neuropeptide Y enhances the release of luteinizing hormone (LH) induced by LH-releasing hormone. *Endocrinology* 120: 941-945.

Dai, X.L., J. Triepel and C. Heym. 1986. Immunohistochemical localization of neuropeptide Y in the guinea pig medulla oblongata. *Histochemistry* 85: 327-334.

Daniel, E.E., M. Costa, J.B. Furness and J.R. Keast. 1985. Peptide neurons in the canine small intestine. *J. Comp. Neurol.* 237: 227-238.

Dawbarn, D., S.P. Hunt and P.C. Emson. 1984. Neuropeptide Y: Regional distribution, chromatographic characterization and immunohistochemical demonstration in post-mortem human brain. *Brain Res.* 296: 168-173.

Dawbarn, D., M.E. de Quidt and P.C. Emson. 1985. Survival of basal ganglia neuropeptide Y-somatostatin neurones in Huntington's disease. *Brain Res.* 340: 251-260.

Dawbarn, D., M.N. Rossor, C.Q. Mountjoy, M. Roth and P.C. Emson. 1986. Decreased somatostatin immunoreactivity but not neuropeptide Y immunoreactivity in cerebral cortex in senile dementia of Alzheimer type. *Neurosci. Lett.* 70: 154-159.

de Quidt, M.E., P.J. Richardson and P.C. Emson. 1985. Subcellular distribution of neuropeptide Y-like immunoreactivity in guinea pig neocortex. *Brain Res.* 335: 354-359.

de Quidt, M.E. and P.C. Emson. 1986a. Distribution of neuropeptide Y-like immunoreactivity in the rat central nervous system-I. Radioimmunoassay and chromatographic characterization. *Neuroscience* 18: 527-543.

de Quidt, M.E. and P.C. Emson. 1986b. Distribution of neuropeptide Y-like immunoreactivity in the rat central nervous system-II. Immunohistochemical analysis. *Neuroscience* 18: 545-618.

DiMaggio, D.A., B.M. Chronwall, K. Buchanan and T.L. O'Donohue. 1985. Pancreatic polypeptide immunoreactivity in rat brain is actually neuropeptide Y. *Neuroscience* 15: 1149-1157.

Edvinsson, L. 1988. The effects of neuropeptide Y on the circulation. In *ISI Atlas of Science: Pharmacology*, 357-361.

Ekblad, E., C. Wahlestedt, M. Ekelund, R. Hakanson and F. Sundler. 1984a. Neuropeptide Y in the gut and pancreas: Distribution and possible vasomotor function. *Front. Horm. Res.* 12: 85-90.

Ekblad, E., R. Hakanson and F. Sundler. 1984b. VIP and PHI coexist with an NPY-like peptide in intramural neurones of the small intestine. *Regul. Peptides* 10: 47-55.

Ekblad, E., M. Ekelund, H. Graffner, R. Hakanson and F. Sundler. 1985. Peptide-containing nerve fibers in the stomach wall of rat and mouse. *Gastroenterology* 89: 73-89.

Ekblad, E., C. Winther, R. Ekman, R. Hakanson and F. Sundler. 1987. Projections of peptide-containing neurons in rat small intestine. *Neuroscience* 20: 169-188.

Emson, P.C. and M.E. de Quidt. 1984. NPY - A new member of the pancreatic polypeptide family. *Trends Neurosci.* 7: 31-35.

Everitt, B.J., T. Hökfelt, L. Terenius, K. Tatemoto, V. Mutt and M. Goldstein. 1984. Differential co-existence of neuropeptide Y (NPY)-like immunoreactivity with catecholamines in the central nervous system of the rat. *Neuroscience* 11: 443-462.

Falkmer, S., E. Dafgard, M. El-Salhy, W. Engstrom, L. Grimelius and A. Zetterberg. 1985. Phylogenetical aspects on islet hormone families: A minireview with particular reference to insulin as a growth factor and to the phylogeny of PYY and NPY immunoreactive cells and nerves in the endocrine and exocrine pancreas. *Peptides* 6 [Suppl 3]: 315-320.

Feher, E. and G. Burnstock. 1986a. Ultrastructural localisation of substance P, vasoactive intestinal polypeptide, somatostatin and neuropeptide Y immunoreactivity in perivascular nerve plexuses of the gut. *Blood Vessels* 23: 125-136.

Feher, E. and G. Burnstock. 1986b. Electron microscopic study of neuropeptide Y-containing nerve elements of the guinea pig small intestine. *Gastroenterology* 91: 956-965.

Foster, G.A. and M. Schultzberg. 1984. Immunohistochemical analysis of the ontogeny of neuropeptide Y immunoreactive neurons in foetal rat brain. *Int. J. Devel. Neurosci.* 2: 387-407.

Foster, G.A., M. Schultzberg and M. Goldstein. 1984. Differential and independent manifestation within co-containing neurones of neuropeptide Y and tyrosine hydroxylase during ontogeny of the rat central nervous system. *Neurochem. Int.* 6: 761-771.

Foster, N.L., C.A. Tamminga, T.L. O'Donohue, K. Tanimoto, E.D. Bird and T.N. Chase. 1986. Brain choline acetyltransferase activity and neuropeptide Y concentrations in Alzheimer's disease. *Neurosci. Lett.* 63: 71-75.

Fried, G., T. Hökfelt, L. Terenius and M. Goldstein. 1985. Neuropeptide Y-(NPY)-like immunoreactivity in guinea pig uterus is reduced during pregnancy in parallel with noradrenergic nerves. *Histochemistry* 83: 437-442.

Fujii, S., S. Baba and T. Fujita. 1982. Pancreatic polypeptide immunoreactive cells and nerves in the canine pituitary. *Biomed. Res.* 3: 525-533.

Furness, J.B., M. Costa, P.C. Emson, R. Hakanson, E. Moghimzadeh, F. Sundler, I.L. Taylor and R.E. Chance. 1983. Distribution, pathways and reactions to drug treatment of nerves with neuropeptide Y- and pancreatic polypeptide-like immunoreactivity in the guinea-pig digestive tract. *Cell Tissue Res.* 234: 71-92.

Furness, J.B., M. Costa and J.R. Keast. 1984. Choline acetyltransferase- and peptide-immunoreactivity of submucous neurons in the small intestine of the guinea-pig. *Cell Tissue Res.* 237: 329-336.

Furness, J.B., M. Costa, I.L. Gibbins, I.J. Llewellyn-Smith and J.R. Oliver. 1985. Neurochemically similar myenteric and submucous neurons directly traced to the mucosa of the small intestine. *Cell Tissue Res.* 241: 155-163.

Fuxe, K., L.F. Agnati, A. Härfstrand, A.M. Janson, A. Neumyer, K. Andersson, M. Ruggeri, M. Zoli and M. Goldstein. 1986. Morphofunctional studies on the neuropeptide Y/adrenaline costoring terminal systems in the dorsal cardiovascular region of the medulla oblongata. Focus on receptor-receptor interactions in cotransmission. *Prog. Brain Res.* 68: 303-320.

Fuxe, K., A. Härfstrand, P. Eneroth, M. Zoli and L.F. Agnati. 1987. Neuropeptide Y mechanisms in neuroendocrine regulation. Focus on Neuropeptide Y-catecholamine interactions in regulation of LH and prolactin secretion. In *The Brain and Female Reproduction Function*, 45-55. Lecture 1987, Capri, Italy, May, 1987. Internat. Pub in Sci. and Tech., New Jersey.

Gall, C., K.B. Seroogy and N. Brecha. 1986. Distribution of VIP- and NPY-like immunoreactivities in rat main olfactory bulb. *Brain Res.* 374: 389-394.

Gehlert, D.R., B.M. Chronwall, M.P. Schafer and T.L. O'Donohue. 1987. Localization of neuropeptide Y messenger ribonucleic acid in rat and mouse brain by *in situ* hybridization. *Synapse* 1: 25-31.

Gibson, S.J., J.M. Polak, J.M. Allen, T.E. Adrian, J.S. Kelly and S.R. Bloom. 1984. The distribution and origin of a novel brain peptide, neuropeptide Y, in the spinal cord of several mammals. *J. Comp. Neurol.* 227: 78-91.

Gray, T.S. and J.E. Morley. 1986. Neuropeptide Y: Anatomical distribution and possible function in mammalian nervous system. *Life Sci.* 38: 389-401.

Gray, T.S., T.L. O'Donohue and D.J. Magnuson. 1986. Neuropeptide Y innervation of amygdaloid and hypothalamic neurons that project to the dorsal vagal complex in rat. *Peptides* 7: 341-349.

Gu, J., J.M. Polak, L. Probert, K.N. Islam, P.J. Marangos, S. Mina, T.E. Adrian, G.P. McGregor, D.J. O'Shaughnessy and S.R. Bloom. 1983. Peptidergic innervation of the human male genital tract. *J. Urology* 130: 386-391.

Gulbenkian, S., J. Wharton, G. W. Hacker, I.M. Varndell, S.R. Bloom and J.M. Polak. 1985. Co-localization of neuropeptide tyrosine (NPY) and its C-terminal flanking peptide (C-PON). *Peptides* 6: 1237-1243.

Gustafson, E.L., J.P. Card and R.Y. Moore. 1986. Neuropeptide Y localization in the rat amygdaloid complex. *J. Comp. Neurol.* 251: 349-362.

Guy, J. and G. Pelletier. 1988. Neuronal interactions between neuropeptide Y (NPY) and catecholaminergic systems in the rat arcuate nucleus as shown by dual immunocytochemistry. *Peptides* 9: 567-570.

Guy, J., S. Li and G. Pelletier. 1988. Studies on the physiological role and mechanism of action of neuropeptide Y in the regulation of luteinizing hormone secretion in the rat. *Regul. Pept.* 23: 209-216.

Härfstrand, A., K. Fuxe, L.F. Agnati, F. Benfenati and M. Goldstein. 1986a. Receptor autoradiographical evidence for high densities of ^{125}I-neuropeptide Y binding sites in the nucleus tractus solitarius of the normal male rat. *Acta Physiol. Scand.* 128: 195-200.

Härfstrand, A., K. Fuxe, L.F. Agnati, P. Eneroth, I. Zini, M. Zoli, K. Andersson, G. Von Euler, L. Terenius, V. Mutt and M. Goldstein. 1986b. Studies on neuropeptide Y-catecholamine interactions in the hypothalamus and in the forebrain of the male rat. Relationship to neuroendocrine function. *Neurochem. Int.* 8: 355-376.

Härfstrand, A., K. Fuxe, L. Terenius and M. Kalia. 1987a. Neuropeptide Y immunoreactive perikarya and nerve terminals in the rat medulla oblongata: Relationship to cytoarchitecture and catecholaminergic cell groups. *J. Comp. Neurol.* 260: 20-35.

Härfstrand, A., P. Eneroth, L. Agnati and K. Fuxe. 1987b. Further studies on the effects of central administration of neuropeptide Y on neuroendocrine function in the male rat: Relationship to hypothalamic catecholamines. *Regul. Peptides* 17: 167-179.

Harrington, M.E., D.M. Nance and B. Rusak. 1985. Neuropeptide Y immunoreactivity in the hamster geniculo-suprachiasmatic tract. *Brain Res. Bull.* 15: 465-472.

Hazelwood, R.L. 1981. Synthesis, storage, secretion, and significance of pancreatic polypeptide in vertebrates. In *The Islet of Langerhans*, eds. S.J. Cooperstein and D. Watkins, 275-318. Academic Press, New York.

Hedlund, H. and K.-E. Andersson. 1985. Effects of some peptides on isolated human penile erectile tissue and cavernous artery. *Acta. Physiol. Scand.* 124: 413-419.

Heinrich, D., M. Reinecke and W.G. Forssmann. 1986. Peptidergic innervation of the human and guinea pig uterus. *Arch. Gynecol.* 237: 213-219.

Heilig, M., C. Wahlestedt and E. Widerlov. 1988. Neuropeptide Y (NPY)-induced suppression of activity in the rat: Evidence for NPY receptor heterogeneity and for interaction with alpha-adrenoreceptors. *Eur. J. Pharmacol.* 157: 205-213.

Hendry, S.H.C., E.G. Jones, J. DeFelipe, J., D. Schmechel, C. Brandon and P.C. Emson. 1984a. Neuropeptide-containing neurons of the cerebral cortex are also GABAergic. *Proc. Natl. Acad. Sci. USA* 81: 6526-6530.

Hendry, S.H.C., E.G. Jones and P.C. Emson. 1984b. Morphology, distribution, and synaptic relations of somatostatin- and neuropeptide Y-immunoreactive neurons in rat and monkey neocortex. *J. Neurosci.* 4: 2497-2517.

Hisano, S., Y. Kagotani, Y. Tsuruo, S. Daikoku, K. Chihara and M.K. Whitnall. 1988. Localization of glucocorticoid receptor in neuropeptide Y-containing neurons in the arcuate nucleus of the rat hypothalamus. *Neurosci. Lett.* 95: 13-18.

Hökfelt, T., J.M. Lundberg, H. Lagercrantz, K. Tatemoto, V. Mutt, J. Lindberg, L. Terenius, B.J. Everitt, K. Fuxe, L. Agnati and M. Goldstein. 1983a. Occurrence of neuropeptide Y (NPY)-like immunoreactivity in catecholamine neurons in the human medulla oblongata. *Neurosci. Lett.* 36: 217-222.

Hökfelt, T., J.M. Lundberg, K. Tatemoto, V. Mutt, L. Terenius, J. Polak, S. Bloom, C. Sasek, R. Elde and M. Goldstein. 1983b. Neuropeptide Y (NPY)- and FMRFamide neuropeptide-like immunoreactivities in catecholamine neurons of the rat medulla oblongata. *Acta Physiol. Scand.* 117: 315-318.

Hökfelt, T., J.M. Lundberg, L. Terenius, G. Jancso and J. Kimmel. 1981. Avian pancreatic polypeptide (APP) immunoreactive neurons in the spinal cord and spinal trigeminal nucleus. *Peptides* 2: 81-87.

Holets, V.R., T. Hökfelt, Å. Rokaeus, L. Terenius and M. Goldstein. 1988. Locus coeruleus neurons in the rat containing neuropeptide Y, tyrosine hydroxylase or galanin and their efferent projections to the spinal cord, cerebral cortex and hypothalamus. *Neuroscience* 24:893-906.

Hooi, S.C., G.S. Richardson, J.K. McDonald, J.M. Allen, J.B. Martin and J.I. Koenig. 1989. Neuropeptide Y (NPY) and vasopressin (AVP) in the hypothalamo-neurohypophysial axis of salt-loaded and/or Brattleboro rats. *Brain Res.* 486: 214-220.

Huang, W.M., J. Gu, M.A. Blank, J.M. Allen, S.R. Bloom and J.M. Polak. 1984. Peptide-immunoreactive nerves in the mammalian female genital tract. *Histochem. J.* 16: 1297-1310.

Huidoboro-Toro, J.P. 1985. Reserpine-induced potentiation of the inhibitory action of neuropeptide Y on the rat vas deferens neurotransmission. *Neurosci. Lett.* 59: 247-252.

Humphreys, G.A., J.S. Davison and W.L. Veale. 1988. Injection of neuropeptide Y into the paraventricular nucleus of the hypothalamus inhibits gastric acid secretion in the rat. *Brain Res.* 456: 241-248.

Hunt, S.P., P.C. Emson, R. Gilbert, M. Goldstein and J.R. Kimmel. 1981. Presence of avian pancreatic polypeptide-like immunoreactivity in catecholamine and methionine-enkephalin-containing neurons within the central nervous system. *Neurosci. Lett.* 21: 125-130.

Jacobowitz, D.M. and J.A. Olschowka. 1982a. Coexistence of bovine pancreatic polypeptide-like immunoreactivity and catecholamine in neurons of the ventral aminergic pathway of the rat brain. *Brain Res. Bull.* 9: 391-406.

Jacobowitz, D.M. and J.A. Olschowka. 1982b. Bovine pancreatic polypeptide-like immunoreactivity in brain and peripheral nervous system: Coexistence with catecholaminergic nerves. *Peptides* 3: 569-590.

Jones, P.M., M.A. Ghatei, J. Steel, D. O'Halloran, G. Gon, S. Legon, J.M. Burrin, U. Leonhardt, J.M. Polak and S.R. Bloom. 1989. Evidence for neuropeptide Y synthesis in the rat anterior pituitary and the influence of thyroid hormone status: Comparison with vasoactive intestinal peptide, substance P, and neurotensin. *Endocrinology* 125: 334-341.

Kalra, S.P. and W.R. Crowley. 1984a. Differential effects of pancreatic polypeptide on luteinizing hormone release in female rats. *Neuroendocrinology* 38: 511-513.

Kalra, S.P. and W. R. Crowley. 1984b. Norepinephrine-like effects of neuropeptide Y (NPY) and human pancreatic polypeptide (hPP) on LH release. *Soc. Neurosci. Abstr.* 10: 1119.

Kalra, S.P. and W.R. Crowley. 1984c. Norepinephrine-like effects of neuropeptide Y on LH release in the rat. *Life Sci.* 35: 1173-1176.

Kalra, S.P., P.S. Kalra, A. Sahu and W.R. Crowley. 1987. Gonadal steroids and neurosecretion: Facilitatory influence on LHRH and neuropeptide Y. *J. Steroid Biochem.* 27: 677-681.

Kalra, S.P., J.T. Clark, A. Sahu, M.G. Dube and P.S. Kalra. 1988. Control of feeding and sexual behaviors by neuropeptide Y: Physiological implications. *Synapse* 2: 254-257.

Kannisto, P., E. Ekblad, G. Helm, Ch. Owman, N.O. Sjöberg, M. Stjernquist, F. Sundler and B. Walles. 1986. Existence and coexistence of peptides in nerves of the mammalian ovary and oviduct demonstrated by immunohistochemistry. *Histochemistry* 86: 25-34.

Katagiri, T., S.J. Gibson, H.C. Su and J.M. Polak. 1986. Composition and central projections of the pudendal nerve in the rat investigated by combined peptide immunocytochemistry and retrograde fluorescent labeling. *Brain Res.* 372: 313-322.

Keast, J.R., J.B. Furness and M. Costa. 1985. Distribution of certain peptide-containing nerve fibres and endocrine cells in the gastrointestinal mucosa in five mammalian species. *J. Comp. Neurol.* 236: 403-422.

Kerkerian, L., J. Guy, G. Lefevre and G. Pelletier. 1985. Effects of neuropeptide Y (NPY) on the release of anterior pituitary hormones in the rat. *Peptides* 6: 1201-1204.

Kerkerian, L., O. Bosler, G. Pelletier and A. Nieoullon. 1986. Striatal neuropeptide Y neurones are under the influence of the nigrostriatal dopaminergic pathway: Immunohistochemical evidence. *Neurosci. Lett.* 66: 106-112.

Kerkerian, L. and G. Pelletier. 1986. Effects of monosodium L-glutamate administration on neuropeptide Y-containing neurons in the rat hypothalamus. *Brain Res.* 369: 388-390.

Khorram, O., K.Y.F. Pau and H.G. Spies. 1987a. Bimodal effects of neuropeptide Y on hypothalamic release of gonadotropin-releasing hormone in conscious rabbits. *Neuroendocrinology* 45: 290-297.

Khorram, O., C.E. Roselli, W.E. Ellinwood and H.G. Spies. 1987b. The measurement of neuropeptide Y in discrete hypothalamic and limbic regions of male rhesus macaques with a human NPY-directed antiserum. *Peptides* 8: 159-163.

Khorram O., K.Y.F. Pau and H.G. Spies. 1988. Release of hypothalamic neuropeptide Y and effects of exogenous NPY on the release of hypothalamic GnRH and pituitary gonadotropins in intact and ovariectomized does *in vitro*. *Peptides* 9: 411-417.

Kimmel, J.R., H.G. Pollock and R.L. Hazelwood. 1968. Isolation and characterization of chicken insulin. *Endocrinology* 83: 1323-1330.

Kimmel, J.R., J. Hayden and H.G. Pollock. 1975. Isolation and characterization of a new pancreatic polypeptide hormone. *J. Biol. Chem.* 250: 9369-9376.

Kimmel, J.R., H.G. Pollock, R.E. Chance, M.G. Johnson, J.R. Reeve, I.L. Taylor, C. Miller and J.E. Shively. 1984. Pancreatic polypeptide from rat pancreas. *Endocrinology* 114: 1725-1731.

Kimmel, J.R., E.M. Plisetskaya, H.G. Pollock, J.W. Hamilton, J.B. Rouse, K.E. Ebner and A.B. Rawitch. 1986. Structure of a peptide from coho salmon endocrine pancreas with homology to neuropeptide Y. *Biochem. Biophys. Res. Comm.* 141: 1084-1091.

Kobayashi, S., J.A. Olschowka and D.M. Jacobowitz. 1983. Bovine pancreatic polypeptide-like immunoreactive nerves in the rat major cerebral arteries. *Brain Res. Bull.* 10: 373-376.

Kohler, C., M. Smialowska, L.G. Eriksson, V. Chan-Palay and S. Davies. 1986. Origin of the neuropeptide Y innervation of the rat retrohippocampal region. *Neurosci. Lett.* 65: 287-292.

Kumoi, Y.K., H. Kiyama, P.C. Emson, J.R. Kimmel and M. Tohyama. 1985. Coexistence of pancreatic polypeptide and substance P in the chicken retina. *Brain Res.* 361: 25-35.

Lee, Y., S. Shiosaka, P.C. Emson, J.F. Powell, A.D. Smith and M. Tohyama. 1985. Neuropeptide Y-like immunoreactive structures in the rat stomach with special reference to the noradrenaline neuron system. *Gastroenterology* 89: 118-126.

Levine, A.S. and J.E. Morley. 1984. Neuropeptide Y: A potent inducer of consummatory behavior in rats. *Peptides* 5: 1025-1029.

Li, S. and G. Pelletier. 1986. The role of dopamine in the control of neuropeptide Y neurons in the rat arcuate nucleus. *Neurosci. Lett.* 69: 74-77.

Lin, T.M. 1980. Pancreatic polypeptide: Isolation, chemistry, and biological function. In *Gastrointestinal Hormones*, ed. G.B. Glass, 276-306. Raven Press, New York.

Lindh, B., T. Hökfelt, L.-G. Elfvin, L. Terenius, J. Fahrenkrug, R. Elde and M. Goldstein. 1986. Topography of NPY-, somatostatin-, and VIP-immunoreactive, neuronal subpopulations in the guinea pig celiac-superior mesenteric ganglion and their projection to the pylorus. *J. Neurosci.* 6: 2371-2383.

Liposits, Zs., L. Sievers and W.K. Paull. 1988. Neuropeptide-Y and ACTH-immunoreactive innervation of corticotropin releasing factor (CRF)-synthesizing neurons in the hypothalamus of the rat. *Histochemistry* 88: 227-234.

Loren, I., J. Alumets, R. Hakanson and F. Sundler. 1979. Immunoreactive pancreatic polypeptide (PP) occurs in the central and peripheral nervous system: Preliminary immunocytochemical observations. *Cell Tissue Res.* 200: 179-186.

Lundberg, J.M., T. Hökfelt, A. Anggard, J. Kimmel, M. Goldstein and K. Markey. 1980a. Coexistence of an avian pancreatic polypeptide (APP) immunoreactive substance and catecholamines in some peripheral and central neurons. *Acta Physiol. Scand.* 110: 107-109.

Lundberg, J.M., A. Anggard, T. Hökfelt and J. Kimmel. 1980b. Avian pancreatic polypeptide (APP) inhibits atropine resistant vasodilation in cat submandibular salivary gland and nasal mucosa: Possible interaction with VIP. *Acta Physiol. Scand.* 110: 199-201.

Lundberg, J.M. and K. Tatemoto. 1982. Pancreatic polypeptide family (APP,BPP,NPY and PYY) in relation to sympathetic vasoconstriction resistant to alpha-adrenoceptor blockade. *Acta Physiol. Scand.* 116: 393-402.

Lundberg, J.M., T. Hökfelt, A. Anggard, L. Terenius, R. Elde, K. Markey, M. Goldstein and J. Kimmel. 1982a. Organizational principles in the peripheral sympathetic nervous system: Subdivision by coexisting peptides (somatostatin-, avian pancreatic polypeptide-, and vasoactive intestinal polypeptide-like immunoreactive materials). *Proc. Natl. Acad. Sci. USA* 79: 1303-1307.

Lundberg, J.M., K. Tatemoto, L. Terenius, P.M. Hellstrom, V. Mutt, T. Hökfelt and B. Hamberger. 1982b. Localization of peptide YY (PYY) in gastrointestinal endocrine cells and effects on intestinal blood flow and motility. *Proc. Natl. Acad. Sci. USA* 79: 4471.

Lundberg, J.M. and L. Stjarne. 1984. Neuropeptide Y (NPY) depresses the secretion of H-noradrenaline and the contractile response evoked by field stimulation, in rat vas deferens. *Acta Physiol. Scand.* 120: 477-479.

Lundberg, J.M., X.-Y. Hua and A. Franco-Cereceda. 1984a. Effects of neuropeptide Y (NPY) on mechanical activity and neurotransmission in the heart, vas deferens and urinary bladder of the guinea-pig. *Acta Physiol. Scand.* 121: 325-332.

Lundberg, J.M., L. Terenius, T. Hökfelt and K. Tatemoto. 1984b. Comparative immunohistochemical and biochemical analysis of pancreatic polypeptide-like peptides with special reference to presence of neuropeptide Y in central and peripheral neurons. *J. Neurosci.* 4: 2376-2386.

Maccarrone, C. and B. Jarrott. 1985. Differences in regional brain concentrations of neuropeptide Y in spontaneously hypertensive (SH) and Wistar-Kyoto (WKY) rats. *Brain Res.* 345: 165-169.

Maccarrone, C. and B. Jarrott. 1986a. Neuropeptide Y: A putative neurotransmitter. *Neurochem. Int.* 8: 13-22.

Maccarrone, C. and B. Jarrott. 1986b. Age-related changes in neuropeptide Y immunoreactivity (NPY-ir) in the cortex and spinal cord of spontaneously hypertensive (SHR) and normotensive Wistar-Kyoto (WKY) rats. *J. Hypertension* 4: 471-475.

Maccarrone, C., B. Jarrott and E.L. Conway. 1986. Comparison of neuropeptide Y immunoreactivity in hypothalamic and brainstem nuclei of young and mature spontaneously hypertensive and normotensive Wistar-Kyoto rats. *Neurosci. Lett.* 68: 232-238.

Mantyh, P.W. and J.A. Kemp. 1983. The distribution of putative neurotransmitters in the lateral geniculate nucleus of the rat. *Brain Res.* 288: 344-348.

Martel, J.-C., S. St-Pierre and R. Quirion. 1986. Neuropeptide Y receptors in rat brain: Autoradiographic localization. *Peptides* 7: 55-60.

Massari, V.J., J. Chan, B.M. Chronwall, T.L. O'Donohue, W.H. Oertel and V.M. Pickel. 1988. Neuropeptide Y in the rat nucleus accumbens: Ultrastructural localization in aspiny neurons receiving synaptic input from GABAergic terminals. *J. Neurosci. Res.* 19: 171-186.

McCann, S.M., V. Rettori, L. Milenkovic, M. Riedel, C. Aguila and J.K. McDonald. 1989. The role of neuropeptide Y (NPY) in the control of anterior pituitary hormone release in the rat. In *Nobel Conference on NPY*, ed. V. Mutt, K. Fuxe, T. HöKfelt, and J. Lüdberg, 215-228. Raven Press, New York.

McDonald, J.K., J.G. Parnavelas, A.N. Karamanlidis and N. Brecha. 1982. The morphology and distribution of peptide-containing neurons in the adult and developing visual cortex of the rat. IV. Avian pancreatic polypeptide. *J. Neurocytol.* 11: 985-995.

McDonald, J.K. and M.D. Lumpkin. 1983. Third ventricular injections of pancreatic polypeptides decrease LH and growth hormone secretion in ovariectomized rats. *65th Annu. Mtng. Endocrine Soc.*, 152.

McDonald, J.K., M.D. Lumpkin, W.K. Samson and S.M. McCann. 1984. Third ventricular injections of neuropeptide Y decrease LH and growth hormone secretion in ovariectomized rats. *Soc. Neurosci. Abstr.* 10: 1214.

McDonald, J.K., M.D. Lumpkin, W.K. Samson and S.M. McCann. 1985a. Neuropeptide Y affects secretion of luteinizing hormone and growth hormone in ovariectomized rats. *Proc. Natl. Acad. Sci. USA* 82: 561-564.

McDonald, J.K., M.D. Lumpkin, W.K. Samson and S.M. McCann. 1985b. Pancreatic polypeptides affect luteinizing and growth hormone secretion in rats. *Peptides* 6: 79-84.

McDonald, J.K., P. Collins and C.A. Reich. 1986. Neonatal injections of MSG inhibit the development of neuropeptide Y in the rat hypothalamus and posterior pituitary. *Soc. Neurosci. Abstr.* 12: 1523.

McDonald, J.K., J.I. Koenig, D.M. Gibbs, P. Collins and B.D. Noe. 1987a. High concentrations of neuropeptide Y in pituitary portal blood of rats. *Neuroendocrinology* 46: 538-541.

McDonald, J.K., J.I. Koenig, D.M. Gibbs, P. Collins and B.D. Noe. 1987b. High concentration of neuropeptide Y in pituitary portal blood of rats. *69th Annu. Mtng. Endocrine Soc.*, 86.

McDonald, J.K., W.L. Dees, C.E. Ahmed, B.D. Noe and S.R. Ojeda. 1987c. Biochemical and immunocytochemical characterization of neuropeptide Y in the immature rat ovary. *Endocrinology* 120: 1703-1710.

McDonald, J.K. 1988. NPY and related substances. In *CRC Critical Reviews in Neurobiology*, Vol. 4, ed. J. Nelson, 97-135. CRC Press, Boca Raton, FL.

McDonald, J.K., C. Han, B.D. Noe and P.W. Abel. 1988a. High levels of NPY in rabbit cerebrospinal fluid and immuno-histochemical analysis of possible sources. *Brain Res.* 463: 259-267.

McDonald J.K., F.D. Sabatino and J.C. Calka. 1988b. Neuropeptide Y and GnRH in the rat hypothalamus: Physiological and anatomical relationships. In *Proc. Int. Symp. Frontiers Reprod. Res.*, 102. Beijing, Peoples Republic of China.

McDonald J.K., J. Tigges, M. Tigges and C. Reich. 1988c. Developmental study of neuropeptide Y-like immunoreactivity in the neurohypophysis and intermediate lobe of the rhesus monkey (Macaca mulatta). *Cell Tissue Res.* 254: 499-509.

McDonald, J.K, M.D. Lumpkin and L.V. DePaolo. 1989. Neuropeptide Y suppresses pulsatile secretion of luteinizing hormone in ovariectomized rats: Possible site of action. *Endocrinology* 125: 186-191.

McNeill, D.L. and H.W. Burden. 1986. Neuropeptide Y and somatostatin immunoreactive perikarya in preaortic ganglia projecting to the rat ovary. *J. Reprod. Fertil.* 78: 727-732.

Milgram, S.L., A. Balasubramaniam, P.C. Andrews, J.K. McDonald and B.D. Noe. 1989. Characterization of aPY-like peptides in anglerfish brain using a novel radioimmunoassay for aPY-Gly. *Peptides* 10: 1013-1017.

Minth, C.D., S.R. Bloom, J.M. Polak and J.E. Dixon. 1984. Cloning, characterization, and DNA sequence of a human cDNA encoding neuropeptide tyrosine. *Proc. Natl. Acad. Sci. USA* 81: 4577-4581.

Minth, C.D., P.C. Andrews and J.E. Dixon. 1986. Characterization, sequence, and expression of the cloned human neuropeptide Y gene. *J. Biol. Chem.* 261: 11974-11979.

Moltz, J.H. and J.K. McDonald. 1985. Neuropeptide Y: Direct and indirect action on insulin secretion in the rat. *Peptides* 6: 1155-1159.

Moore, R.Y., E.L. Gustafson and J.P. Card. 1984. Identical immunoreactivity of afferents to the rat suprachiasmatic nucleus with antisera against avian pancreatic polypeptide, molluscan cardioexcitatory peptide and neuropeptide Y. *Cell Tissue Res.* 236: 41-46.

Morley, J.E., A.S. Levine, B.A. Gosnell, J.E. Mitchell, D.D. Krahn and S.E. Nizielski. 1985. Peptides and feeding, *Peptides* [Suppl. 2] 6: 181-192.

Morris, J.L., I.L. Gibbins, J.B. Furness, M. Costa and R. Murphy. 1985. Colocalization of neuropeptide Y, vasoactive intestinal polypeptide and dynorphin in non-noradrenergic axons of the guinea pig uterine artery. *Neurosci. Lett.* 62: 31-37.

Muske, L.E., G.J. Dockray, K.S. Chohan and W.K. Stell. 1987. Segregation of FMRF amide-immunoreactive efferent fibers from NPY-immunoreactive amacrine cells in goldfish retina. *Cell Tissue Res.* 247: 299-307.

Nakagawa, Y., S. Shiosaka, P.C. Emson and M. Tohyama. 1985. Distribution of neuropeptide Y in the forebrain and diencephalon: An immunohistochemical analysis. *Brain Res.* 361: 52-60.

Nakajima, T., Y. Yashima and K. Nakamura. 1986. Quantitative autoradiographic localization of neuropeptide Y receptors in the rat lower brainstem. *Brain Res.* 380: 144-150.

Nakamura, S. and S.R. Vincent. 1986. Somatostatin- and neuropeptide Y-immunoreactive neurons in the neocortex in senile dementia of Alzheimer's type. *Brain Res.* 370: 11-20.

Noe, B.D., J.K. McDonald, F. Greiner, J.G. Wood and P.C. Andrews. 1986. Anglerfish islets contain NPY immunoreactive nerves and produce the NPY analog aPY. *Peptides* 7: 147-154.

Noe, B.D., S.L. Milgram, B. Balasubramaniam, P.C. Andrews, J. Calka and J.K. McDonald. 1989a. Localization and characterization of NPY-like peptides in the brain and islet organ of anglerfish. *71st Annu. Mtng. Endocrine Soc.*, p. 398.

Noe, B.D., S.L. Milgram, A. Balasubramaniam, P.C. Andrews, P.C., J. Calka and J.K. McDonald. 1989b. Localization and characterization of NPY-like peptides in the brain and islet organ of the anglerfish (*Lophius Americanus*). *Cell Tissue Res.* 257: 303-311.

O'Donohue, T.L., B.M. Chronwall, R.M. Pruss, E. Mezey, J.Z. Kiss, L.E. Eiden, V.J. Massari, R.E. Tessel, V.M. Pickel, D.A. DiMaggio, A.J. Hotchkiss, W.R. Crowley and Z. Zukowska-Grojec. 1985. Neuropeptide Y and peptide YY neuronal and endocrine systems. *Peptides* 6: 755-768.

Ohhashi, T. and D.M. Jacobowitz. 1983. The effects of pancreatic polypeptides and neuropeptide Y on the rat vas deferens. *Peptides* 4: 381-386.

Olschowka, J.A., T.L. O'Donohue and D.M. Jacobowitz. 1981. The distribution of bovine pancreatic polypeptide-like immunoreactive neurons in rat brain. *Peptides* 2: 309-331.

Olschowka, J.A. and D.M. Jacobowitz. 1983. The coexistence and release of bovine pancreatic polypeptide-like immunoreactivity from noradrenergic superior cervical ganglia neurons. *Peptides* 4: 231-238.

Osborne, N.N., S. Patel, G. Terenghi, J.M. Allen, J.M. Polak and S.R. Bloom. 1985. Neuropeptide Y (NPY)-like immunoreactive amacrine cells in retinas of frog and goldfish. *Cell Tissue Res.* 241: 651-656.

Papka, R.E., J.P. Cotton and H.H. Traurig. 1985. Comparative distribution of neuropeptide tyrosine-, vasoactive intestinal polypeptide-, substance P-immunoreactive, acetylcholinesterase-positive and noradrenergic nerves in the reproductive tract of the female rat. *Cell Tissue Res.* 242: 475-490.

Papka, R.E. and H.H. Traurig. 1988. Distribution of subgroups of neuropeptide Y-immunoreactive and noradrenergic nerves in the female rat uterine cervix. *Cell Tissue Res.* 252: 533-541.

Pau, K.Y.F., O. Khorram, A.H. Kaynard and H.G. Spies. 1989. Simultaneous induction of neuropeptide Y and gonadotropin-releasing hormone release in the rabbit hypothalamus. *Neuroendocrinology* 49: 197-201.

Pelletier, G., L. Desy, L. Kerkerian and J. Cote. 1984. Immunocytochemical localization of neuropeptide Y (NPY) in the human hypothalamus. *Cell Tissue Res.* 238: 203-205.

Pettersson, M., I. Lundquist and B. Ahren. 1987a. Neuropeptide Y and calcitonin gene-related peptide: Effects on glucagon and insulin secretion in the mouse. *Endocr. Res.* 13: 407-417.

Pettersson, M., B. Ahren, I. Lundquist, G. Bottcher and F. Sundler. 1987b. Neuropeptide Y: Intrapancreatic neuronal localization and effects on insulin secretion in the mouse. *Cell Tissue Res.* 248: 43-48.

Polak, J.M. and S.R. Bloom. 1984. Regulatory peptides - the distribution of two newly discovered peptides: PHI and NPY. *Peptides* 5: 79-89.

Potter, E.K. 1988. Neuropeptide Y as an autonomic neurotransmitter. *Pharmac. Ther.* 37: 251-273.

Rodriguez-Sierra, J.F., D.M. Jacobowitz and C.A. Blake. 1987. Effects of neuropeptide Y on LH, FSH and TSH release in male rats. *Peptides* 8: 539-542.

Rossor, M., P. Emson, D. Dawbarn, G. Dockray, C. Mountjoy and M. Roth. 1986. Postmortem studies of peptides in Alzheimer's disease and Huntington's disease. In *Neuropeptides in Neurologic and Psychiatric Disease*, eds. J.B. Martin and J.D. Barchas, 259-277. Raven Press, New York.

Saavedra, J.M. and R. Cruciani. 1988. Quantitative autoradiographic localization of neuropeptide Y (NPY) binding sites in rat posterior pituitary lobe. *Cell. Molec. Neurobiology* 8: 333-338.

Sabatino, F.D. and J.K. McDonald. 1987. Neuropeptide Y affects the release of LHRH from the median eminence *in vitro*. *Soc. Neurosci. Abstr*. 13: 1310.

Sabatino, F.D., J.M. Murnane, R.A. Hoffman and J.K. McDonald. 1987. The distribution of neuropeptide Y-like immunoreactivity in the hypothalamus of the adult golden hamster. *J. Comp. Neurol*. 257: 93-104.

Sabatino, F.D. and J.K. McDonald. 1988. NPY stimulation of LHRH release from rat median eminence: Effects of estrogen and calcium. *Soc. Neurosci. Abstr*. 14: 144.

Sabatino, F.D., P. Collins and J.K. McDonald. 1989a. Effects of progesterone on neuropeptide Y stimulated LHRH release from rat median eminence. *71st Annu. Mtng. Endocrine Soc*., 76.

Sabatino F.D., P. Collins and J.K. McDonald. 1989b. Neuropeptide Y stimulation of luteinizing hormone-releasing hormone secretion from the median eminence *in vitro* by estrogen dependent and extracellular Ca^{2+} independent mechanisms. *Endocrinology* 124: 2089-2098.

Sahu, A., S.P. Kalra, W.R. Crowley, T.L. O'Donohue and P.S. Kalra. 1987. Neuropeptide Y levels in microdissected regions of the hypothalamus and *in vitro* release in response to KCl and prostaglandin E_2: Effects of castration. *Endocrinology* 120: 1831-1836.

Sahu, A., S.P. Kalra, W.R. Crowley and P.S. Kalra. 1988. Evidence that NPY-containing neurons in the brainstem project into selected hypothalamic nuclei: Implication in feeding behavior. *Brain Res*. 457: 376-378.

Sahu, A., S.P. Kalra, W.R. Crowley and P.S. Kalra. 1989. Testosterone raises neuropeptide-Y concentration in selected hypothalamic sites and *in vitro* release from the medial basal hypothalamus of castrated male rats. *Endocrinology* 124: 410-414.

Samuelson, U.E. and D.-J. Dalsgaard. 1985. Action and localization of neuropeptide Y in the human fallopian tube. *Neurosci. Lett*. 58: 49-54.

Saria, A., E. Theodorsson-Norheim and J.M. Lundberg. 1985. Evidence for specific neuropeptide Y-binding sites in rat brain synaptosomes. *Eur. J. Pharmacol*. 107: 105-107.

Sasek, C.A. and R.P. Elde. 1985. Distribution of neuropeptide Y-like immunoreactivity and its relationship to FMRF-amide-like immunoreactivity in the sixth lumbar and first sacral spinal cord segments of the rat. *J. Neurosci*. 5: 1729-1739.

Sawchenko, P.E., L.W. Swanson, R. Grzanna, P.R.C. Howe, S.R. Bloom and J.M. Polak. 1985. Colocalization of neuropeptide Y immunoreactivity in brainstem catecholaminergic neurons that project to the paraventricular nucleus of the hypothalamus. *J. Comp. Neurol*. 241: 138-153.

Sawchenko, P.E. and S.W. Pfeiffer. 1988. Ultrastructural localization of neuropeptide Y and galanin immunoreactivity in the paraventricular nucleus of the hypothalamus in the rat. *Brain Res*. 474: 231-245.

Scott, J.W., J.K. McDonald and J.L. Pemberton. 1987. Short axon cells of the rat olfactory bulb display NADPH-diaphorase activity, neuropeptide Y-like immunoreactivity and somatostatin-like immunoreactivity. *J. Comp. Neurol*. 260: 378-391.

Senba, E., M. Tohyama, Y. Shiotani, Y. Kawasaki, T. Kubo, T. Matsunaga, P.C. Emson and H.W.M. Steinbusch. 1985. Peptidergic and aminergic innervation of the facial nucleus of the rat with special reference to ontogenetic development. *J. Comp. Neurol*. 238: 429-439.

Simerly, R.B. and L.W. Swanson. 1987. The distribution of neurotransmitter-specific cells and fibers in the anteroventral periventricular nucleus: Implications for the control of gonadotropin secretion in the rat. *Brain Res.* 400: 11-34.

Smialowska, M. 1988. Neuropeptide Y immunoreactivity in the locus coeruleus of the rat brain. *Neuroscience* 25: 123-131.

Smith, Y., A. Parent, L. Kerkerian and G. Pelletier. 1985. Distribution of neuropeptide Y immunoreactivity in the basal forebrain and upper brainstem of the squirrel monkey (Saimiri sciureus). *J. Comp. Neurol.* 236: 71-89.

Smith, Y. and A. Parent. 1986. Neuropeptide Y-immunoreactive neurons in the striatum of cat and monkey: Morphological characteristics, intrinsic organization and co-localization with somatostatin. *Brain Res.* 372: 241-252.

Solomon, T.E. 1985. Pancreatic polypeptide, peptide YY, and neuropeptide Y family of regulatory peptides. *Gastroenterology* 88: 838-841.

Stanley, B.G. and S.F. Leibowitz. 1984. Neuropeptide Y: Stimulation of feeding and drinking by injection into the paraventricular nucleus. *Life Sci.* 35: 2635-2642.

Stanley, B.G. and S.F. Leibowitz. 1985. Neuropeptide Y injected in the paraventricular hypothalamus: A powerful stimulant of feeding behavior. *Proc. Natl. Acad. Sci. USA* 82: 3940-3943.

Stanley, B.G., A.S. Chin and S.F. Leibowitz. 1985a. Feeding and drinking elicited by central injection of neuropeptide Y: Evidence for a hypothalamic site(s) of action. *Brain Res. Bull.* 14: 521-524.

Stanley, B.G., D.R. Daniel, A.S. Chin and S.F. Leibowitz. 1985b. Paraventricular nucleus injections of peptide YY and neuropeptide Y preferentially enhance carbohydrate ingestion. *Peptides* 6: 1205-1211.

Stanley, B.G., S.E. Kyrkouli, S. Lampert and S.F. Leibowitz. 1986. Neuropeptide Y chronically injected into the hypothalamus: A powerful neurochemical inducer of hyperphagia and obesity. *Peptides* 7: 1189-1192.

Stjarne, L. and J.M. Lundberg. 1986. On the possible roles of noradrenaline, adenosine 5'-triphosphate and neuropeptide Y as sympathetic cotransmitters in the mouse vas deferens. *Prog. Brain Res.* 68: 263-278.

Stjarne, L., J.M. Lundberg and P. Astrand. 1986. Neuropeptide Y - a cotransmitter with noradrenaline and adenosine 5'-triphosphate in the sympathetic nerves of the mouse vas deferens? A biochemical, physiological and electropharmacological study. *Neuroscience* 18: 151-166.

Stjernquist, M., P. Emson, C.H. Owman, N.-O. Sjoberg, F. Sundler and K. Tatemoto. 1983. Neuropeptide Y in the female reproductive tract of the rat. Distribution of nerve fibres and motor effects. *Neurosci. Lett.* 39: 279-284.

Stone, R.A. 1986. Neuropeptide Y and the innervation of the human eye. *Exp. Eye Res.* 42: 349-355.

Stone, R.A., A.M. Laties and P.C. Emson. 1986. Neuropeptide Y and the ocular innervation of rat, guinea pig, cat and monkey. *Neuroscience* 17: 1207-1216.

Sundler, F., E. Moghimzadeh, R. Hakanson, M. Ekelund and P. Emson. 1983. Nerve fibers in the gut and pancreas of the rat displaying neuropeptide-Y immunoreactivity. *Cell Tissue Res.* 230: 487-493.

Sundler, F., R. Hakanson, E. Ekblad, R. Uddman and C. Wahlestedt. 1986. Neuropeptide Y in the peripheral adrenergic and enteric nervous systems. *Int. Review Cytol.* 102: 243-269.

Sutin, E.L. and D.M. Jacobowitz. 1988. Immunocytochemical localization of peptides and other neurochemicals in the rat laterodorsal tegmental nucleus and adjacent area. *J. Comp. Neurol.* 270: 243-270.

McDonald *153*

Sutton, S., T. Toyama, S. Otto and P. Plotsky. 1987. Presence of neuropeptide Y (NPY) in hypophysial-portal plasma: Analysis of putative central and adenohypophysial actions. *69th Annu. Mtng. Endocrine Soc.*, 204.
Sutton, S.W., N. Mitsugi, P.M. Plotsky and D.K. Sarkar. 1988a. Neuropeptide Y (NPY): A possible role in the initiation of puberty. *Endocrinology* 123: 2152-2154.
Sutton S.W., T.T. Toyama, S. Otto and P.M. Plotsky. 1988b. Evidence that neuropeptide Y (NPY) released into the hypophysial-portal circulation participates in priming gonadotrophs to the effects of gonadotropin releasing hormone (GnRH). *Endocrinology* 123: 1208-1210.
Takeuchi, T., D.L. Gumucio, T. Yamada, M.H. Meisler, C.D. Minth, J.E. Dixon, R.E. Eddy and T.B. Shows. 1986. Genes encoding pancreatic polypeptide and neuropeptide Y are on human chromosomes 17 and 7. *J. Clin. Invest.* 77: 1038-1041.
Tatemoto, K. 1982a. Isolation and characterization of peptide YY (PYY), a candidate gut hormone that inhibits pancreatic exocrine secretion. *Proc. Natl. Acad. Sci. USA* 79: 2514-2518.
Tatemoto, K. 1982b. Neuropeptide Y: Complete amino acid sequence of the brain peptide. *Proc. Natl. Acad. Sci. USA* 79: 5485-5489.
Tatemoto, K., M. Carlquist and V. Mutt. 1982. Neuropeptide Y - a novel brain peptide with structural similarities to peptide YY and pancreatic polypeptide. *Nature* 296: 659-660.
Taylor, I.L. 1985. Distribution and release of peptide YY in dog measured by specific radioimmunoassay. *Gastroenterology* 88: 731-737.
Tenmoku, S., B. Ottesen, M.M.T. O'Hare, S. Sheikh, B. Bardrum, B. Hansen, B. Walker, R.F. Murphy and T.W. Schwartz. 1988. Interaction of NPY and VIP in regulation of myometrial blood flow and mechanical activity. *Peptides* 9: 269-275.
Terenghi, G., J.M. Polak, J.M. Allen, S.Q. Zhang, W.G. Unger and S.R. Bloom. 1983. Neuropeptide Y-immunoreactive nerves in the uvea of guinea pig and rat. *Neurosci. Lett.* 42: 33-38.
Tigges, J., J.K. McDonald and M. Tigges. 1987. Neuropeptide Y-like immunoreactivity in the hypothalamus of the young rhesus monkey (Macaca mulatta). *Soc. Neurosci. Abstr.* 13: 1579.
Tigges, M., J. Tigges, J.K. McDonald, M. Slattery and A. Fernandes. 1989. Postnatal development of neuropeptide Y-like immunoreactivity in area 17 of normal and visually deprived rhesus monkeys. *Visual Neuroscience* 2: 315-328.
Uddman, R., L. Edvinsson, R. Hakanson, C. Owman and F. Sundler. 1982. Immunohistochemical demonstration of APP (avian pancreatic polypeptide)-immunoreactive nerve fibres around cerebral blood vessels. *Brain Res. Bull.* 9: 715-718.
Ueda, S., M. Kawata and Y. Sano. 1986. Identification of neuropeptide Y immunoreactivity in the suprachiasmatic nucleus and the lateral geniculate nucleus of some mammals. *Neurosci. Lett.* 68: 7-10.
Unden, A. and T. Bartfai. 1984. Regulation of neuropeptide Y (NPY) binding by guanine nucleotides in the rat cerebral cortex. *FEBS Lett.* 177: 125-128.
Unden, A., K. Tatemoto, V. Mutt and T. Bartfai. 1984. Neuropeptide Y receptor in the rat brain. *Eur. J. Biochem.* 145: 525-530.
VanDongen, P.A.M., T. Hökfelt, S. Grillner, A.A.J. Verhofstad, H.W.M. Steinbusch, A.C. Cuello and L. Terenius. 1985. Immunohistochemical demonstration of some putative neurotransmitters in the lamprey spinal cord and spinal ganglia: 5-hydroxytryptamine-, tachykinin-, and neuropeptide-Y-immunoreactive neurons and fibers. *J. Comp. Neurol.* 234: 501-522.

Vincent, S.R., L. Skirboll, T. Hökfelt, O. Johansson, J.M. Lundberg, R.P. Elde, L. Terenius and J. Kimmel. 1982a. Coexistence of somatostatin- and avian pancreatic polypeptide (APP)-like immunoreactivity in some forebrain neurons. *Neuroscience* 7: 439-446.

Vincent, S.R., O. Johansson, T. Hökfelt, B. Meyerson, C. Sachs, R.P. Elde, L. Terenius and J. Kimmel. 1982b. Neuropeptide coexistence in human cortical neurones. *Nature* 298: 65-67.

Vincent, S.R. and O. Johansson. 1983. Striatal neurons containing both somatostatin- and avian pancreatic polypeptide (APP)-like immunoreactivities and NADPH-diaphorase activity: A light and electron microscopic study. *J. Comp. Neurol.* 217: 264-270.

Vincent, S.R., O. Johansson, T. Hökfelt, L. Skirboll, R.P. Elde, L. Terenius, J. Kimmel and M. Goldstein. 1983. NADPH-diaphorase: A selective histochemical marker for striatal neurons containing both somatostatin- and avian pancreatic polypeptide (APP)-like immunoreactivities. *J. Comp. Neurol.* 217: 252-263.

Wahle, P., G. Meyer and K. Albus. 1986. Localization of NPY-immunoreactivity in the cat's visual cortex. *Exp. Brain Res.* 61: 364-374.

Wahlestedt, C., N. Yanaihara and R. Hakanson. 1986. Evidence for different pre- and post-junctional receptors for neuropeptide Y and related peptides. *Regul. Pept.* 13: 307-318.

Wahlestedt, C., G. Skagerberg, R. Ekman, M. Heilig, F. Sundler and R. Hakanson. 1987. Neuropeptide Y (NPY) in the area of the hypothalamic paraventricular nucleus activates the pituitary-adrenocortical axis in the rat. *Brain Res.* 417: 33-38.

Wahlestedt, C., R. Ekman and E. Widerlov. 1989. Neuropeptide Y (NPY) and the central nervous system: Distribution effects and possible relationship to neurological and psychiatric disorders. *Prog. Neuro-Psychopharmacol. Biol. Psychiat.* 13: 31-54.

Wang, Y.-N., J.K. McDonald and R.J. Wyatt. 1987. Immunocytochemical localization of neuropeptide Y-like immunoreactivity in adrenergic and non-adrenergic neurons of the rat gastrointestinal tract. *Peptides* 8: 145-151.

Wehrenberg, W.B., R. Corder and R.C. Gaillard. 1988. A physiological role for neuropeptide Y in regulating the estrogen progesterone induced LH surge in ovariectomized rats. *70th Annu. Mtng. Endocrine Soc.*, 311.

Woller, M.J., J.K. McDonald and E. Terasawa. 1989. *In vivo* release of neuropeptide-Y in the stalk-median eminence of castrated rhesus monkeys is pulsatile. *Soc. Neurosci. Abstr.* 15: 188.

Yamazoe, M., S. Shiosaka, P.C. Emson and M. Tohyama. 1985. Distribution of neuropeptide Y in the lower brainstem: An immunohistochemical analysis. *Brain Res.* 335: 109-120.

Yui, R., T. Iwanaga, H. Kuramoto and T. Fujita. 1985. Neuropeptide immunocytochemistry in protostomian invertebrates, with special reference to insects and molluscs. *Peptides* 6 [Suppl. 3]: 411-415.

Zhang, S.Q., G. Terenghi, W.G. Unger, K.W. Ennis and J. Polak. 1984. Changes in substance P and neuropeptide Y-immunoreactive fibres in rat and guinea pig irides following unilateral sympathectomy. *Exp. Eye Res.* 39: 365-372.

Zukowska-Grojec, Z. and A.C. Vaz. 1988. Role of neuropeptide Y (NPY) in cardiovascular responses to stress. *Synapse* 2: 293-298.

OPIOIDS AND INDUCTION OF OVULATION: MEDIATION BY NEUROPEPTIDE Y

S.P. Kalra, L.G. Allen*, A. Sahu, and W.R. Crowley**
R.W. Johnson Medical School*
Department of Anatomy
Piscataway, NJ
and
Department of Pharmacology**
University of Tennessee
College of Medicine
Memphis, TN

I.   Introduction

II.   EOP in the Induction Of LH Surge

    A.   The Preovulatory LH Surge

    B.   The E_2-induced LH Surge in Ovx Rats

III.   Where and How do EOP Neurons Communicate with LHRH Neurons?

IV.   The Adrenergic Connection

V.   The Neuropeptide Y Connection

VI.   Summary

Introduction

The concept of opiate involvement in the control of reproduction by the brain stems from clinical observations that female narcotic addicts suffer from severe menstrual cycle disorders and that morphine, a synthetic alkaloid, readily inhibited ovulation and the preovulatory release of gonadotropins (Santen *et al.*, 1975; Pang *et al.*, 1977; Ieiri *et al.*, 1980). We have previously reviewed the anatomical distribution of the three endogenous opioid peptides (EOP), namely β-endorphin (βE), dynorphin and enkephalins in the hypothalamus, the general effects of each on LH release and the opiate receptor subtype (μ, κ and γ) each EOP interacts with to inhibit LH release (Kalra and Kalra, 1983; Kalra, 1986; Kalra *et al.*, 1989a). Also, the two views on the mode of EOP participation, that is, whether they mediate or monitor the inhibitory feedback effects of gonadal steroids on LH release in the rat, have been critically evaluated (Kalra *et al.*, 1989a). This knowledge of the physiology and pharmacology of the EOP and of their receptors in close proximity to luteinizing hormone-releasing hormone (LHRH) neurons has prompted us to define more carefully the part played by EOP in the preovulatory release of LH in the rat. More than 40 years ago, Everett, Sawyer and associates demonstrated both that the preovulatory release of gonadotropins was triggered by neurogenic stimuli during a well-defined "critical period" on the afternoon of proestrus of rats displaying regular estrous cycles and that this neural event was tightly entrained to the circadian periodicity (Everett and Sawyer, 1950; Everett *et al.*, 1949; Sawyer *et al.*, 1949). Subsequent research demonstrated that an interaction of the ovarian steroid milieu, dominated by estradiol 17β (E$_2$), with the circadian clock, facilitated the onset, duration and magnitude of LH release (Ferin *et al.*, 1969; Caligaris *et al.*, 1971), and that the facilitatory action of E$_2$ on the central nervous system was completed by 3 AM of proestrus (Kalra, 1975). Thereafter, the neural clock (NC) triggered the release of LHRH at an accelerated rate during the critical period to precipitate a massive discharge of LH from the pituitary lasting for 6-8 h (Kalra *et al.*, 1971; Sarkar *et al.*, 1976; Ching, 1982). Despite intensive research over the years, the origin and route of transmission of the neurogenic stimuli responsible for LHRH hypersecretion during the critical period are not fully characterized.

Our recent studies showed that a complex hypothalamic neural circuitry, composed of a multitude of inhibitory and excitatory monoaminergic and neuropeptidergic circuits, participates in the induction of ovulation in the rat (Kalra and Kalra, 1983; Kalra, 1986). Consequently, the goal of this communication is two fold: first, to define precisely the role of opioids in the normal physiological regulation of preovulatory secretion of LH and, second, to establish the functional connectivity of opioids with other neural systems engaged in the origin and transmission of neurogenic stimuli from the NC to the final destination of the median eminence (ME), wherein LHRH is discharged into the hypophyseal portal veins for stimulation of the pituitary. For these investigations we have employed a general opiate receptor antagonist, naloxone (NAL), which readily antagonizes the tonic inhibitory influence of EOP on the episodic secretion of LHRH under a variety of experimental conditions (Kalra, 1983).

EOP in the Induction of LH Surge

The Preovulatory LH Surge

The evidence that stimulation of opiate receptors before the critical period on proestrus with either an EOP, βE (Leadem and Kalra, 1985a), or an opiate agonist, morphine, blocked the preovulatory LH release and ovulation and suppressed the ovarian steroid-induced LH surge in ovariectomized (ovx) rats (Pang *et al.*, 1977; Ieiri *et al.*, 1980; Kalra and Simpkins, 1981) suggests that the hypothalamic EOP network normally exerts an inhibitory influence and that experimental sustenance of this inhibition counters the expression of neurogenic stimuli responsible for the preovulatory discharge of HL. Further studies from our laboratory (Gabriel *et al.*, 1983) and those of Piva *et al.* (1985) verified this assumption. It was observed that although blockade of opiate receptors with NAL injection readily stimulated LH release on each day of the estrous cycle, it was completely ineffective during the period of LH surge either on proestrus or during that induced by sequential estrogen-progesterone treatment on ovx rats. Additionally, NAL was rendered ineffective, after administration of progesterone on the morning of proestrus, in advancing the preovulatory LH surge. If NAL stimulated LH secretion by restraining the prevailing inhibitory influence after displacing EOP from the receptor sites, then its ineffectiveness during the LH surge period suggested to us that the inhibitory EOP effects are either absent or curtailed drastically to allow LHRH and LH hypersecretion to occur during the critical period. This line of reasoning and the findings that intraventricular βE blocked the preovulatory LH surge and ovulation and decreased the high-amplitude, high-frequency LH pulses in ovx rats (Leadem and Kalra, 1985a,b), led us to postulate that some time prior to and during the critical period a decrease in opioid tone in the preoptic-tuberal pathway may trigger the preovulatory LH surge.

This hypothesis, originally advanced in 1983 (Kalra and Gallo, 1983; Kalra, 1983), was tested by examining the effects on LH secretion of a sustained decrease of opioid tone with NAL infusion prior to the LH surge in the afternoon of proestrus (Allen and Kalra, 1986; Allen *et al.*, 1988). The results of these studies showed the following (Figs. 1 and 2): 1) NAL infusion on the morning of proestrus can prematurely evoke the preovulatory LH surge; 2) Normally LH is secreted as low-amplitude pulses occurring at 50-75 min intervals prior to the critical period on proestrus. NAL infusion during this period, at the rate of 0.5 mg/h, evoked discrete episodes of greater amplitude at a mean interval of 37 min. Infusion of NAL, at a higher rate of 2 mg/h, accelerated the frequency of LH episodes further to 30-37 min, and the magnitude of the LH response increased correspondingly. As evident in figure 1, in some instances the LH episodes were progressively more robust and rapidly

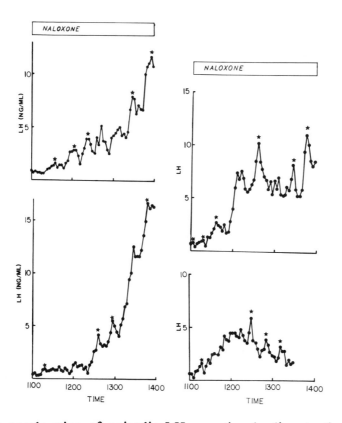

Fig. 1. Shows acceleration of episodic LH secretion leading to the induction of the LH surge in four rats during infusion of 2 mg NAL/h between 1100-1400 h on proestrus. Note a progressive rise in LH characterized by discrete LH episodes during the ascending phase (right panel). Asterisks denote the identified LH peaks.

merged into a steep ascending phase, whereas in other, discrete LH peaks of progressively increasing strength were clearly observed during the ascending phase. Thus, it was evident that a decrease in opioid tone on proestrus accelerated the episodic LH discharge and a dose-dependent relationship existed between the degree of restraint of the inhibitory opioid tone imposed by NAL and the magnitude of LH response. 3) Since NAL readily stimulated LHRH release from the hypothalamus in a dose-related fashion (Leadem *et al.*, 1985; Kalra *et al.*, 1987), it suggested that an accelerated rate of episodic LHRH discharge and, perhaps, an increase in the amount of LHRH per episode may be responsible for the premature onset of the LH surge in NAL-infused rats. Indeed, infusion of an extremely small, invariant amount of LHRH at a frequency of 30 min, and not at 60 min, reproduced a proestrous-type LH surge in pentobarbital-blocked proestrous rats (Allen *et al.*, 1988). 4) A closer scrutiny of these results revealed an intriguing effect of the decrease in opioid tone on episodic LH discharge on proestrus. The augmentation of episodic LH secretion by NAL was reminiscent of the episodic pattern of LH hypersecretion in the afternoon of proestrus and in ovx rats (Gallo, 1980; Leadem and Kalra, 1985b).

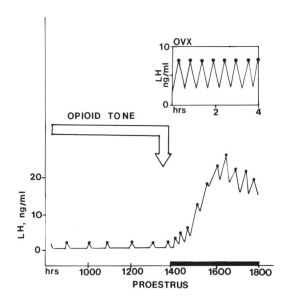

Fig. 2. A diagrammatic summary of the effects of NAL infusion prior to the critical period on proestrus on the preovulatory LH surge. Our thesis states that the NC may decrease the opioid tone (indicated by the arrow) to evoke a preovulatory LH rise during the afternoon of proestrus. A diminution in the inhibitory tone may gradually accelerate the frequency of LH discharge to the frequency normally observed in ovx rats. For details, see text.

The preovulatory LH surge is composed of robust episodes occurring at a frequency varying from 16-35 min (Gallo, 1981; Allen *et al.*, 1988) and, similarly, after gonadectomy LH is secreted episodically at a frequency ranging between 20-40 min (Gallo, 1980, Leadem and Kalra, 1985b). As presented in figure 2, a decrease in inhibitory opioid tone by NAL accelerated the frequency of LH discharge that approached the frequency normally seen after ovariectomy and in the afternoon of proestrus. On the basis of these observations, we proposed the following mode of EOP involvement in initiation of the preovulatory release of LH: The opioid systems are interposed between the NC and LHRH neurons and to evoke preovulatory LH hypersecretion on the afternoon of proestrus, the NC curtails, perhaps, the steroid-independent (Gabriel *et al.*, 1983; Gallo *et al.*, 1987; Kalra *et al.*, 1989a) inhibitory tone to permit accelerated episodic LHRH discharge at a rate close to that normally occurring in ovx rats. As a consequence of this sustained hypothalamic activation of the pituitary, there is a massive discharge of LH lasting for 6-8 h. Whether the NC communicates directly with the EOP neurons or whether other neural circuits intervene to decrease the opioid tone remains to be elucidated.

The E_2-induced LH Surge in Ovx Rats

This concept of a decrease in opioid tone, as a crucial, early step in the sequence of events that triggers the LH surge, was recently verified in experiments designed to account for the diminished LH surge response in E_2-treated ovx rats (Masotto *et al.*, 1989). It is well known that the E_2-induced LH surge in ovx rats is invariably smaller in magnitude than the preovulatory LH surge or that induced by progesterone (P) in estrogen-primed ovx rats (Caligaris *et al.*, 1971; Kalra and McCann, 1973; Hiemke and Ghraf, 1984; Hiemke *et al.*, 1987). Remarkably, the facilitatory effects of estrogen wane after ovariectomy in a time-dependent manner so that in chronically ovx rats, physiological as well as pharmacological doses of E_2 fail to evoke an appreciable LH rise (Hiemke *et al.*, 1987; Caligaris *et al.*, 1971; Kalra and McCann, 1973). Evidence showed that this loss of responsivity occurred centrally because the pituitary response to LHRH was not adversely altered (McGinnis *et al.*, 1981; Camp *et al.*, 1985). Furthermore, LHRH accumulation in the ME, normally seen antecedent to the preovulatory LH surge on proestrus, is absent in rats treated with estrogen alone, but not in rats treated sequentially with E_2 and P (Kalra, *et al.*, 1973; Kalra and Kalra, 1977, 1979; Kalra and Simpkins, 1981). It remains to be discovered how the underlying neuroendocrine factors that impart loss of responsivity to estrogen in ovx rats or how P overcomes this loss. In view of the crucial role of a

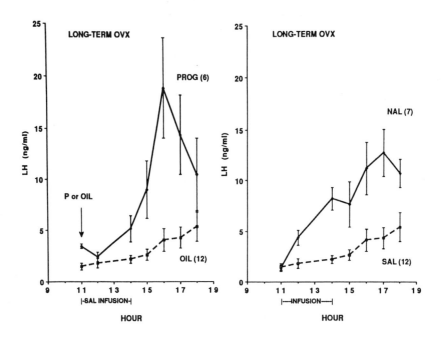

Fig. 3. The effects of progesterone (P) or NAL infusion between 1100-1400 h on the LH surge in long-term ovx rats. Estrogen alone evoked a small rise in plasma LH in the afternoon (broken lines). P injection (left) or NAL infusion (right) were equally effective in amplifying the LH surge response.

decrease in opioid tone in the preovulatory discharge of LH surge, we speculated that the reduction in opioid tone in estrogen-treated rats may be inadequate to allow a proestrus-type LH surge. In a series of recent studies we observed that an additional transient decrease in opioid tone, induced experimentally with NAL infusion prior to the onset of spontaneous LH surge, advanced the onset and magnitude of the LH surge so that it resembled the preovulatory LH surge on proestrus or that induced by P in these estrogen-treated rats (Fig. 3; Masotto *et al.*, 1990). In addition, an analysis of the episodic LH secretory pattern showed that NAL infusion accelerated not only the frequency and amplitude of LH discharge but significantly changed the contour of LH episodes by decreasing the interpeak interval and pulse width accompanied by increments in pulse height. Consequently, LH episodes were sharper and leaner during NAL infusion (Fig. 4). Thus, consistent with the findings on proestrus, it was obvious that an additional decrease in opioid tone markedly changed the configuration and rate of hypothalamic activation of the pituitary to enhance the magnitude of the LH

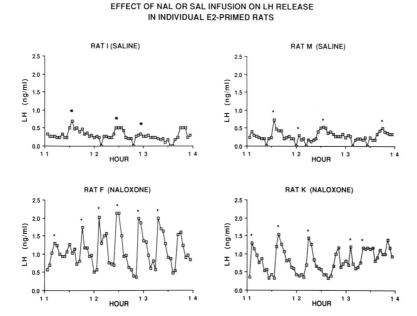

Fig. 4. Effect of NAL or saline infusion on LH release in individual E_2-primed ovx rats. NAL infusion accelerated the frequency and markedly changed the contour of LH episodes.

surge in E_2-treated rats. Further, the striking similarity of the effects of NAL infusion and P treatment in advancing the onset and in amplifying the LH surge (Fig. 3) suggested that as with NAL, one of the neuroendocrine sequelae of central P action is to reduce the central inhibitory opioid tone. This concept is supported by evidence of the ineffectiveness of NAL after P treatment (Gabriel *et al.*, 1983), the decrease in ^{3}H-NAL binding in hypothalamic slices of estrogen-treated ovx rats (Jacobson and Kalra, 1989), and the disclosures that opiate receptors engaged in the NAL-induced LH surge reside in the basal hypothalamus where βE-containing cells concentrate P (Kalra, 1981; Fox *et al.*, 1986).

In summary, these studies are in line with our hypothesis that normally the NC may initially decrease the EOP inhibitory tone in the hypothalamus to trigger a chain of events that result in the preovulatory LH surge, and that inadequate restraint of the opioid tone may underlie the diminution in LH response in estrogen-treated rats. NAL infusion and P treatment intensify the NC-induced decrements in EOP tone to recuperate the afternoon LH response in E_2-treated rats.

Where and How Do EOP Neurons Communicate with LHRH Neurons?

Our results suggested that the opioid system is interposed between the NC and LHRH neurons. The opioid-LHRH functional link appears to be confined to the preoptic-tuberal pathway, primarily in the medial preoptic area (MPOA) and ME-arcuate nucleus (ME-ARC). Since NAL implantation in the ME-ARC stimulated LH release *in vivo* and NAL stimulated LHRH release *in vitro* from the medial basal hypothalamus (MBH) or the ME-ARC (Kalra, 1981; Leadem *et al.*, 1985; Negro-Vilar *et al.*, 1985; Masotto *et al.*, 1989), it is reasonable to assume that the relevant link between the two systems resides in the proximity of LHRH nerve terminals in the ME. On the other hand, immunohistochemical studies demonstrate a few synaptic connections between βE immunoreactive axons and LHRH perikarya in the MPOA (Leranth *et al.*, 1988). Whereas the physiological significance of the axo-somatic connections between the two peptidergic systems in the MPOA remains to be understood, the possibility that this relationship may regulate the biosynthesis of the decapeptide in LHRH cell bodies in the MPOA is favored by the findings that stimulation of opiate receptors with morphine completely blocked the P-induced increments in MBH LHRH levels associated with the afternoon LH surge and that morphine causes a decrease in LHRH levels in the MPOA (Kalra and Simpkins, 1981; Negro-Vilar *et al.*, 1985). Therefore, it is conceivable that more than one site of interaction between EOP and LHRH neurons may exist in the preoptic-tuberal pathway. Presumably then, whereas the inhibitory control on biosynthesis of LHRH may manifest itself through a direct EOP-LHRH neuronal link in the MPOA, the inhibition exercised by EOP on the release of LHRH may be mediated by multiple neural systems (see below).

The Adrenergic Connection

These formulations paved the way for identification of the hypothalamic neuronal systems likely to relay the EOP inhibitory signals to LHRH nerve terminals in the ME. Our endeavors have identified an excitatory circuit consisting of adrenergic and neuropeptide Y (NPY) neurons as an essential component of the transmission line between the EOP and LHRH neurons (Kalra, 1986; Crowley, 1987). The mediation by adrenergic systems of the effects of EOP on LH release are based on the following: 1) a decrease in hypothalamic adrenergic transmission, produced either by administration of α-adrenergic receptor antagonists or specific norepinephrine (NE) or epinephrine (EPI) synthesis inhibitors, blocked the NAL-induced LH release and conversely, adrenergic receptor activation by clonidine countered this pharmacologic blockade (Kalra and Simpkins, 1981; Kalra and Crowley, 1983; Kalra et al., 1972); 2) NAL increased and morphine decreased hypothalamic adrenergic turnover (Adler and Crowley, 1984); 3) NAL stimulated the concomitant release of NE, EPI and LHRH from the hypothalamus *in vitro* (Leadem *et al.*, 1985; Kalra *et al.*, 1987), and α-adrenergic receptor antagonists blocked the *in vitro* NAL-induced LHRH release from the ME (Negro-Vilar *et al.*, 1985; Masotto *et al.*, 1989). Thus, these results concur with the idea that EOP receptors are present on the hypothalamic adrenergic systems and this axo-axonic interaction regulates the output of excitatory adrenergic signals to LHRH nerve terminals locally in the ME.

The Neuropeptide Y Connection

The proposal of adrenergic mediation of EOP effects on LHRH release failed to accommodate the findings that a drastic reduction in adrenergic inputs to the hypothalamus on a chronic basis did not impair normal ovulatory cycles in the rat (Clifton and Sawyer, 1979; Nicholson *et al.*, 1978). As a consequence we envisioned that additional excitatory circuits, possibly localized within the hypothalamus, may play a crucial and, perhaps, a mandatory role in the control of pituitary gonadotropin secretion (Kalra, 1986). With these possibilities in mind, in 1982 we tested the effects of human pancreatic peptide (hPP) on LH release (Kalra and Crowley, 1984a). The basis of these new undertakings was that since PP was shown to coexist with adrenergic transmitters in the brain stem and PP immunoreactivity was localized in the hypothalamus (Hökfelt *et al.*, 1984; Jacobowitz and Olschowka, 1982; Lundberg *et al.*, 1980; Olschowka *et al.*, 1981), it is possible that PP, either alone or in concert with adrenergic transmitters, may affect LH release. In late 1982 and early 1983 we observed that hPP, like NE (Gallo and Drouva, 1979), inhibited LH release in ovx rats but stimulated LH release in steroid-primed ovx rats (Kalra and Crowley, 1984a). This unexpected discovery of the dual effects of PP acting as an inhibitory peptide in an ovarian steroid-free environment and as an excitatory peptide in the ovarian steroid-primed environment provided a new avenue for exploration of the hypothalamic control of pituitary LH release. Additionally, these findings disclosed that the excitatory properties of PP are likely to be of greater physiological significance in gonad-intact rats. McDonald and Lumpkin (1983) reported that PP of avian origin inhibited LH release in ovx rats, a finding that advocated an inhibitory role of aPP in the hypothalamic control of LH release.

Subsequently, in August–September 1983, we became aware of another member of the PP family, NPY, a biologically active peptide of neural origin with a distribution pattern in the rat brain similar to that of PP (Allen *et al.*, 1983; Guy *et al.*, 1983; Tatemoto, 1982; Tatemoto *et al.*, 1982). Consequently, from December 1983 onwards we studied the effects of intraventricular NPY on LH release. Similar to NE and hPP, NPY inhibited LH release in ovx rats and stimulated LH release in a dose-related fashion in ovarian steroid-primed ovx rats (Kalra and Crowley, 1984b). Thus, the concept that NPY may indeed be an endogenous excitatory neuropeptidergic system, functionally linked with the LHRH-LH axis in gonad-intact rats, was advanced. These formulations were validated directly in gonad-intact rats (Allen *et al.*, 1985; Rodriguez-Sierra *et al.*, 1987). In line with our findings, McDonald *et al.* (1985) reported an inhibition of LH release by NPY in ovx rats, but contrary to our hypothesis (Kalra and Crowley, 1984a,b), implied that like avian PP, NPY may normally act as an inhibitory hypothalamic peptide in the regulation of LH release.

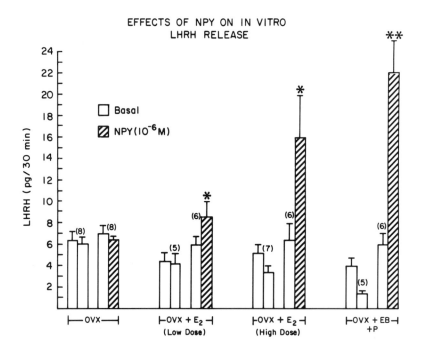

Fig. 5.   Effects of NPY on *in vitro* LHRH release from the MBH of ovx rats primed with either subphysiological (10-15 pg/ml, low dose) or physiological E_2 (30 pg/ml, high dose) or estrogen and progesterone.   Asterisks denote significantly different from respective basal levels.

Two additional studies from our laboratory further affirmed the notion that NPY acted as an excitatory peptide in the rat. First, Crowley et al. (1985) observed that NPY concentrations in the ME fluctuated in parallel with LHRH in association with the P-induced LH hypersecretion in estrogen-primed ovx rats, thereby suggesting for the first time that augmented NPY release activated LHRH secretion to precipitate the LH surge. Recent studies utilizing passive immunization against NPY have fully affirmed the involvement of NPY in the preovulatory release of LH in the rat (Sutton et al., 1988; Wherenberg et al., 1989). Second, Crowley and Kalra (1987; Fig. 5) reported that whereas NPY failed to alter the basal LHRH release in vitro from the MBH of ovx rats, it readily stimulated LHRH release from the MBH of rats pretreated with either subphysiological (10-15 pg/ml, low dose) or physiological (30 pg/ml, high dose) doses of E_2. Further, combined treatment with E_2 and P allowed a relatively higher release of LHRH from the MBH as compared to that from rats treated with E_2 alone. These studies were extended to show that stimulation of LHRH release by NPY may not require entry of extracellular Ca^{++} through voltage-sensitive Ca^{++} channels (Fig. 6; Crowley and Kalra, 1988). Khorram and co-workers (1987, 1988) similarly showed that intrahypothalamic NPY infusion in E_2-primed rabbits stimulated LHRH release, but in gonadectomized rabbits it inhibited LHRH release. Recently, Sabatino et al. (1989) reported similar stimulatory effects of NPY on LHRH release from the ME of E_2-primed ovxrats and concurred with the excitatory role originally assigned to NPY in the

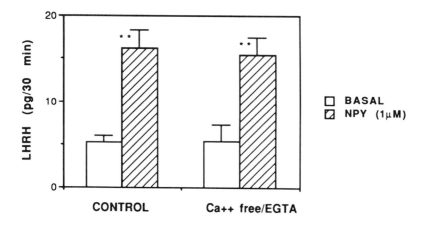

Fig. 6. Effects of NPY (1 μM) on the in vitro release of LHRH from the MBH incubated in media with and without Ca^{++}. **$p < 0.01$ vs. respective basal.

control of LHRH release in gonad-intact rats (Kalra and Crowley, 1984a,b; Kalra, 1986; Allen *et al.*, 1985; Crowley and Kalra, 1987).

The EOP-NPY Transmission Line

Because NPY and LHRH levels in the ME fluctuated in parallel in association with the LH surge induced by P in estrogen-primed ovx rats (Crowley *et al.*, 1985), it seemed likely that NPY may similarly be involved during the preovulatory discharge of LHRH on proestrus. Indeed, Sahu *et al.* (1989a; Fig. 7) observed that of the five hypothalamic sites examined for changes in NPY concentration, only the ME NPY levels displayed marked fluctuations in close association with the LH surge. NPY concentrations were low between 1000-1300 h and rose abruptly at 1400 h preceding the onset of LH rise at 1500 h. These

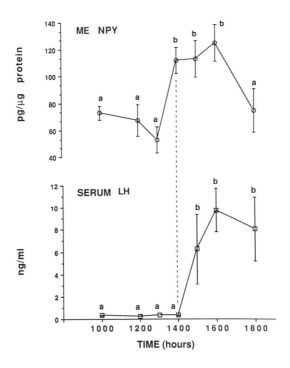

Fig. 7. Dynamic change in ME NPY concentrations in association with the preovulatory LH surge on proestrus afternoon. The values that share the same superscript, in this and subsequent figures, are not statistically different (p > 0.05).

elevated levels were maintained until 1600 h during the plateau phase of the LH surge. Thereafter, ME NPY levels fell at 1800 h to the low, early morning levels. In contrast, the pattern of changes in NPY levels in the ARC, supra-chiasmatic nucleus and MPOA, the sites known to participate in the preovulatory discharge of LH, were markedly different and failed to correlate with the LH surge. These findings of site-specific, dynamic changes in NPY levels on proestrus, and not on diestrus II, in a manner documented previously for LHRH (Kalra *et al.*, 1973; Kalra and Kalra, 1977), supported the hypothesis that a subset of NPY neurons terminating in the ME may represent a component of an

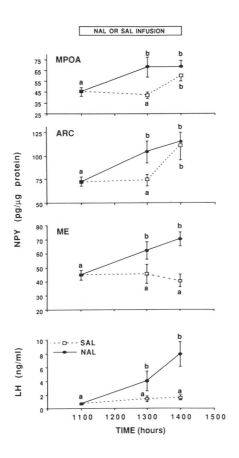

Fig. 8. Effect of NAL or SAL (saline) infusion on NPY concentrations in microdissected hypothalamic sites in association with increase in LH release in E_2-primed ovx rats. MPOA = medial preoptic area; ARC = arcuate nucleus; ME = median eminence.

excitatory neural circuitry that either independently, or in conjunction with adrenergic systems, is responsible for induction of the preovulatory LH release.

When one considers the observations that 1) NPY and LHRH concentrations in the ME fluctuated in parallel in association with the LH surges on proestrus, or that induced by P in E_2-primed rats; 2) a spontaneous increase in hypothalamic adrenergic turnover and ME LHRH occurred prior to the onset of the LH surge (Kalra, 1986; Crowley, 1987); 3) NAL infusion advanced the onset and magnitude of the LH surge (Allen and Kalra, 1986; Allen *et al.*, 1988) and evoked increments in ME LHRH levels (Jacobson *et al.*, 1988) and in adrenergic turnover in the hypothalamus (Adler and Crowley, 1984; Crowley, 1987), it appears likely that the hypothalamic NPY system may also be activated by a decrease in opioid tone induced by NAL infusion. Our recent studies confirmed this suspicion (Sahu *et al.*, 1989b).

Two days after E_2-priming, rats were infused with either saline (SAL, control) or NAL between 1100-1400 h. In association with a moderate LH surge beginning at 1500 h in SAL-infused rats, NPY levels increased in the ARC,

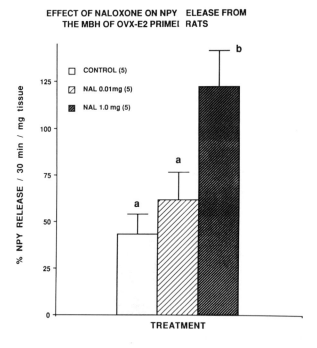

EFFECT OF NALOXONE ON NPY ELEASE FROM THE MBH OF OVX-E2 PRIMEI RATS

Fig. 9. Effect of NAL on NPY release from the MBH of E_2-primed ovx rats. Values are represented as a percentage of the basal release in the first 30 min period.

ventromedial hypothalamus and MPOA at 1400 h. In contrast, NAL infusion advanced the onset to 1300 h and markedly enhanced the magnitude of LH stimulation and concurrently activated NPY neurons. NPY levels were elevated significantly at 1300 h in the ME, ARC and MPOA (Fig. 8). These effects of NAL or SAL on NPY levels in other hypothalamic sites were not seen. These studies showed for the first time that a decrease in opioid tone stimulated NPY levels in the preoptic-tuberal pathway in association with LH stimulation. More recently, in a preliminary study we observed that NAL stimulated the *in vitro* release of NPY from the MBH of E_2-primed rats (Fig. 9). Since NPY readily stimulated LHRH release, these new results indicated that a decrease in opioid

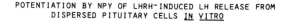

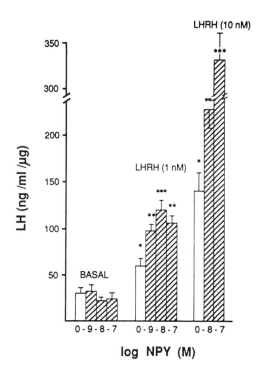

POTENTIATION BY NPY OF LHRH-INDUCED LH RELEASE FROM DISPERSED PITUITARY CELLS IN VITRO

Fig. 10. Effect of NPY and LHRH on LH release from cultured anterior pituitary cells during a 3 h incubation. Cells were exposed to NPY alone in the of LHRH (basal) or to 1 or 10 nM LHRH alone or with NPY. *p < 0.01 vs. respective basal; **p < 0.05 vs. respective LHRH alone; ***p < 0.01 vs. respective LHRH alone. In addition, LH release induced by LHRH (10 nM) plus NPY (10^{-7} M) was significantly (p < 0.01) greater than that induced by LHRH (10 nM) plus NPY (10^{-8} M).

tone may activate NPY neurosecretion and an obvious consequence of the absence increased NPY secretion would be to evoke LHRH release. When these results are considered in the context of the key role of EOP in the induction of the preovulatory LH surge on proestrus, we suggest that NPY, along with adrenergic systems (Allen *et al.*, 1987), constitute an excitatory neural link in the transmission line between the EOP and LHRH neurons.

Crowley *et al.* (1987) also explored the possibility that NPY may be released into the hypophyseal portal system for an effect at the level of the pituitary. Although a slight stimulatory effect of NPY on LH release from hemipituitaries in culture was observed, unlike the results of McDonald *et al.* (1985), we failed to observe any effect of NPY on LH release from dispersed pituitary cells (Fig. 10). However, an unexpected, novel outcome of these studies was a robust dose-dependent potentiation by NPY of the effects of LHRH on LH release from hemipituitaries and dispersed pituitary cells in culture (Fig. 10).

Cumulatively, it seems likely that in response to a decrease in opioid tone, the increased release of NPY from afferents in the ME acts in at least two ways: locally in the ME in concert with adrenergic transmitters to potentiate the release of LHRH (Allen *et al.*, 1987; Crowley and Kalra, 1989; Kalra *et al.*, 1989b), and after reaching the pituitary, to amplify the response of gonadotrophs to LHRH.

SEQUENTIAL NEUROENDOCRINE INTERACTIONS
PRECEDING PREOVULATORY LH SURGE

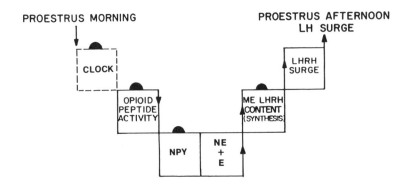

Fig. 11. Shows the sequential neuroendocrine events associated with the preovulatory LH surge. For details see text.

An underlying, two-pronged cooperative action of NPY--one in the hypothalamus with NE to stimulate LHRH release and the other with LHRH at the level of the pituitary--appears to be crucial for expression of the massive preovulatory LH discharge which lasts for several hours.

Summary

Although the formulation that neurogenic stimuli during the critical period on preoestrus evoke the preovulatory gonadotropin surge has been known for a long time, only recently have the neuroendocrine factors responsible for the origin and route of transmission of these neurogenic stimuli become apparent. The precise location of the NC in the preoptic area is still unknown. Our systematic research over the years has allowed us not only to trace the transmission line from the NC to LHRH neurons in the preoptic-tuberal pathway but also to delineate the sequence of neurosecretory events in components of this transmission line (Fig. 11). The research summarized herein strongly suggests that EOP play a pivotal role in the clock-induced preovulatory LH surge. As depicted in figure 11, we propose that the NC, under the facilitatory influence of the ovarian steroid milieu, curtails the tonic, steroid-independent, influence of EOP locally in the preoptic-tuberal pathway in advance of the critical period on proestrus. The gradual decrease of the opioid restraint augments the rate of adrenergic turnover and NPY and LHRH elaboration in the ME nerve terminals in preparation for the trigger for LHRH release during the critical period. Further, either the continued intense restraint on the opioid tone near the time of the critical period itself or an additional input signal increases the intermittent efflux of adrenergic transmitters and NPY in the ME. In the ME, NPY and adrenergic transmitters, either synergistically or in an additive manner, accelerate the episodic LHRH discharge to its innate maximal frequency rate, as found in ovx rats. As a consequence, the sustained hypothalamic activation of pituitary gonadotrophs by LHRH, in synergism with NPY, results in a massive discharge of LH on the afternoon of proestrus. This dynamic progression of neural events underscores the fact that the NC acts at an early stage of signal propagation sequence to restrain EOP influence and thereby allow an unfolding, in a time-dependent manner, of the core excitatory adrenergic, NPY and LHRH circuits. While these studies envision EOP systems as a paramount intermediate step between the NC and LHRH neurons, the morphological organizations of EOP with the NC on one hand, and with the excitatory adrenergic and NPY circuits on the other, remain to be defined.

ACKNOWLEDGMENTS

Supported by grants from NIH, HD 08634 (SPK) and HD 13703 (WRC).
Thanks are due to Ms. Sally McDonell for secretarial assistance.

REFERENCES

Adler, B.A. and W.R. Crowley. 1984. Modulation of luteinizing hormone release
     and catecholamine activity by opiate in the female rat. *Neuroendocrinology*
     38: 248-253.
Allen, L.G. and S.P. Kalra. 1986. Evidence that a decrease in opioid influence
     evokes preovulatory LH release. *Endocrinology* 118: 1275-1281.
Allen, L.G., W.R. Crowley and S.P. Kalra. 1987. Interactions between
     Neuropeptide Y and adrenergic systems in stimulation of LH release in
     steroid-primed ovariectomized rats. *Endocrinology* 121: 1953-1959.
Allen, L.G., E. Hahn, D. Caton and S.P. Kalra. 1988. Evidence that a decrease
     in opioid tone on proestrus changes the episodic pattern of LH secretion:
     Implications in the preovulatory LH hypersecretion. *Endocrinology* 122:
     1004-1013.
Allen, L.G., P.S. Kalra, W.R. Crowley and S.P. Kalra. 1985. Comparison of the
     effect of neuropeptide Y and adrenergic transmitters on LH release and food
     intake in male rats. *Life Sci.* 37: 617-623.
Allen, Y.S., T.E. Adrian, J.M. Allen, K. Tatemato, T.J. Crow, S.R. Bloom and
     J.M. Polak. 1983. Neuropeptide Y distribution in the rat brain. *Science*
     221: 877-879.
Caligaris, L., J. Astrada and S. Taleisnik. 1971. Release of luteinizing hormone
     induced by estrogen injection into ovariectomized rats. *Endocrinology* 88:
     910-915.
Camp, P., A. Akabori and C.A. Barraclough. 1985. Correlation of luteinizing
     hormone surge with estrogen nuclear and progesterone cytosol receptors in
     the hypothalamus and pituitary gland. II. Temporal estradiol effect.
     Neuroendocrinology 40: 54-62.
Ching, M. 1982. Correlative surges of LHRH, LH and FSH in pituitary stalk
     plasma and systemic plasma of rat during proestrus: Effects of anesthetics.
     Neuroendocrinology 34: 279-285.
Clifton, D.K. and C.H. Sawyer. 1979. LH release and ovulation in the rat
     following depletion of hypothalamic norepinephrine: Chronic vs. acute
     effects. *Neuroendocrinology* 28: 441-449.
Crowley, W.R. 1987. Control of luteinizing hormone secretion by ovarian
     hormone-monoamine-neuropeptide interaction. In *Integrative
     Neuroendocrinology: Molecular, Cellular and Clinical Aspects*, eds. S.M.
     McCann and R.I. Weiner, 54-69. S. Karger, Basel.
Crowley, W.R. and S.P. Kalra. 1987. Neuropeptide Y stimulates the release of
     luteinizing hormone-releasing hormone from medial basal hypothalamus *in
     vitro*: Modulation by ovarian hormones. *Neuroendocrinology* 46: 97-103.
Crowley, W.R. and S.P. Kalra. 1988. Regulation of luteinizing hormone
     secretion by neuropeptide Y in rats: Hypothalamic and pituitary actions.
     Synapse 2: 276-281.

Crowley, W.R. and S.P. Kalra. 1989. Regulation of preovulatory luteinizing hormone secretion by ovarian hormone-monoamine-neuropeptide interaction in the rat. In *Neural Control of Reproductive Function*, eds. J.M. Lakoski, J.R. Perez-Palo and D.K. Rassin, 79-93. Alan R. Liss, New York.

Crowley, W.R., A. Hassid and S.P. Kalra. 1987. Neuropeptide Y enhances the release of luteinizing hormone induced by luteinizing hormone-releasing hormone. *Endocrinology* 120: 941-945.

Crowley, W.R., R.E. Tessel, T.L. O'Donohue, B.A. Adler and S.P. Kalra. 1985. Effects of ovarian hormones on the concentrations of immunoreactive neuropeptide Y in discrete brain regions of the female rat: Correlation with serum LH and median eminence LHRH. *Endocrinology* 117: 1151-1155.

Everett, J.W. and C.H. Sawyer. 1950. A 24 h periodicity in the "LH-release apparatus" of female rats disclosed by barbiturate sedation. *Endocrinology* 47: 198-218.

Everett, J.W., C.H. Sawyer and J.E. Markee. 1949. A neurogenic timing factor in control of the ovulatory discharge of luteinizing hormone in the cyclic rat. *Endocrinology* 44: 234-250.

Ferin, M., A. Tempone, P. Simmering and R.L. VandeWiele. 1969. Effect of antibodies to 17B-estradiol and progesterone on the estrous cycle of the rat. *Endocrinology* 85: 1070-1078.

Fox, S.R., B.D. Shivers, R.E. Harland and D.W. Pfaff. 1986. Gonadotrophs and β-endorphin-immunoreactive neurons contain progesterone receptors, but luteinizing hormone-releasing hormone immunoreactive neurons do not. *19th Ann. Mtg. SSR*, p. 62.

Gabriel, S.M., J.W. Simpkins and S.P. Kalra. 1983. Modulation of endogenous opioid influence on luteinizing hormone secretion by progesterone and estradiol. *Endocrinology* 113: 1806-1811.

Gallo, R.V. 1980. Neuroendocrine regulation of pulsatile luteinizing hormone release in the rat. *Neuroendocrinology* 30: 122-131.

Gallo, R.V. 1981. Pulsatile LH release during the ovulatory LH surge on proestrus in the rat. *Biol. Reprod.* 24: 100-104.

Gallo, R.V. and S.V. Drouva. 1979. Effect of intraventricular infusion of catecholamines on luteinizing hormone release in ovariectomized and ovariectomized steroid-primed rats. *Neuroendocrinology* 29: 149-162.

Gallo, R.V., G.N. Babu, A. Bona-Gallo, E. DeVorshak-Harvey, R.E. Leipheimer and J. Marco. 1987. In *Regulation of Ovarian and Testicular Function*, eds. V.B. Mahesh, D.S. Dhindsa, E. Andersen and S.P. Kalra, 109-130. Plenum Publishing Corp., New York.

Guy, J., Y.S. Allen, J.M. Polak and G. Pelletier. 1983. Immunocytochemical localization of neuropeptide Y (NPY) in the rat brain. *13th Ann. Mtg. Soc. Neurosci.* 9: 291.

Hiemke, C. and R. Ghraf. 1984. Re-establishment of stimulatory estradiol effects on luteinizing hormone secretion in long-term ovariectomized rats. *Brain Res.* 294: 182-185.

Hiemke, C., A. Schmid, D. Buttner and R. Ghraf. 1987. Improved model for the induction of proestrous-like gonadotropin surges in the long-term ovariectomized rats. *J. Ster. Biochem.* 28: 357.

Hökfelt, T., O. Johansson and M. Goldstein. 1984. Chemical anatomy of the brain. *Science* 225: 1326-1334.

Ieiri, T., H.T. Chen, G.A. Campbell and J. Meites. 1980. Effects of naloxone and morphine on proestrous surge of prolactin and gonadotropin in the rat. *Endocrinology* 106: 1568-1570.

Jacobowitz, D.M. and J.A. Olschowka. 1982. Bovine pancreatic polypeptide-like immunoreactivity in brain and peripheral nervous systems: Coexistence with catecholaminergic nerves. *Peptides* 3: 569-590.

Jacobson, W. and S.P. Kalra. 1989. Decreases in mediobasal hypothalamic and preoptic area opioid (^{3}H naloxone) binding are associated with the progesterone-induced LH surge. *Endocrinology* 124: 199-205.

Jacobson, W., A. Sahu and S.P. Kalra. 1988. Dynamic changes in LHRH concentration in hypothalamic sites following premature induction of the proestrus LH surge with an opioid antagonist. *18th Ann. Mtg. Soc. Neurosci.* 14: 438.

Kalra, P.S. and S.P. Kalra. 1977. Temporal changes in the hypothalamic and serum luteinizing hormone-releasing hormone (LH-RH) levels and the circulating ovarian steroids during the rat estrous cycle. *Acta Endocrinol.* 85: 449-455.

Kalra, P.S. and S.M. McCann. 1973. Involvement of catecholamines in feedback mechanisms. *Progr. Brain Res.* 39: 185-198.

Kalra, P.S., W.R. Crowley and S.P. Kalra. 1987. Differential *in vitro* stimulation by naloxone of LHRH and catecholamine release from the hypothalami of intact and castrated rats. *Endocrinology* 120: 178-185.

Kalra, P.S., S.P. Kalra, L. Krulich, C.P. Fawcett and S.M. McCann. 1972. Involvement of norepinephrine in transmission of the stimulatory influence of progesterone on gonadotropin release. *Endocrinology* 90: 1168-1176.

Kalra, S.P. 1975. Observation on facilitation of the preovulatory rise of LH by estrogen. *Endocrinology* 96: 23-28.

Kalra, S.P. 1981. Neural loci involved in naloxone-induced luteinizing hormone release: Effects of a norepinephrine synthesis inhibitor. *Endocrinology* 109: 1805-1810.

Kalra, S.P. 1983. Opioid peptides - Inhibitory neuronal systems in regulation of gonadotropin secretion. In *Role of Peptides and Proteins in Control of Reproduction*, eds. S.M. McCann and D.S. Dhindsa, 63-87. Elsevier Biomedical, New York.

Kalra, S.P. 1986. Neural circuitry involved in the control of LHRH secretion: A model for preovulatory LH release. In *Frontiers in Neuroendocrinology*, eds. L. Martini and W.F. Ganong, 31-75. Raven Press, New York.

Kalra, S.P. and W.R. Crowley. 1983. Epinephrine synthesis inhibitors block naloxone-induced LH release. *Endocrinology* 82: 1403-1405.

Kalra, S.P. and W.R. Crowley. 1984a. Differential effects of pancreatic polypeptide on luteinizing hormone release in female rats. *Neuroendocrinology* 38: 511-513.

Kalra, S.P. and W.R. Crowley. 1984b. Norepinephrine-like effects of neuropeptide Y on LH release in the rat. *Life Sci.* 35: 1173-1176.

Kalra, S.P. and R.V. Gallo. 1983. Effects of intraventricular catecholamines on luteinizing hormone release in morphine-treated rats. *Endocrinology* 113: 23-28.

Kalra, S.P. and P.S. Kalra. 1979. Dynamic changes in hypothalamic LHRH levels associated with the ovarian steroid-induced gonadotropin surge. *Acta Endocrinol.* 92: 1-7.

Kalra, S.P. and P.S. Kalra. 1983. Neural regulation of luteinizing hormone secretion in the rat. *Endocrinol. Rev.* 4: 311-351.

Kalra, S.P. and J.W. Simpkins. 1981. Evidence for noradrenergic mediation of opioid effects on luteinizing hormone secretion. *Endocrinology* 109: 776-782.

Kalra, S.P., K. Ajika, L. Krulich, C.P. Fawcett, M. Quijada and S.M. McCann. 1971. Effects of hypothalamic and preoptic electrochemical stimulation on gonadotropins and prolactin release in proestrous rats. *Endocrinology* 8: 1150-1158.

Kalra, S.P., L.G. Allen and P.S. Kalra. 1989a. Opioids in the steroid-adrenergic circuit regulating LH secretion: Dynamics and diversities. In *Brain Opioid Systems in Reproduction*, eds. R.G. Dyer and R.J. Bicknell, 95-111. Cambridge University Press, Cambridge.

Kalra, S.P., J.T. Clark, A. Sahu, P.S. Kalra and W.R. Crowley. 1989b. Hypothalamic NPY: A local circuit in the control of reproduction and behavior. In *Nobel Conference on NPY*, ed. V. Mutt, 229-241. Raven Press, New York.

Kalra, S.P., L. Krulich and S.M. McCann. 1973. Changes in gonadotropin-releasing factor content in the rat hypothalamus following electrochemical stimulation of anterior hypothalamic area and during the estrous cycle. *Neuroendocrinology* 12: 321-333.

Khorram, O., K-Y.F. Pau and H.G. Spies. 1987. Biomodal effects of Neuropeptide Y on hypothalamic release of gonadotropin-releasing hormone in conscious rabbits. *Neuroendocrinology* 45: 290-297.

Khorram, O., K-Y.F. Pau and H.G. Spies. 1988. Release of hypothalamic Neuropeptide Y and effects of exogenous NPY on the release of hypothalamic GnRH and pituitary gonadotropins in intact and ovariectomized does *in vitro*. *Peptides* 9: 411-417.

Leadem, C.A. and S.P. Kalra. 1985a. Reversal of β-endorphin-induced blockade of ovulation and LH surge with prostaglandin E_2. *Endocrinology* 117: 684-689.

Leadem, C.A. and S.P. Kalra. 1985b. Effects of endogenous opioid peptides and opiates on luteinizing hormone and prolactin secretion in ovariectomized rats. *Neuroendocrinology* 41: 342-352.

Leadem, C.A., W.R. Crowley, J.W. Simpkins and S.P. Kalra. 1985. Effects of naloxone on catecholamine and LHRH release from the perifused hypothalamus of the steroid-primed rat. *Neuroendocrinology* 40: 497-500.

Leranth, C., N.J. MacLusky, M. Shanabrough and F. Naftolin. 1988. Immunohistochemical evidence for synaptic connections between pro-opioimelanocortin-immunoreactive axons and LHRH neurons by the preoptic area of the rat. *Brain Res.* 449: 167-176.

Lundberg, J.M., T. Hökfelt, A. Anggard, J. Kimmel, M. Goldstein and K. Markey. 1980. Coexistence of avian pancreatic polypeptide (APP) immunoreactive substance and catecholamines in some peripheral and central neurons. *Acta Physiol. Scand.* 110: 107-109.

Masotto, C., G. Wisnewski and A. Negro-Vilar. 1989. Different γ-aminobutyric acid receptor subtypes are involved in the regulation of opiate-dependent and independent luteinizing hormone-releasing hormone secretion. *Endocrinology* 125: 548-553.

Masotto, C., A. Sahu, M.G. Dube and S.P. Kalra. 1990. A decrease in opioid tone amplifies the LH surge in estrogen-treated ovariectomized rats: Comparisons with progesterone effects. *Endocrinology* 126: 18-25.

McDonald, J.K. and M.D. Lumpkin. 1983. Third ventricular injection of pancreatic polypeptides decreases LH and growth hormone secretion in ovariectomized rats. *65th Ann. Mtg. Endocr. Soc.*, p. 152.

McDonald, J., M.D. Lumpkin, W. Samson and S.M. McCann. 1985. Neuropeptide Y affects secretion of luteinizing hormone and growth hormone in ovariectomized rats. *Proc. Natl. Acad. Sci.* 82: 561-564.

McGinnis, M.Y., L.C. Krey, W. MaClusky and B.S. McEwen. 1981. Steroid receptor levels in intact and ovariectomized estrogen-treated rats: An examination of quantitative temporal and endocrine factors influencing the efficacy of an estradiol stimulus. *Neuroendocrinology* 33: 158-165.

Negro-Vilar, A., M.M. Valenca and C. Masotto. 1985. Tonic inhibitory effects of endogenous opiates on gonadotropin secretion: Site and mechanism of action. *J. Androl.* 107: 17-25.

Nicholson, G., G. Greeley, J. Humm, W. Youngblood and J.S. Kizer. 1978. Lack of effect of a noradrenergic denervation of the hypothalamus and medial preoptic area on the feedback regulation of gonadotropin secretion and the estrous cycle of the rat. *Endocrinology* 103: 559-566.

Olschowka, J.A., T.L. O'Donohue and D.M. Jacobowitz. 1981. The distribution of bovine pancreatic polypeptide-like immunoreactive neurons in rat brain. *Peptides* 2: 309-331.

Pang, C.N., E. Zimmerman and C.H. Sawyer. 1977. Morphine inhibition of preovulatory surges of plasma luteinizing hormone and follicle-stimulating hormone in rat. *Endocrinology* 101: 1726-1732.

Piva, F., R. Maggi, P. Limonta, M. Motta and L. Martini. 1985. Effects of naloxone on luteinizing hormone, follicle-stimulating hormone and prolactin secretion in the different phases of the estrous cycle. *Endocrinology* 117: 766-772.

Rodriguez-Sierra, J., D. Jacobowitz and C.A. Blake. 1987. Effects of neuropeptide Y on LH, FSH and TSH in male rats. *Peptides* 8: 539-542.

Sabatino, F.D., P. Collins and J.K. McDonald. 1989. Neuropeptide Y stimulation of luteinizing hormone releasing hormone secretion from the median eminence *in vitro* by estrogen-dependent and extracellular Ca^{++}-independent mechanism. *Endocrinology* 124: 2089-2098.

Sahu, A., W.R. Crowley and S.P. Kalra. 1989b. Naloxone stimulates hypothalamic neuropeptide Y (NPY): Evidence for an opioid-NPY link in the estrogen-induced LH hypersecretion. *71st Ann. Mtg. Endocr. Soc.*, p. 77.

Sahu, A., W. Jacobson, W.R. Crowley and S.P. Kalra. 1989a. Dynamic changes in neuropeptide Y concentrations in the median eminence in association with preovulatory luteinizing hormone (LH) release in the rats. *J. Neuroendocrinol.* 1: 83-87.

Santen, R.J., J. Sofsky, N. Bilic and K. Lippert. 1975. Mechanism of action of narcotics in the production of menstrual dysfunction in women. *Fertil. Steril.* 26: 538-548.

Sarkar, D.K., S.A. Chiappa and G. Fink. 1976. Gonadotropin-releasing hormone surge in pro-oestrous rats. *Nature* 264: 461-463.

Sawyer, C.H., J.W. Everett and J.E. Markee. 1949. A neural factor in the mechanisms by which estrogen induces the release of luteinizing hormone in the rat. *Endocrinology* 44: 218-233.

Sutton, S., T. Toyama, S. Otto and P. Plotsky. 1988. Evidence that neuropeptide Y (NPY) released into the hypophyseal-portal circulation participates in priming gonadotrops to the effects of gonadotropin releasing hormone (GnRH). *Endocrinology* 123: 1208-1210.

Tatemoto, K. 1982. Neuropeptide Y: Complete amino acid sequence of the brain peptide. *Proc. Natl. Acad. Sci. U.S.A.* 79: 5485-5489.

Tatemoto, K., M. Carlquist and V. Mutt. 1982. Neuropeptide Y-a novel brain peptide with structural similarities to peptide YY and pancreatic polypeptide. *Nature* 296: 659-660.

Wherenberg, W.B., R. Cordor and R.C. Gaillard. 1989. A physiologic role for Neuropeptide Y in regulating the estrogen/progesterone-induced luteinizing hormone surge in ovariectomized rats. *Neuroendocrinology* 46: 680-682.

THE ROLE OF NEUROTENSIN IN CONTROL OF ANTERIOR PITUITARY HORMONE SECRETION

S.M. McCann[1] and E. Vijayan[2]

[1]Department of Physiology, Neuropeptide Division
The University of Texas Southwestern
Medical Center at Dallas
Dallas, Texas
and
[2]Department of Biological Sciences
Pondicherry University
Pondicherry, India

I. Introduction

II. Role of Neurotensin in Control of Anterior Pituitary Hormone Release

    A. Prolactin

    B. FSH and LH

    C. TSH

    D. Growth Hormone

III. Discussion

Introduction

While attempting to purify corticotropin-releasing factor from hypothalamic extracts, Susan Leeman noticed that the rats salivated following injection of certain fractions (Leeman and Carraway, 1982). She and her associates characterized the molecule responsible for this sialogogic activity. It was substance P (Chang and Leeman, 1970). While screening columns to locate sialogogic activity, she noticed a characteristic vasodilatation of exposed cutaneous regions, particularly around the face and ears of the assay rats. She and her associates went on to isolate and determine the structure of this vasodilatory peptide which was named neurotensin (Carraway and Leeman, 1973). The structure of the tridecapeptide (Fig. 1) has pyroglutamic acid at the N-terminal and a carboxyl group at the C-terminal end of the molecule.

In the rat the peptide is widely distributed in the central nervous system particularly to limbic system structures. There are cell bodies in the nucleus interstialis stria terminalis, medial preoptic area, periventricular region, paraventricular nucleus, arcuate nucleus and lateral hypothalamus, areas which also contain fibers of these neurons. Some of these axons, particularly those arising in the arcuate nucleus, project to the lateral aspect of the external layer of the median eminence (Jennes *et al.*, 1982; Kahn *et al.*, 1980; Kahn *et al.*, 1982). Neurotensin may be colocalized with dopamine (Hökfelt *et al.*, 1984; Ibata *et al.*, 1983) and CRF (Sawchenko *et al.*, 1984). This distribution clearly suggests an important role for the peptide in hypothalamic-pituitary function. Such a role is further supported by the presence of binding sites for the peptide in a variety of brain sites, including those related to areas where terminals or perikarya of neurotensin neurons are found (Goedert *et al.*, 1985).

Anterior pituitary cells have been shown, by immunocytochemistry, to contain neurotensin (Goedert *et al.*, 1982). Levels of neurotensin in hypophyseal portal blood, collected from anesthetized rats by a modification of the technique of Worthington and Fink, were higher than those in peripheral blood, indicating that the peptide is secreted from the median eminence into portal blood and passes to the anterior pituitary (Eckland and Lightman, 1987). There, it may

<Glu–Leu–Tyr–Glu–Asn–Lys–Pro–Arg–Arg–Pro–Tyr–ILeu–Leu

Fig. 1. The structure of neurotensin.

well have direct actions to alter pituitary hormone release. Since the hormone persists in the gland following pituitary stalk section, it not only reaches the pituitary via the portal vessels but actually may be synthesized within pituitary cells (Goedert *et al.*, 1984).

In fact, considerable evidence has now accumulated to indicate that the peptide can alter the release of most anterior pituitary hormones by hypothalamic and/or direct pituitary effects. The remainder of this chapter will be devoted to examining its effects on the various pituitary hormones, determining whether the actions are on the brain or the pituitary directly, evaluating the possible role of other neuronal systems in mediating the effects, evaluating the mechanism of action of the effects and determining their physiological significance.

Role of Neurotensin in Control of Anterior Pituitary Hormone Release

Prolactin

Hypothalamic action. The largest amount of work has been done to clarify the role of neurotensin in the control of prolactin release. Following injection into the third ventricle, neurotensin can reduce plasma levels of prolactin in conscious ovariectomized, ovariectomized estrogen progesterone-primed (OEP), or male rats (Vijayan and McCann, 1979). It was difficult to observe this lowering of prolactin in conscious male rats because of the already low levels of prolactin; however, when these levels were elevated by exposing the animals to ether there was a dramatic lowering of plasma prolactin induced by intraventricular injection of neurotensin (Koenig *et al.*, 1982). Consequently, it is clear that the action of neurotensin on structures adjacent to the ventricle is to alter the balance of prolactin releasing and prolactin inhibiting factor release so as to suppress the release of prolactin from the adenohypophysis.

Next, an attempt was made to determine which neurotransmitters might be involved in inducing the suppressive action of neurotensin. Fluoxetine, a blocker of reuptake of serotonin (5-hydroxytryptamine, 5HT) and 5-hydroxytryptophan (5HTP), a precursor of 5HT, produced marked elevations in plasma prolactin, presumably by increasing the amounts of 5HT in the synaptic clefts adjacent to the hypothalamic mechanisms controlling prolactin release. This elevation was dramatically reduced by intraventricular injection of neurotensin, which indicates that the suppressive action of neurotensin on prolactin release is not mediated directly by 5HT (Koenig *et al.*, 1982).

On the other hand, when adrenergic transmission was blocked by the inhibitor of tyrosine hydroxylase (α-methyl para-tyrosine; αMT)--to halt catecholamine synthesis--prolactin levels also rose by elimination of catecholaminergic inhibitory control. Under these circumstances the prolactin lowering ability of intraventricular neurotensin was blocked. This was true whether or not plasma prolactin was elevated even further by ether anesthesia in the αMT-treated animals. These results indicate that the effect of neurotensin is mediated by catecholamines, which bring about the prolactin-lowering action of the peptide, but do not identify the catecholamine involved. Spiroperidol, the dopamine receptor blocker, elevates plasma prolactin by elimination of dopaminergic inhibitory control and in these animals neurotensin had no effect on plasma prolactin, which indicates that the catecholamine, dopamine, is involved (Koenig *et al.*, 1982).

The evidence indicates that intraventricular neurotensin stimulates the release of dopamine from dopaminergic neurons, presumably those of the tuberoinfundibular dopaminergic system. This dopamine then enters hypophyseal portal vessels and passes down to the pituitary to suppress prolactin release from the lactotrophs (Koenig *et al.*, 1982). *In vitro* incubation experiments with hypothalamic fragments indicate that neurotensin can indeed release dopamine, so this mechanism appears to be fairly well established (Vijayan and McCann, 1979; McCann *et al.*, 1982).

Direct pituitary action. In contrast to the prolactin-lowering activity of intraventricularly injected neurotensin, prolactin levels in plasma can be elevated by intravenous injection of the peptide. This action is almost certainly due to a direct stimulatory effect of the peptide on the anterior pituitary gland since *in vitro* incubation of hemipituitaries from ovariectomized female rats with neurotensin produced a dose-related stimulation of prolactin release (Vijayan and McCann, 1979). The direct stimulatory effect of neurotensin on the lactotrophs has been amply confirmed in subsequent work (MacLeod *et al.*, 1988).

The mechanism of stimulation appears to involve membrane polyphosphoinositide degradation. The inositol triphosphate (IP3) liberates Ca^{++} from intracellular stores. Diacylglycerol (DAG) is also released which activates protein kinase C. This acts synergistically with the liberated Ca^{++}. DAG as well as IP2, by the action of phospholipase A_2, liberates archidonate which is metabolized to leukotrienes, prostaglandins and epoxides which may also stimulate prolactin release (MacLeod *et al.*, 1988).

Physiological role of neurotensin to alter prolactin release. To examine the possible role of endogenous neurotensin to alter prolactin release, highly specific antiserum to neurotensin (NT-AS) was injected either into the third ventricle or intravenously into unanesthetized intact male, ovariectomized female or OEP rats.

Intraventricular injection of 1 or 3 μl of NT-AS significantly increased plasma prolactin levels in ovariectomized as well as OEP rats on the first measurement 1 h after injection and the effect was dose-related at early times after the injection.

In intact male rats intraventricular injection of NT-AS also produced an elevation of plasma prolactin, but the elevations were lower than those seen in ovariectomized or OEP rats. These results are opposite to those obtained from injection of the peptide itself and support the physiological significance of the intrahypothalamic inhibitory action of neurotensin in all these types of rats. Furthermore, they strongly suggest that neurotensin is involved in tonically controlling the basal secretion of prolactin.

On the other hand, intravenous injection of a dose of antiserum which had been previously shown to block the hypotensive effect of neurotensin produced the opposite result and suppressed plasma prolactin concentrations in ovariectomized as well as OEP animals. The effect was more pronounced in OEP rats, which demonstrate elevated plasma prolactin levels compared to their ovariectomized controls. There was no effect of intravenous injection of the antiserum in intact males. The direct stimulatory action of the peptide on prolactin release appears to be physiologically and tonically significant in females. Apparently in the male the direct action of the peptide is not being exerted under resting conditions. The lack of physiological significance of neurotensin at the pituitary level in the male is also supported by the failure of the antiserum to alter prolactin release from dispersed anterior pituitary cells of males. Whether it would be demonstrable under conditions of enhanced release of prolactin requires further experimentation (Vijayan *et al.*, 1988).

Thus, the evidence is extremely good that neurotensin has a suppressive action of physiological significance on the release of prolactin via the hypothalamus, which is probably mediated by release of dopamine that passes down the portal vessels and directly inhibits the lactotrophs. On the other hand, there is an opposite stimulatory action of physiological significance directly on the lactotrophs, at least in the female. This situation is quite similar to that of many brain peptides which appear to have opposite actions, at hypothalamic and pituitary levels, to alter pituitary hormone release.

FSH and LH

In ovariectomized animals in which plasma FSH and LH values were elevated, because of removal of ovarian steroid negative feedback, intraventricular injection of neurotensin lowered plasma LH within 5 min and the effect persisted for 30 min following the 0.5 μg dose, but a progressive lowering occurred throughout the 1 h course of the experiment following the 2 μg dose. That this was a central effect is suggested by the fact that a dose of 1 μg given intravenously had no effect on plasma LH. In contrast to the effect on LH, there was no effect of neurotensin on plasma FSH. Since the peptide had no effect on the release of LH by hemipituitaries incubated *in vitro*, it was concluded that it exerted an inhibitory action on LHRH release in the castrate rat (Vijayan and McCann, 1979).

Subsequently, Ferris *et al.* (1984) microinjected neurotensin into the medial preoptic area of anesthetized, ovariectomized animals and found an elevation instead of a suppression of plasma LH but found, as we had, that injecting it into the ventricle lowered plasma LH levels. These observations suggest that the action of the peptide may be different depending on the locus in the hypothalamus affected.

To determine the physiological significance of the effects of neurotensin on LH release, we injected purified neurotensin antiserum into the third ventricle of ovariectomized rats. There was no effect from the control injection of normal rabbit serum; however, the microinjection of 1 μl of neurotensin antiserum produced a slight increase in plasma LH at 1 h following injection. Values remained elevated, but not significantly so, during the remaining 4 h of observation. When the dose of antiserum was raised to 3 μl, a highly significant increase in plasma LH occurred, peaking at 2 and 3 h, this was followed by a precipitous fall back to control values at 4 and 5 h after injection. This observation supports the physiological significance of the plasma LH-lowering effect that occurred following intraventricular injection of the peptide.

That the action of the neurotensin antiserum was exerted within the brain is suggested by the fact that there was no effect from peripheral injection of the antiserum in ovariectomized animals just as there had been no effect of i.v. injection of the peptide itself (Vijayan *et al.*, 1990c).

The plasma LH levels were elevated following intraventricular injection of the antiserum in ovariectomized, estrogen progesterone-primed animals at 3, 4 and 5 h after the intraventricular injection of the lower, but not the higher, dose. This supports the physiological significance of the inhibitory action of endogenous neurotensin in this type of rat as well (Vijayan *et al.*, 1990c).

Recently, Leeman and associates (Alexander *et al.*, 1989) have microinjected antiserum against neurotensin into the medial preoptic area (in the afternoon) in ovariectomized, estrogen progesterone-treated animals which were undergoing a proestrus-like surge of LH. The antiserum blocked the LH surge, indicating that the stimulatory action of the peptide in the preoptic area, found earlier by this group, may have physiological significance in augmenting the proestrus surge of LH, and further supports the idea of opposite actions of the peptide at rostral and more caudal sites in the hypothalamus. Perhaps the failure of the higher intraventricular dose of NT-AS to affect plasma LH in our experiments is related to the spread of the antiserum from the area in which neurotensin inhibits LHRH release to the area in which it stimulates it.

It is now apparent that blockade of the action of a number of transmitters or neuromodulators can interfere with the proestrus surge of LH. Blockade of α-adrenergic, dopaminergic (McCann, 1982), neuropeptide Y (Wehrenberg *et al.*, 1989), angiotensin II (Steele *et al.*, 1983), and oxytocin (Johnston and Negro-Vilar, 1988), transmission, as well as the release or action of LHRH, blocks the surge (McCann, 1982). Furthermore, LHRH apparently exerts a positive feedback at this time to augment its own release during the surge (Hiruma *et al.*, 1989). Thus, the proestrus surge of LH is probably due to a stimulatory action of a variety of factors that augment the release of LHRH. At this time it may be further augmented by direct action on the pituitary from transmitters released into the hypophyseal portal vessels; these transmitters could include not only LHRH but also neuropeptide Y (McDonald *et al.*, 1985) and even epinephrine and norepinephrine (Petrovic *et al.*, 1984).

TSH

Intraventricular injection of either 0.5 or 2 μg of neurotensin failed to modify plasma levels of TSH; however, intravenous injection of 1 μg of the peptide dramatically elevated plasma TSH within 5 min and this effect persisted for 30 min after injection. The effect of the intravenous injection of neurotensin was probably mediated at the pituitary level since incubation of hemipituitaries from ovariectomized animals revealed a stimulatory effect of neurotensin on TSH release at a minimal effective dose of 4 nM of the peptide. There was a suggestion of a dose-response relationship (Vijayan and McCann, 1980).

This pituitary action of neurotensin may have physiological significance since intravenous injection of neurotensin antiserum lowered plasma TSH in both ovariectomized and ovariectomized, estrogen progesterone-treated rats. There was also a suppression of plasma TSH following intraventricular injection of the antiserum in males, as well as the ovariectomized female groups. Since there was no effect of intraventricular injection of the peptide itself, these effects may be related to spread of the antiserum to the pituitary with suppression of the stimulation of TSH release induced by endogenous neurotensin (Vijayan *et al.*, 1990a).

Growth Hormone

Third ventricular injection of 2 µg of neurotensin induced dramatic elevations in plasma growth hormone within 5 min of injection, and values remained significantly elevated for the 60 min duration of the experiment. They were also elevated following injection of a lower, 0.5 µg dose, but declined to control values within 60 min. There was no significant difference between the response to the 2 doses except at this latter time. When hemipituitaries of ovariectomized animals were incubated with the peptide, there was no effect on growth hormone release indicating that the actions observed following intraventricular injection were mediated at the hypothalamic level (Vijayan and McCann, 1980).

In the experiments with the intraventricular injection of neurotensin antiserum, it was found that the antiserum lowered growth hormone in ovariectomized females, OEP females and also male rats. Thus, the peptide appears to have a physiologically significant action within the hypothalamus to stimulate growth hormone secretion either via activation of growth hormone-releasing factor release, inhibition of somatostatin release, or by both actions (Vijayan *et al.*, 1990b).

Surprisingly, the intravenous injection of the antiserum produced the opposite effect and an elevation in growth hormone levels in both types of ovariectomized rats. We have no explanation for this latter result since we found no effect of the peptide on growth hormone release *in vitro*. Perhaps there is a direct inhibitory effect on the release of growth hormone that went undetected.

Table 1. Summary of Results with Neurotensin Antiserum.

Type of Rat	Route Injection	Effect on Plasma Hormone Concentration				
		FSH	LH	Prl	GH	TSH
Male	3V	o	o	+	-	-
OVX	3V	o	+	+	-	-
	IV	o	o	-	+	-
OEP	3V	o	+	+	-	-
	IV	o	o	-	+	-

Discussion

Thus, neurotensin alters pituitary hormone secretion by both hypothalamic and pituitary actions. The best established effects are those on prolactin in which it is clear that the peptide suppresses prolactin release by stimulating dopamine release which then directly inhibits secretion at the lactotrophs. This effect appears to be of physiological significance, based on the antiserum studies (Table 1). In addition, there is a direct stimulatory effect of neurotensin on prolactin release and this is of physiologic significance, at least in the female (Fig. 2).

In the case of LH, there may be dual actions of the peptide to stimulate the release of LHRH, either directly or indirectly, in the medial preoptic area and to inhibit the release of LHRH more caudally by actions on the median eminence LHRH neuronal axons, or interneurons connected to these. The possible transmitters involved in these putative interneuronal connections have yet to be determined, but these actions appear to be of physiological significance based on the antibody studies. There appears to be no direct action of the peptide on the

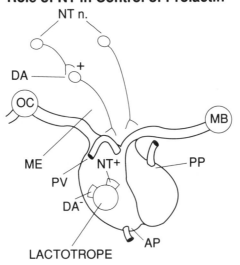

Fig. 2. The Diagrammatic Illustration of the Role of Neurotensin in Control of Prolactin Release. Key to abbreviations in this and Fig. 3: OC = optic chiasm; MB = mammillary bodies; ME = median eminence; PV = portal vessel; PP = posterior pituitary; AP = anterior pituitary; NTn. = neurotensin neuron; DA = dopamine; + = stimulation; - = inhibition.

Role of NT in Control of LH

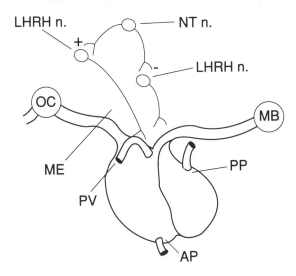

Fig. 3. The Diagrammatic Representation of the Role of Neurotensin in Control of LH Release. Abbreviations as in Fig. 2 except that LHRHn. = luteinizing hormone releasing hormone neuron.

gonadotrophs (Fig. 3). As in the case of many other brain peptides, although there are clear effects of the peptide on LHRH release, there is no effect on FSH release, providing yet another example of dissociation of the hypothalamic control of these two gonadotropins (McCann, 1983; Lumpkin *et al.*, 1989).

We have found no intrahypothalamic action of the peptide to alter the release of the third glycoprotein hormone, TSH, but instead neurotensin appears to have a physiologically significant stimulatory action on the thyrotrophs, on the basis of the antibody studies reported here.

Lastly, the peptide appears to have a physiologically significant intrahypothalamic action to stimulate growth hormone release either by inhibition of somatostatin release, stimulation of growth hormone releasing factor or by a combination of these two actions either directly or indirectly via interneurons. In this case as well, we found no effect of the peptide directly on the somatotrophs.

We have not cited work done in anesthetized animals since it is quite obvious that anesthesia profoundly modifies the neuronal function. Results can often be considerably different in anesthetized animals and bear little relevance to the situation in the unanesthetized freely moving animals such as were used in these studies (Pass and Ondo, 1977).

Table 2. Actions of neurotensin.

Hormone	Hypothalamus	Pituitary	Physiol. Significance
Prolactin	-	+	+
LH	+,-	o	+
FSH	o	o	o
GH	+	o	+
TSH	o	+	+

Although it is now clear that neurotensin has powerful actions intrahypothalamically, and sometimes on the pituitary, to alter pituitary hormone secretion (Table 2), much work remains to be done. For example, the synaptic transmitters which may mediate the actions of the peptide on the brain have yet to be fully elucidated. More work needs to be done with localization of the sites of action. It would be very interesting to do studies to evaluate the effect of the peptide on the release of the various releasing hormones and inhibitory hormones *in vitro* and, furthermore, it would be of great interest to evaluate the effects of various stimuli on the release of neurotensin into hypophyseal portal blood or into pituitary push-pull cannulae. Only by a thorough study using all of these methodologies can a full understanding of the role of neurotensin in control of anterior pituitary hormone secretion be elucidated.

REFERENCES

Alexander, M.J., P.D. Mahoney, C.F. Ferris, R.E. Carraway and S.E. Leeman. 1989. Evidence that neurotensin participates in the central regulation of the preovulatory surge of luteinizing hormone in the rat. *Endocrinology* 124: 783-788.
Carraway, R.E. and S.E. Leeman. 1973. The isolation of a new hypotensive peptide, neurotensin, from bovine hypothalami. *J. Biol. Chem.* 248: 6854-6861.
Chang, M.M. and S.E. Leeman. 1970. Isolation of a sialogogic peptide from bovine hypothalamic tissue and its characterization as substance P. *J. Biol. Chem.* 245: 4787-4790.
Eckland, D.J.A. and S.L. Lightman. 1987. Neurotensin in hypothalamo-hypophyseal portal blood. *Brain Res.* 421: 161-166.

Ferris, C.F., J.X. Pan, E.A. Singer, N.D. Boyd, R.E. Carraway and S.E. Leeman. 1984. Stimulation of luteinizing hormone release after stereotaxic microinjection of neurotensin into the medial preoptic area of rats. *Neuroendocrinology* 38: 145-151.

Goedert, M., S.L. Lightman and P.C. Emson. 1984. Neurotensin in the rat anterior pituitary gland: Effects of endocrinological manipulations. *Brain Res.* 299: 160-163.

Goedert, M., S.L. Lightman, P.W. Mantyhi, S.P. Hunt and P.C. Emson. 1985. Neurotensin-like immunoreactivity and neurotensin receptors in the rat hypothalamus and in the neurointermediate lobe of the pituitary gland. *Brain Res.* 358: 59-69.

Goedert, M., S.L. Lightman, J.I. Nagy, P.D. Marley and P.C. Emson. 1982. Neurotensin in the rat anterior pituitary gland. *Nature* 298: 163-165.

Hiruma, H., T. Funabashi and F. Kimura. 1989. LHRH injected into the medial preoptic area potentiates LH secretion in ovariectomized estrogen-primed and proestrous rats. *Neuroendocrinology* 50: 421-426.

Hökfelt, T., B.-J. Everitt, E. Theodorsson-Norheim and M. Goldstein. 1984. Occurrence of neurotensin-like immunoreactivity in subpopulations of hypothalamic, mesencephalic, and medullary catecholamine neurons. *J. Comp. Neurol.* 222: 543-559.

Ibata, Y., K. Fukui, H. Okamura, T. Kawakami, M. Tanaka, H.L. Obata, T. Tsuto, H. Terubayashi, C. Yanaihara and N. Yanaihara. 1983. Coexistence of dopamine and neurotensin in hypothalamic arcuate and periventricular neurons. *Brain Res.* 269: 177-179.

Jennes, L., W.E. Stumpf and P.W. Kalivas. 1982. Neurotensin: Topographical distribution in rat brain by immunohistochemistry. *J. Comp. Neurol.* 210: 211-224.

Johnston, C.A. and A. Negro-Vilar. 1988. Role of oxytocin on prolactin secretion and in different physiological or pharmacological paradigms. *Endocrinology* 122: 341-350.

Kahn, D., G.M. Adams, E.A. Zimmerman, R.E. Carraway and S.E. Leeman. 1980. Neurotensin neurons in the rat hypothalamus: An immunocytochemical study. *Endocrinology* 107: 47-54.

Kahn, D., A. Hou-You and E.A. Zimmerman. 1982. Localization of neurotensin in the hypothalamus. *Ann. N.Y. Acad. Sci.* 400: 117-131.

Koenig, J.I., M.A. Mayfield, S.M. McCann and L. Krulich. 1982. On the prolactin-inhibiting effect of neurotensin. *Neuroendocrinology* 35: 277-281.

Leeman, S.E. and R.E. Carraway. 1982. Neurotensin: Discovery isolation, characterization, synthesis and possible physiological roles. In *Neurotensin, a Brain and Gastrointestinal Peptide*, Vol. 400, ed. C.B. Nemeroff, A.J. Prange Jr., 1-15. Ann. N.Y. Acad. Sci., New York.

Lumpkin, M.D., J.K. McDonald, W.K. Samson and S.M. McCann. 1989. Destruction of the dorsal anterior hypothalamic region suppresses pulsatile release of follicle stimulating hormone but not luteinizing hormone. *Neuroendocrinology* 50: 229-235.

MacLeod, R.M., A.M. Judd, B.L. Spangelo, P.C. Ross, W.D. Jarvis and I.S. Login. 1988. Systems that regulate prolactin release in prolactin gene family and its receptors. In *Excerpta Medica International Congress Series 819*, ed. K. Hoshino, 13-28. Excerpta Medica, New York.

McCann, S.M. 1982. Physiology and pharmacology of LHRH and somatostatin. *Ann. Rev. Pharmacol. Toxicol.* 22: 491-515.

McCann, S.M. 1983. Differential hypothalamic control of FSH secretion. A review. *Psychoneuroendocrinology* 8: 299-308.

McCann, S.M., E. Vijayan, J. Koenig and L. Krulich. 1982. The effects of neurotensin on anterior pituitary hormone secretion. In *Neurotensin, a Brain and Gastrointestinal Peptide*, Vol. 400, ed. C.B. Nemeroff, A.J. Prange Jr., 160-171. Ann. N.Y. Acad. Sci., New York.

McDonald, J., M.D. Lumpkin, W.K. Samson and S.M. McCann. 1985. Neuropeptide Y affects luteinizing hormone secretion in ovariectomized rats. *Proc. Natl. Acad. Sci.* 82: 561-564.

Pass, K.A. and J.G. Ondo. 1977. Effect of gamma amino butyric acid on prolactin and gonadotropin secretion in unanesthetized rats. *Endocrinology* 100: 1437-1442.

Petrovic, S.L., J.K. McDonald, G.D. Snyder and S.M. McCann. 1984. Testosterone control of brain and anterior pituitary β-adrenergic receptors. *Life Sci.* 34: 2399-2406.

Sawchenko, P.E., L.W. Swanson and W.W. Vale. 1984. Corticotropin-releasing factor: Co-expression within distinct subsets of oxytocin-, vasopressin-, and neurotensin-immunoreactive neurons in the hypothalamus of the male rat. *J. Neurosci.* 4: 1118-1129.

Steele, M.A., R.V. Gallo and W.F. Ganong. 1983. Possible role for the brain renin-angiotensin system in the regulation of LH secretion. *Am. J. Physiol.* 245: R805-R810.

Vijayan, E., R.E. Carraway, S.E. Leeman and S.M. McCann. 1988. Use of antiserum to neurotensin reveals a physiological role for the peptide in rat prolactin release. *Proc. Natl. Acad. Sci.* 85: 9866-9869.

Vijayan, E., R.E. Carraway, S.E. Leeman and S.M. McCann. 1990a. Effect of antineurotensin serum on the release of TSH in the rat. Submitted.

Vijayan, E., R.E. Carraway, S.E. Leeman and S.M. McCann. 1990b. Effect of antisera against neurotensin on growth hormone release in the rat. Submitted.

Vijayan, E., R.E. Carraway, S.E. Leeman and S.M. McCann. 1990c. Effect of intraventricular injection of neurotensin antiserum on LH release. Submitted.

Vijayan, E. and S.M. McCann. 1979. *In vivo* and *in vitro* effects of substance P and neurotensin on gonadotropin and prolactin release. *Endocrinology* 105: 64-68.

Vijayan, E. and S.M. McCann. 1980. Effects of substance P and neurotensin on growth hormone and thyrotropin release *in vivo* and *in vitro*. *Life Sci.* 26: 321-327.

Wehrenberg, W.B., R. Corder and R.C. Gaillard. 1989. A physiological role for neuropeptide Y in regulating the estrogen/progesterone induced luteinizing hormone surge in ovariectomized rats. *Neuroendocrinology* 49: 680-682.

GALANIN: A POTENTIALLY SIGNIFICANT NEUROENDOCRINE MODULATOR

J.I. Koenig[1], S.M. Gabriel[2], and L.M. Kaplan[3]
Departments of Neurology[1] and Medicine[3]
Massachusetts General Hospital
and Harvard Medical School
Boston, MA
and
Department of Psychiatry[2]
Mount Sinai Medical School
New York, NY

I.   Introduction

II.   Neuroendocrine Effects of Galanin

        A.   Growth Hormone

        B.   Prolactin

        C.   Luteinizing Hormone

III.   Galanin Expression in the Anterior Pituitary

IV.   Hormonal Regulation of CNS Galanin Expression

V.   Summary

Introduction

In 1983, Tatemoto and co-workers isolated and characterized a novel 29-amino acid peptide from the porcine intestine on the basis of its unusual carboxy-terminal alanine amide residue. Its name, galanin, derives from the N-terminal glycine and C-terminal alanine moieties (Tatemoto *et al.*, 1983). In early studies, galanin was found to stimulate smooth muscle contraction and to produce a hyperglycemic response by blunting insulin secretion. These activities served as the basis for bioassays that facilitated purification of the peptide. Although the amino acid sequence of porcine galanin suggested 2-3 residue homologies to several other regulatory peptides, these similarities are not present in galanin peptides from other species. Thus, galanin appears unrelated to any known peptide families.

Using mixed oligonucleotide probes, Rokaeus and Brownstein (1986) isolated cDNAs encoding bovine galanin. Analysis of these cDNAs revealed that porcine galanin is encoded as a 123-amino acid precursor peptide. The predicted structure of preprogalanin includes a secretory signal peptide sequence, galanin itself, and a 59-amino acid sequence of currently unknown significance (galanin mRNA-associated peptide). Within the precursor, the galanin residues are flanked by pairs of basic amino acids. A C-terminal glycine appears to serve as the amide donor. cDNAs encoding rat preprogalanin were isolated from pituitary (Vrontakis *et al.*, 1987) and hypothalamic (Kaplan *et al.*, 1988a) tissues. These clones predict that rat galanin is also a 29-amino acid amidated peptide cleaved from a 124-amino acid precursor. Rat and porcine galanin share 26 of 29 amino acids. The three variant residues are located within the carboxy-terminal heptapeptide. The predicted C-terminal residue in the rat molecule is threonine. More recently, the structure of bovine galanin and its mRNA have been determined (Rokaeus and Carlquist, 1988), revealing a similar structure. Once again, all variations from the rat and porcine sequences are located in the C-terminal region. The complete conservation of sequence in the N-terminal 15 amino acids of the galanin peptides from these three species suggests that this region is important for biological activity. Consistent with the lengths of the reported cDNA clones, Northern blot analysis reveals a single mRNA species of approximately 900 bases in all three species.

Immunohistochemical studies have shown the presence of galanin-containing cell bodies throughout the rat brain, with the highest concentration observed in the hypothalamus. Galanin immunoreactive cell bodies are abundant in the cerebral cortex, nucleus basalis, locus coeruleus, raphe nuclei and hippocampus

(Melander *et al.*, 1986b; Ch'ng *et al.*, 1985; Skofitsch and Jacobowitz, 1985; Rokaeus *et al.*, 1984; Levin *et al*, 1987). In the hypothalamic area, cell bodies are concentrated in the arcuate, paraventricular (PVN), supraoptic (SON) and dorsomedial nuclei (Melander *et al.*, 1986b; Levin *et al.*, 1987; Palkovits *et al.*, 1987; Meister *et al.*, 1987). Scattered cells also are present lateral to the suprachiasmatic nucleus of the monkey (Gabriel *et al.*, 1988c). Galanin immunoreactive nerve terminals and fibers are concentrated in the median eminence (ME) (Skofitsch and Jacobowitz, 1986a; Palkovits *et al.*, 1987; Gabriel *et al.*, 1988a), an area also known to contain dense receptors for galanin (Melander *et al.*, 1986a, 1987; Skofitsch *et al.*, 1986b). Furthermore, galanin-immunoreactive nerve fibers are present in the PVN, preoptic region, periventricular and dorsomedial nuclei. Outside of the hypothalamus, galanin fibers can be seen in the septal region and in the medial forebrain bundle (Melander *et al.*, 1986b; Skofitsch and Jacobowitz, 1985; Ch'ng *et al.*, 1985).

Throughout the CNS, galanin has been found to coexist with numerous other peptides and monoamine neurotransmitters. One of the most striking examples of colocalization occurs in the locus coeruleus between galanin and dopamine β-hydroxylase or tyrosine hydroxylase (Melander *et al.*, 1986c; Levin *et al.*, 1987). Many of these cells are thought to project to the oxytocin-rich regions of the hypothalamic paraventricular nucleus. In addition, neurons of the dorsomedial nucleus containing both galanin and dopamine β-hydroxylase project to the parvocellular regions of the PVN (Melander *et al.*, 1986c). Galanin also is localized in the serotonergic neurons of the brainstem raphe nuclei (Melander *et al.*, 1986c), and with acetylcholine in the nucleus basalis (Melander *et al.* 1985). Interestingly, galanin appears to modulate the activity of the serotonergic neurons (Sundstrom and Melander, 1988) and selectively decreases the K_D of the serotonin 1-A receptors in limbic cortex areas (Fuxe *et al.*, 1988). The relationship between the biological actions of galanin and its coexistence in serotonergic neurons is currently unknown.

In the hypothalamus, galanin-containing cell bodies in the magnocellular regions of the PVN and SON also stain with antisera directed against vasopressin (Melander *et al.*, 1986c). Several recent studies have shown that galanin immunoreactivity in these neurons vary with the hydration state of the animal (Rokaeus *et al.*, 1988b; Koenig *et al.*, 1989; Gaymann and Martin 1989; Skofitsch *et al.*, 1989). Furthermore, these changes are consistent with reported alterations in the concentration of vasopressin in the posterior pituitary (Hooi *et al.*, 1989). Brattleboro rats, characterized by the lack of vasopressin and increased plasma osmolality, have elevated amounts of galanin mRNA in PVN neurons (Rokaeus *et al.*, 1988b), consistent with activation of these neurons and a potential modulatory role for galanin in regulating vasopressin release.

In parvocellular regions of the PVN, galanin coexists with corticotropin-releasing hormone (CRH) and neurotensin (Ceccatelli *et al.*, 1989). Galanin and neurotensin also coexist in neurons of the arcuate nucleus (Meister *et al.*, 1987). Colocalization of galanin with tyrosine hydroxylase immunoreactivity has been demonstrated in the arcuate nucleus and PVN (Melander *et al.*, 1986c; Meister *et al.*, 1987) and there is a striking colocalization of growth hormone releasing hormone (GRH) and γ-aminobutyric acid (GABA) in galanin-containing cell bodies of the arcuate nucleus (Melander *et al.*, 1986c; Meister *et al.*, 1987). Further studies are required to determine the functional significance of the coexistence of galanin and these other neuroactive substances.

The anatomical and cellular localization of galanin within the hypothalamus suggests an important neuromodulatory role for this peptide. We have embarked on a number of studies to determine the importance of galanin in regulating neuroendocrine function in the rat.

For many of these studies, we have used a locally prepared antiserum to porcine galanin which cross-reacts with rat galanin. In our initial studies using this antiserum in a radioimmunoassay, we observed that the highest concentrations of galanin in the rat brain are found in the ME, followed by several regions within the hypothalamus. Treatment of male rats with monosodium glutamate (MSG) every other day for the first ten days of life markedly depletes numerous peptidergic systems, including GRH (Gabriel *et al.*, 1988c). Galanin concentrations in the ME of adult MSG-treated rats are reduced approximately 50% from 124 ± 11 ng/mg protein. Furthermore, mediobasal hypothalamic galanin concentrations are reduced 15% from 10.4 ± 0.4 ng/mg protein to 8.9 ± 0.4 ng/mg protein and significant reductions also occur in the septum (Gabriel *et al.*, 1988a). These studies suggest that galanin-containing cell bodies in several areas of the hypothalamus project to the ME. Additionally, the arcuate nucleus appears to contain subsets of galanin neurons which may be differentiated by the presence or absence of glutamate receptors.

Neuroendocrine Effects of Galanin

Growth Hormone

The initial effects of galanin on anterior pituitary function were reported by Ottlecz *et al.* (1986). They demonstrated that intraventricular infusion of 50-400 pmol of the peptide stimulated growth hormone (GH) secretion in ovariectomized female rats. This finding was confirmed by Melander *et al.* (1987) in a study using male rats. In normal human volunteers, the intravenous infusion of galanin raised serum GH levels (Davis *et al.*, 1987). The physiological role of galanin in controlling GH secretion has been implicated in a study by Ottlecz *et al.* (1988). Intraventricular infusion of galanin antisera (obtained from Peninsula Laboratories) markedly lowered plasma levels of GH in male rats. In a more extensive analysis of the effect of galanin in regulating episodic GH secretion, we have observed that intraventricular infusion of galanin antiserum significantly decreases the amplitude of GH secretory events and increases the GH pulse frequency (Maiter *et al.*, 1990). In short, galanin immunoneutralization converted male GH sectretory patterns into patterns more closely resembling those seen in females.

The mechanism of the observed effect on pulsatile GH secretion is unclear. However, several investigators have begun to unravel the means by which galanin may control GH secretion. The stimulatory effect of galanin on GH release after intraventricular infusions appears to be mediated by GRH (Cella *et al.*, 1987, Murakami *et al.*, 1989) as GRH immunoneutralization attenuates this response. In human studies, galanin tripled GRH-induced GH release suggesting that in normal men, galanin either potentiates the action of GRH or reduces the inhibitory tone of somatostatin (Davis *et al.*, 1987). A similar potentiation of GRH-mediated stimulation of GH release in rats has been reported by Gabriel *et al.* (1988b) using dispersed pituitary cells in culture. It is likely, however, that the mechanism by which galanin affects GH release may be more complex than initially thought. Chatterjee *et al.* (1988) have demonstrated that the GH response to iv galanin infusion in man is mediated by a cholinergic mechanism. In addition, studies in the rat also support a role for α-adrenergic and γ-aminobutyric acid (GABA) pathways in the GH-releasing effect of galanin (Murakami *et al.*, 1989; Cella *et al.*, 1987).

A direct effect of galanin on pituitary GH secretion remains controversial. Ottlecz *et al.* (1986) failed to stimulate GH release from short-term dispersed pituitary cells obtained from ovariectomized female rats. These workers subsequently confirmed their negative findings using dispersed pituitary cells obtained from male pituitaries. In these studies galanin also failed to augment GRH-induced GH release. Meister and Hulting (1987) also failed to induce GH secretion with galanin from dispersed pituitary cells derived from male rats. Interestingly, these authors did demonstrate that galanin and dopamine coincubated with GRH could inhibit GRH-induced GH release. This effect was reversed by neurotensin. The physiological significance of these observations are unknown.

In contrast to these negative reports, we have reported that 10^{-6} M porcine galanin stimulates GH release under *in vitro* conditions (Gabriel *et al.*, 1988b). Furthermore, galanin at these doses augmented GRH-induced GH release. Torsello *et al.* (1988) also reported a stimulatory effect of galanin on GH release from dispersed pituitary cells obtained from immature female, but not male, rats. They also observed additive effects of galanin and GRH on GH release from these cells. These studies suggest that under certain conditions, galanin may act directly on the pituitary to stimulate GH secretion. However, due to the differences among *in vitro* systems, absolute comparisons between reports are difficult and varying observations may result from differences in factors present in the media, some of which may sensitize the cells to the actions of galanin. Whether galanin acts directly on the somatotropic cell in the hypophysis remains to be clearly established. However, these studies clearly demonstrate that galanin exerts significant actions on hypothalamic neurons regulating GH secretion.

Prolactin

Ottlecz *et al.* (1986), reported that galanin injected centrally was unable to alter prolactin (PRL) secretion in the ovariectomized female rat. Subsequently, however, it was reported by Melander *et al.* (1987) and Koshiyama *et al.* (1987) that intraventricular injections of galanin in conscious or urethane-anesthetized male rats elevated plasma concentrations of PRL. The action of galanin on PRL is 4-fold less potent than its effects on GH (Ottlecz *et al.*, 1988). Immunoneutralization of endogenous galanin by icv infusion of galanin antiserum does not alter PRL secretion in the rat (Ottlecz *et al.*, 1988; Maiter *et al.*, 1990). These studies imply the absence of a tonic galaninergic influence on the tuberoinfundibular dopamine and VIP-ergic neurons that regulate rat PRL secretion and suggest that galanin is involved in the hypothalamic control of phasic PRL release. Further studies will be required to explore this possibility.

Direct effects of galanin on PRL secretion from dispersed rat pituitary cells have not been observed (Ottlecz *et al.*, 1986; 1988; Inoue *et al.*, 1988). However Ottlecz *et al.* (1988) reported that galanin potentiates TRH-induced PRL and thyroid stimulating hormone (TSH) secretion *in vitro*. Vrontakis and Neill (1989) have shown that immunoneutralization of galanin inhibits PRL secretion from dispersed pituitary cells. Therefore, it appears that exogenous galanin has only minimal effects on pituitary function but endogenous galanin may modulate PRL release from lactotrophs.

The effects of galanin on the hypothalamic control of PRL secretion have been analyzed in more detail. There appears to be a strong influence of galanin on the tuberoinfundibular dopaminergic neurons. Melander *et al.* (1987), using quantitative histofluorescence, reported a depletion of the catecholamines from the medial palisade zone of the ME of male rats. Upon further investigation, Nordstrom and co-workers (1987) found that the release of ^{3}H-dopamine induced by K$^+$-depolarization from the basal hypothalamus was reduced by galanin. The IC_{50} for this effect was in the nanomolar range consistent with the known K_D of galanin binding to its receptor in the basal hypothalamus (Skofitsch and Jacobowitz, 1986). Jansson *et al.* (1989) confirmed these findings by demonstrating a decreased turnover of catecholamines in the ME of rats treated with galanin and α-methyl-para-tyrosine.

In addition to its effects on dopaminergic systems, galanin may also regulate PRL secretion by a VIP-mediated pathway. Koshiyama *et al.* (1987) found that PRL secretion induced by central galanin administration was attentuated by systemic VIP immunoneutralization. Furthermore, Inoue *et al.* (1988) reported that cerebrospinal fluid concentrations of VIP increased 2-fold after the infusion of galanin and incubation of isolated hypothalamic tissue with galanin (10^{-7}M) *in vitro* increased the concentration of VIP in the media to a level similar to that observed after K+ depolarization. These studies suggest that galanin-induced PRL secretion may occur by complex mechanisms that include the inhibition of dopamine secretion from tuberoinfundibular neurons or by increasing the secretion of VIP into the hypophysial portal blood. With respect to the effect of galanin on VIP, Sundstrom and Melander (1988) showed that galanin activates serotonergic neurons in the hypothalamus, which in turn, could modulate the activity of the hypothalamic VIP neurons. Thus, galanin appears to modulate dopamine, serotonin and VIP systems central to regulating PRL secretion in mammals.

Luteinizing Hormone (LH)

Melander *et al.* (1987) have reported that infusions of galanin into the cerebral ventricles of male rats did not change plasma concentrations of LH in male rats. Sahu *et al.* (1987) investigated the actions of galanin in a model system more responsive to changes in LH secretion. These investigators demonstrated that LH secretion in ovariectomized rats was unresponsive to galanin. However, galanin increased plasma concentrations of LH in the ovariectomized estrogen-progesterone-primed female rat. The effect of galanin could be demonstrated at doses as low as 0.5 μg up to 10 μg, intraventricularly.

These observations suggest that galanin may participate in the regulation of LH release but the effects of the peptide are strongly dependent on the hormonal milieu and the resulting physiological state of the pituitary.

Galanin Expression in the Anterior Pituitary

In addition to its widespread distribution in the central nervous system, galanin immunoreactivity and the mRNA encoding preprogalanin are present in endocrine cells of the rat anterior pituitary (Kaplan *et al.*, 1988b; Vrontakis *et al.*, 1987). Moreover, the peptide and mRNA species detected in pituitary tissue appear to be the same as those found in nervous tissue. We have observed that galanin and its mRNA are considerably more abundant in the pituitaries of female rats than male animals (Kaplan *et al.*, 1988b; Gabriel *et al.*, 1989). The observed sex differences develop at the time of sexual maturation (Gabriel *et al.*, 1990), and ovariectomy reduces the levels in females to those observed in males (Kaplan *et al.*, 1988b). Administration of pregnant mare serum to juvenile female rats leads to precocious puberty. Gonadotropins present in this serum induce ovarian maturation and estrogen secretion. Concurrent with these hormonal changes, galanin immunoreactivity in the anterior pituitary increases up to 3-fold (Gabriel *et al.*, 1990). Because these observations suggested that reproductive hormones might regulate galanin gene expression and peptide synthesis, we examined the effects of estrogens and androgens directly. Normal male rats have low concentrations of galanin immunoreactivity (0.5-10.0 ng/mg protein) and a virtual absence of mRNA in the anterior pituitary. Estrogen treatment of both male and ovariectomized female rats increased mRNA levels up to 3000-fold in a dose- and time-dependent manner. In the same groups of animals, galanin immunoreactivity in the anterior pituitary increased up to 50-fold, to a maximum of approximately 65 ng/mg protein. These stimulatory effects were specific to estrogen and were quantitatively similar in male and female animals (Kaplan *et al.*, 1988b). Testosterone, progesterone, thyroxine, and dexamethasone each had no effect on galanin mRNA levels. These studies indicate that pituitary galanin gene expression can be induced by estrogen under pharmacological conditions. We have also shown that variations in normal circulating levels of estrogen regulate galanin gene expression. Over the course of the 4-day estrous cycle in adult female rats, galanin mRNA levels in the anterior pituitary vary 30-fold. Maximal gene expression correlated with circulating estrogen levels and were seen on the evening of proestrus and morning of estrus. Thereafter, they fell rapidly and reached a nadir on diestrus (Kaplan *et al.*, 1988b). Galanin peptide concentrations also varied approximately

10-fold over the estrous cycle.   Levels of immunoreactive galanin followed mRNA levels with a delay of approximately 24 h and were maximal on diestrus day 1 (Fig. 1).   Galanin gene expression also increases approximately 100-fold during the last trimester of pregnancy, consistent with the grossly elevated levels of estrogen observed during this period.   Thus, observed physiologic variations in pituitary galanin expression, including differences between expression in male and female rats, appear to be due to differences in the circulating estrogen concentrations, rather than an intrinsic sexual dimorphism.

Although the precise distribution of galanin-expressing cell bodies in the pituitary is currently undefined, lactotrophs and somatotrophs contain high levels of galanin mRNA and peptide under conditions where circulating estrogen levels are elevated.   The implantable MtTW15 pituitary tumor line is derived from a rat lactosomatotrophic cell tumor and secretes high levels of prolactin and growth

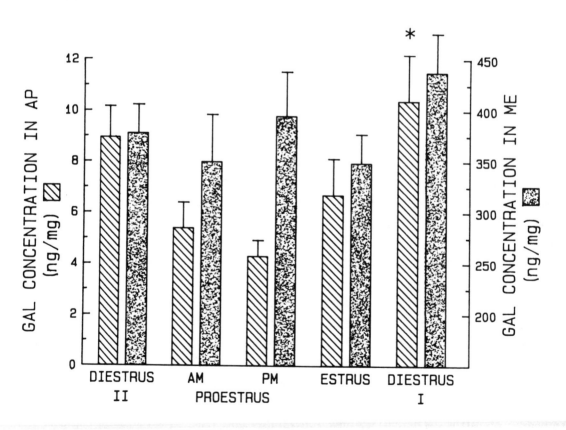

hormone constitutively. We have observed that galanin expression in this tumor is regulated by estrogen in the same manner as in the native pituitary (Kaplan *et al.*, 1988b). Vrontakis and Neill (1989), using dispersed pituitary cultures, demonstrated the coexistence of galanin with both prolactin and growth hormone by immunohistochemical techniques. More recently, we have used *in situ* hybridization techniques to show that galanin gene expression is primarily localized to lactotrophs and somatotrophs. A minority of thyrotrophs and only scattered corticotrophs and gonadotrophs appear to express this gene. Galanin, therefore, appears to be synthesized in a subset of pituitary cells, lactotrophs and somatotrophs whose secretory function is controlled by this peptide. As noted earlier, evidence for a direct effect of galanin on pituitary GH and PRL is inconclusive. However, these observations certainly raise the intriguing possibility that galanin acts as a paracrine or autocrine regulator within the pituitary gland itself. Conversely, we have observed that neither GH nor PRL regulates galanin gene expression in the rat pituitary (Kaplan *et al.*, 1988b).

Hormonal Regulation of CNS Galanin Expression

Pharmacological administration of estrogen also induces the synthesis of immunoreactive galanin in the central nervous system. Under the influence of estradiol valerate (10 mg/100 gm) for seven days, galanin concentrations in the neurointermediate lobe of the male rat pituitary increased from 23.4 ± 2.7 ng/mg protein to 31.8 ± 3.4 ng/mg protein ($P<0.05$). In addition, basal hypothalamic concentrations of galanin also increased by 33% ($P<0.05$) and concentrations in the ME increased 50% (Gabriel *et al.*, 1990).

These observations are extended by studies of the developmental expression of galanin in the hypothalamus (Gabriel *et al.*, 1989). Galanin immunoreactivity appears in the rat hypothalamus by day 15 of gestation and increases steadily until the time of birth. Postpartum hypothalamic galanin immunoreactivity decreases transiently for 10 days after which concentrations increase. Adult levels are attained by postnatal day 45. Adult animals demonstrated significant sex differences in the concentration of galanin in the ME and neurointermediate lobe of the pituitary. In the ME, the concentration of galanin on day 25 is approximately 100 ng/mg protein in both male and female animals. Adult (day 90) concentrations in the male rats (230 ng/mg protein) are reached by day 35. However, concentrations in female rats are nearly 2-fold higher (440 ng/mg protein) around the time of puberty (day 45) and then fall to adult levels (300 ng/mg protein) by day 90. Thus, random cycling adult female rats have significantly greater concentrations of galanin in the ME than their male littermates.   In the neurointermediate lobe, galanin concentrations fall during development from levels of approximately 75 ng/mg protein to 30-40 ng/mg protein by day 90.   In this region as well, female animals have significantly greater concentrations of galanin than their male siblings.

Pregnant mare serum (PMSG) induces precocious puberty in juvenile female rats (Gabriel *et al.*, 1990).   PMSG administration to prepubertal female rats also increased galanin immunoreactivity in several regions of the neuroendocrine axis (Gabriel *et al.*, 1990).   Hypothalamic concentrations of galanin increased 48 h after treatment, whereas ME and NIL concentrations increased after 72 h.   No changes in galanin immunoreactivity in these regions were observed in similarly treated male or ovariectomized female rats.   These studies demonstrate that increases in circulating estrogen levels produced by either pharmacological or physiological manipulations cause increased tissue galanin concentrations in the brain as well as the pituitary.   In the context of the observed effects of galanin on GH, PRL and LH secretion, they suggest that galanin may participate in the response of these hormones to estrogenic stimuli.

Summary

Galanin is a neuroendocrine peptide that regulates the secretion of several anterior pituitary hormones, including GH, PRL and LH.   Physiologic and pharmacologic studies of the action of this peptide suggest that it participates in a variety of neuroendocrine pathways involved in the regulation of the rat hypothalamic-pituitary axis.   Its distribution within the hypothalamus, ME, neurointermediate lobe, and anterior pituitary is consistent with its predicted regulatory functions.

Galaninergic neurons within the hypothalamus appear to play an important role in the control of basal and episodic GH secretion in the rat.   The effect of local, pituitary galanin on GH secretion is less well established.   The means by which galanin regulates PRL secretion are also poorly characterized. Physiological studies suggest that galanin modulates the tuberoinfundibular dopaminergic system.   However, several studies also suggest a potential for PRL regulation by peptide produced within the pituitary itself.   The regulation of hypothalamic, ME and neurointermediate lobe galanin content by estrogen suggests that this peptide may help to generate sexually dimorphic patterns of hypothalamic activity.   The dramatic regulation of galanin gene expression and peptide synthesis within pituitary lactotrophs and somatotrophs indicate that pituitary gelanin may help to mediate the effects of estrogen on pituitary growth and function.   Together, these observations suggest that galanin plays an important role in the neuroendocrine regulation of growth, lactation and reproductive behavior in the rat. Further studies to elucidate the physiologic functions of this peptide, and its interactions with other neuroendocrine regulatory systems, will likely permit a more sophisticated understanding of the subtle interplay of factors which control reproductive function. Such an understanding may lead to the development of more selective, and effective, agents to promote individual fertility on one hand, or population control on the other.

ACKNOWLEDGEMENTS

The authors wish to acknowledge the support of NIH grants DK40788 and DK26252 for this work and to thank Mrs. Elizabeth Thompson for her excellent secretarial assistance.

REFERENCES

Ceccatelli, S., M. Eriksson and T. Hökfelt. 1989. Distribution and coexistence of corticotropin-releasing factor, neurotensin-, enkephalin-, cholecystokinin-, galanin- and vasoactive intestinal polypeptide/peptide histidine isoleucine-like peptides in the parvocellular part of the paraventricular nucleus. *Neuroendocrinology* 49: 309-323.

Cella, S.G., V. Locatelli, V. De Gennaro, G.P. Bondiolotti, C. Pintor, S. Loche, M. Provezza and E.E. Müller. 1987. Epinephrine mediates the growth hormone-releasing effect of galanin in infant rats. *Endocrinology* 122: 855-859.

Chatterjee, V.K.K., J.A. Ball, T.M.E. Davis, C. Proby, J.M. Burrin and S.R. Bloom. 1988. The effect of cholinergic blockade on the growth hormone response to galanin in humans. *Metabolism* 37: 1089-1091.

Ch'ng, J.L.C., N.D. Christofides, P. Anand, S.J. Gibson, Y.S. Allen, H.C. Su and K. Tatemoto. 1985. Distribution of galanin immunoreactivity in the central nervous system and responses of galanin-containing neuronal pathways to injury. *Neuroscience* 16: 343-354.

Davis, T.M.E., J.M. Burrin and S.R. Bloom. 1987. Growth hormone (GH) release in response to GH-releasing hormone in man is 3-fold enhanced by galanin. *J. Clin. Endocrinol. Metab.* 65: 1248-1252.

Fuxe, F., G. von Euler, L.F. Agnati and S.O. Ogren. 1988. Galanin selectively modulates 5-hydroxytryptamine 1A receptors in the rat ventral limbic cortex. *Neurosci. Lett.* 85: 163-167.

Gabriel, S.M., L.M. Kaplan, J.B. Martin and J.I. Koenig. 1989. Tissue-specific sex differences in galanin-like immunoreactivity and galanin mRNA during development in the rat. *Peptides* 10: 369-374.

Gabriel, S.M., J.I. Koenig and L.M. Kaplan. 1990. Galanin-like immunoreactivity is influenced by estrogen in peripubertal and adult rats. *Neuroendocrinology*. 51: 168-173.

Gabriel, S.M., U.M. MacGarvey, J.I. Koenig, K.J. Swartz, J.B. Martin and M.F. Beal. 1988a. Characterization of galanin-like immunoreactivity in the rat brain: Effects of neonatal glutamate treatment. *Neurosci. Lett.* 87: 114-120.

Gabriel, S.M., P.E. Marshall and J.B. Martin. 1988c. Interactions between growth hormone releasing hormone, somatostatin and galanin in the control of growth hormone secretion. In *Basic and Clinical Aspects of Growth Hormone*, ed. B. Bercu, 73-82. Plenum Press, New York.

Gabriel, S.M., C.M. Milbury, J.A. Nathanson and J.B. Martin. 1988b. Galanin stimulates rat pituitary growth hormone secretion *in vitro*. *Life Sci.* 42: 1981-1986.

Gaymann, W. and R. Martin. 1989. Immunoreactive galanin-like material in magnocellular hypothalamo-neurohypophyseal neurons of the rat. *Cell Tissue Res.* 255: 139-147.

Hooi, S.C., G.S. Richardson, J.K. McDonald, J.M. Allen, J.B. Martin and J.I. Koenig. 1989. Neuropeptide Y and vasopressin in the hypothalamo-neurohypophysial axis of the salt-loaded or Brattleboro rats. *Brain Res.* 486: 214-220.

Inoue, T., Y. Kato, H. Koshiyama, N. Yanaihara and H. Imura. 1988. Galanin stimulates the release of vasoactive intestinal polypeptide from perifused hypothalamic fragments *in vitro* and from periventricular structures into the cerebrospinal fluid *in vivo* in the rat. *Neurosci. Lett.* 85: 95-100.

Jansson, A., K. Fuxe, P. Eneroth and L.F. Agnati. 1989. Centrally administered galanin reduces dopamine utilization in the median eminence and increases dopamine utilization in the medial neostriatum of the male rat. *Acta Physiol. Scand.* 135: 199-200.

Kaplan, L.M., S.M. Gabriel, J.I. Koenig, M.E. Sunday, E.R. Spindel, J.B. Martin and W.W. Chin. 1988. Galanin is an estrogen-inducible, secretory product of the rat anterior pituitary. *Proc. Natl. Acad. Sci. U.S.A.* 85: 7408-7412.

Kaplan, L.M., E.R. Spindel, K.J. Isselbacher and W.W. Chin. 1988. Tissue-specific expression of the rat galanin gene. *Proc. Natl. Acad. Sci. U.S.A.* 85: 1065-1069.

Koenig, J.I., S. Hooi, S.M. Gabriel and J.B. Martin. 1989. Potential involvement of galanin in the regulation of fluid hemostasis in the rat. *Regul. Pept.* 24: 81-86.

Koshiyama, H., Y. Kato, T. Inoue, Y. Murakami, Y. Ishikawa, N. Yanaihara and H. Imura. 1987. Central galanin stimulates pituitary prolactin secretion in rats: Possible involvement of hypothalamic vasoactive intestinal polypeptide. *Neurosci. Lett.* 75: 49-54.

Levin, M.C., P.E. Sawchenko, P.R.C. Howe, S.R. Bloom and J.M. Polak. 1987. Organization of galanin-immunoreactive inputs to the paraventricular nucleus with special reference to their relationship to catecholaminergic afferents. *J. Comp. Neurol.* 261: 562-582.

Maiter, D., S.C. Hooi, J.I. Koenig and J.B. Martin. 1990. Galanin is a physiological regulator of pulsatile growth hormone secretion in the rat. Abstract of the *Endocrinol. Meeting*, 126:1216-1222.

Meister, B., T. Hökfelt, O. Johansson and A.L. Hulting. 1987. Distribution of growth hormone-releasing factor, somatostatin and coexisting messengers in the brain. In *Growth Hormone--Basic and Clinical Aspects*, eds. O. Isaksson *et al.*, 29-52. Elsevier, Amsterdam, The Netherlands.

Meister, B. and A.L. Hulting. 1987. Influence of coexisting hypothalamic messengers on growth hormone secretion from rat anterior pituitary cells *in vitro*. *Neuroendocrinology* 46: 387-394.

Melander, T., K. Fuxe, A. Harfstrand, P. Eneroth and T. Hökfelt. 1987a. Effects of intraventricular injections of galanin on neuroendocrine functions in the male rat. Possible involvement of hypothalamic catecholamine neuronal systems. *Acta Physiol. Scand.* 131: 25-32.

Melander, T., T. Hökfelt, S. Nilsson and E. Brodin. 1986a. Visualization of galanin binding sites in the rat central nervous system. *Eur. J. Pharmacol.* 124: 381-382.

Melander, T., T. Hökfelt and A. Rokaeus. 1986b. Distribution of galanin-like immunoreactivity in the rat central nervous system. *J. Comp. Neurol.* 248: 475-517.

Melander, T., T. Hökfelt, A. Rokaeus, A.C. Cuello, W.H. Oertel, A. Verhofstad and M. Goldstein. 1986c. Coexistence of galanin-like immunoreactivity with catecholamines, 5-hydroxytryptamine, GABA and neuropeptides in the rat CNS. *J. Neurosci.* 6: 3640-3654.

Melander, T., W.A. Staines, T. Hökfelt, A. Rokaeus, F. Eckenstein, P.M. Salvaterra and B.H. Wainer. 1985. Galanin-like immunoreactivity in cholinergic neurons of the septum basal forebrain complex projecting to the hippocampus of the rat. *Brain Res.* 30: 130-138.

Murakami, Y., Y. Kato, A. Shimatsu, H. Koshiyama, N. Hattori, N. Yanaihara and H. Imura. 1989. Possible mechanisms involved in growth hormone secretion induced by galanin in the rat. *Endocrinology* 124: 1224-1229.

Nordstrom, O., T. Melander, T. Hökfelt, T. Bartfai and M. Goldstein. 1987. Evidence for an inhibitory effect of the peptide galanin on dopamine release from the rat median eminence. *Neurosci. Lett.* 73: 21-26.

Ottlecz, A., W.K. Samson and S.M. McCann. 1986. Galanin: Evidence for a hypothalamic site of action to release growth hormone. *Peptides* 7: 51-53.

Ottlecz, A., G.D. Snyder and S.M. McCann. 1988. Regulatory role of galanin in control of hypothalamic-anterior pituitary function. *Proc. Natl. Acad. Sci. U.S.A.* 85: 9861-9865.

Palkovits, M., A. Rokaeus, F.A. Antoni and A. Kiss. 1987. Galanin in the hypothalamo-hypophyseal system. *Neuroendocrinology* 46: 417-423.

Rokaeus, A. and M.J. Brownstein. 1986. Construction of a porcine adrenal medullary cDNA library and nucleotide sequence analysis of two clones encoding a galanin precursor. *Proc. Natl. Acad. Sci. U.S.A.* 83: 6287-6291.

Rokaeus, A. and M. Carlquist. 1988. Neucleotide sequence analysis of cDNAs encoding a bovine galanin precursor protein in the adrenal medulla and chemical isolation of bovine gut galanin. *FEBS Lett.* 234: 400-406.

Rokaeus, A., T. Melander, T. Hökfelt, J.M. Lundberg, K. Tatemoto, M. Carlquist and V. Mutt. 1984. A galanin-like peptide in the central nervous system and intestine of the rat. *Neurosci. Lett.* 47: 161-166.

Rokaeus, A., W.S. Young III and E. Mezey. 1988b. Galanin coexists with vasopressin in the normal rat hypothalamus and galanin's synthesis is increased in the Brattleboro (diabetes insipidus) rat. *Neurosci. Lett.* 90: 45-50.

Sahu, A., W.R. Crowley, K. Tatemoto, A. Balasubramaniam and S.P. Kalra. 1987. Effects of neuropeptide Y, NPY analog (norleucine [4]NPY), galanin and neuropeptide K on LH release in ovariectomized (ovx) and ovx estrogen, progesterone-treated rats. *Peptides* 8: 921-926.

Skofitsch, G. and D.M. Jacobowitz. 1985. Immunohistochemical mapping of galanin-like neurons in the rat central nervous system. *Peptides* 6: 509-546.

Skofitsch, G. and D.M. Jacobowitz. 1986. Quantitative distribution of galanin-like immunoreactivity in the rat central nervous system. *Peptides* 7: 609-623.

Skofitsch, G., D.M. Jacobowitz, R. Amann and F. Lembeck. 1989. Galanin and vasopressin coexist in the rat hypothalamo-neurohypophyseal system. *Neuroendocrinology* 49: 419-427.

Skofitsch, G., M.A. Sills and D.M. Jacobowitz. 1986. Autoradiographic distribution of I-galanin binding sites in the rat central nervous system. *Peptides* 7: 1029-1042.

Sundstrom, E. and T. Melander. 1988. Effects of galanin on 5HT neurons in the rat CNS. *Eur. J. Pharmacol.* 146: 327-329.

Tatemoto, K., A. Rokaeus, H. Jornvall, T.J. McDonald, and V. Mutt. 1983. Galanin--a novel biologically active peptide from the porcine intestine. *FEBS Lett.* 164: 124-128.

Torsello, A., S.G. Cella, M.C. Carolec, V. Locatelli, R. Sellan and E.E. Müller. 1988. Galanin modulates growth hormone secretion from rat anterior pituitary cells. Abstract of the *Endocrine Society Meeting*, p. 312.

Vrontakis, M. and J.D. Neill. 1989. Inhibition of prolactin secretion by galanin antiserum. Abstract of the *Endocrine Society Meeting*, p. 227.

Vrontakis, M.E., L.M. Peden, M.L. Duckworth and H.G. Friesen. 1987. Isolation and characterization of a complementary DNA (galanin) clone from estrogen-induced pituitary tumor messenger RNA. *J. Biol. Chem.* 62: 16755-16758.

CHOLECYSTOKININ AND NEUROENDOCRINE SECRETION

Joseph G. Verbalis and Edward M. Stricker
Departments of Medicine and
Behavioral Neuroscience
University of Pittsburgh
Pittsburgh, PA

I.   Introduction

II.   Neuroanatomy and Neurophysiology

III.   Effect of CCK on Neuroendocrine Secretion

      A.   Gonadotropins

      B.   Oxytocin and Vasopressin

      C.   Prolactin

      D.   CRH and ACTH

IV.   Physiological Functions of CCK

      A.   Grooming

      B.   Reproductive Behavior

      C.   Ingestive Behavior

V.   Future Directions

      A.   Receptor Antagonists

      B.   Functional Neuroanatomical Mapping

VI.   Summary

Introduction

Cholecystokinin (CCK) is one of the most widely distributed and abundant peptides found in the brain, including particularly prominent localization in the cortex, hippocampus, midbrain and hypothalamus (Crawley, 1985a). This widespread distribution implies the interaction/involvement of this peptide in a variety of neural system functions. However, it also suggests that there probably is no simple chemical coding by which CCK mediates physiological functions, whether neuroendocrine or behavioral. In addition, the distribution of CCK throughout the brain hinders investigators from answering functional questions by administering CCK agonists or antagonists into the cerebral ventricles, because the possibility of producing simultaneous effects on multiple brain pathways subserving diverse functions complicates simple interpretations of such data.

This chapter will review briefly the neuroanatomy of brain CCK-containing neuronal systems, and then will discuss more extensively the role of CCK and CCK-activated pathways in the control of neuroendocrine secretion from the hypothalamus, median eminence, and pituitary, as well as their potential involvement in the control of grooming, reproductive, and ingestive behaviors. We conclude with a brief consideration of several possible directions for future research to better understand central actions of this peptide.

Neuroanatomy and Neurophysiology

Within the hypothalamus, CCK-containing neurons have been identified in the paraventricular (PVN) and supraoptic (SON) nuclei (Lorén *et al.*, 1979; Vanderhaeghen *et al.*, 1980), the medial preoptic area, and the dorsomedial hypothalamus (Micevych *et al.*, 1987). Of these areas, the PVN and SON have been best studied, largely because they contain the most dense CCK immunoreactivity and were discovered first. Initial studies appropriately focused on the hypothalamo-neurohypophyseal system after CCK immunoreactivity was detected in the posterior pituitary and was demonstrated to be of hypothalamic origin (Beinfeld *et al.*, 1980). Additional studies suggested that neural lobe CCK was co-secreted with vasopressin (AVP) and/or oxytocin (OT) because treatments which deplete neural lobe content of these hormones, such as salt loading or lactation also cause a similar decrease in neural lobe CCK content. Immunohistochemical studies later confirmed that CCK immunoreactivity was present in the SON and PVN, and also demonstrated that CCK was selectively co-localized in oxytocinergic rather than vasopressinergic magnocellular neurons (Vanderhaeghen *et al.*, 1981). Furthermore, both parvocellular and magnocellular neurons of the PVN which project to the median eminence (Anhut et al., 1983; Deschepper et al., 1983; Palkovits et al., 1984) were found to contain CCK immunoreactivity.

Kiss et al. (1984) quantified hypothalamic efferent projections of PVN CCK-immunoreactive neurons by analyzing CCK-positive cell bodies following selective axotomy of fiber tracts projecting from the PVN. These studies demonstrated that approximately 40% of the CCK-positive PVN neurons were magnocellular in nature and projected to the neural lobe, while 60% were parvocellular and projected mostly to the median eminence. This quantitative predominance of parvocellular over magnocellular hypothalamic neurons is consistent with studies showing a 3:1 excess of median eminence CCK over neural lobe content (Palkovits et al., 1984). In addition to the prominent median eminence and neural lobe projections, a small but significant number of CCK-containing PVN parvocellular neurons were found to descend caudally to the brainstem, as identified by increased immunohistochemical staining following transection of the median forebrain bundle. However, unlike in the magnocellular system, a large number of the CCK-containing parvocellular neurons were found to be co-localized with AVP and corticotropin releasing hormone (CRH) rather than OT (Mezey et al., 1986). These results suggest, therefore, that CCK co-localized in the magnocellular and parvocellular neurons of the PVN may subserve quite different functions, as will be discussed in subsequent sections.

Projections from the preoptic and dorsomedial areas of the hypothalamus have not been as thoroughly identified. However, studies quantifying the distribution of such neurons in rats suggest that a sexual dimorphism exists for some of these neurons and their terminal fields (Micevych et al., 1987). Four such areas were shown to be sexually dimorphic: two in which the number of CCK-immunoreactive cells were greater in males (the posterior magnocellular subdivision of the PVN and the central division of medial preoptic nucleus), and two in which a preponderance of CCK-immunoreactive cells was found in females (the periventricular and dorsal areas of the medial preoptic nucleus). In addition, terminal fields of CCK-immunoreactive neurons in the ventromedial hypothalamus (VMH), presumably representing efferent projections of CCK-containing neurons from either the dorsomedial hypothalamus or the PVN, also were found to be significantly more dense in females as compared to males. The significance of these sexually dimorphic differences in immunohistochemical staining has not yet been determined, though some investigators have suggested that these areas may participate in the control of reproductive behavior, as will be discussed later.

Studies of co-localization of CCK with other peptides have shown that CCK is present in a wide variety of neuronal types within the brain (Hökfelt *et al.*, 1985). Although CCK in the magnocellular neurohypophyseal system is almost exclusively co-localized within oxytocinergic neurons, while that in the parvocellular system appears to be mainly co-localized within AVP- and CRH-containing neurons, the actual amounts of CCK within these systems constitute a small fraction of the major neurosecretory products. For example, in the magnocellular system, the measured concentration of CCK in the neural lobe (approximately 200-300 pg) is several orders of magnitude below the concentration of OT (500-800 ng). Consequently, speculation regarding physiological functions for brain CCK must take into account the fact that, in many neuronal systems, CCK concentrations amount to less than 1% of the peptide(s) with which it is co-localized. In addition to the hypothalamo-neurohypophyseal system, CCK has been co-localized within tyrosine hydroxylase-positive dopaminergic neurons in the midbrain (Fallon and Seroogy, 1985). Such neurons are found predominantly in the ventral tegmental area (VTA), whereas CCK in the substantia nigra appears to be in neurons separate from the dopamine-containing neurons (Crawley, 1985b; Hökfelt *et al.*, 1985). Recently, spinal cord neurons that appear to contain both CCK and substance P have also been described (Hökfelt *et al.*, 1985).

Multiple studies have now confirmed that the predominant component of brain CCK immunoreactivity consists of the octapeptide CCK-8 (Dockray, 1976; Beinfeld *et al.*, 1980). Although the last five amino acids of the carboxy-terminal of CCK-8 are identical to pentagastrin, CCK-8 immunoreactivity in the central nervous system has definitely been distinguished from gastrin immunoreactivity. These studies also have suggested that the octapeptide in brain is largely, if not exclusively, present as the sulfated form of CCK (Dockray *et al.*, 1985). Some smaller forms that, presumably, are degradation products of CCK-8 also have been detected by HPLC studies, but the biological significance of such forms remains uncertain (Brownstein and Rehfeld, 1985).

Given the abundant distribution of CCK-immunoreactive neurons and terminal fields within the central nervous system, it is not surprising that CCK receptors are similarly localized in neural tissue. Initial studies of tissue CCK receptors suggested significant differences in the ligand binding characteristics of brain CCK receptors as compared to peripheral pancreatic CCK receptors. Although receptors from both tissues showed greatest binding affinity for sulfated CCK-8, the brain receptors also demonstrated a smaller binding affinity for desulfated CCK-8 as well as to CCK-4, gastrin, and pentagastrin, in contrast to pancreatic receptors which were almost completely insensitive to these peptides (Innis and Snyder, 1980). On the basis of this characterization, the more specific pancreatic receptors were named CCK A (alimentary) receptors and the less specific brain receptors were named CCK B (brain) receptors. More recently, pancreatic CCK receptors expressed in Xenopus oocytes have similarly shown a greater response to sulfated CCK-8 than to desulfated CCK-8 and CCK-4 (Williams *et al.*, 1988). However, the fact that such receptors still demonstrate some sensitivity to the desulfated and shorter forms of CCK points out that differences in CCK receptor binding affinities are relative rather than absolute. Perhaps even more importantly, it has now been demonstrated that not all of the brain CCK receptors are of the B type. Studies by Moran *et al.* (1986) showed that CCK receptors in the area postrema and nucleus tractus solitarius of the brainstem possess binding characteristics more analogous to A type receptors whereas caudate nucleus CCK receptors had binding characteristics of B type receptors. Other areas in the brain, specifically the posterior hypothalamic nuclei and interpeduncular nucleus, also had CCK receptors demonstrating characteristics more analogous to A type receptors. Consequently, the distribution of CCK receptor types appears to be more complex than originally proposed.

Effects of CCK on Neuroendocrine Secretion

The effects of CCK on several neuroendocrine systems have been studied: gonadotropins, neurohypophyseal hormones (OT and AVP), prolactin, CRH, and ACTH. Each system will be reviewed separately, and in each case the differences between possible direct central actions of hypothalamic CCK versus the effects produced by pathways activated by peripheral CCK injections will be emphasized.

Gonadotropins

Initial studies employing intracerebroventricular (icv) injections of CCK showed no significant acute effects on plasma follicle stimulating hormone levels but a significant suppression of plasma luteinizing hormone (LH) levels (Vijayan *et al.*, 1979). The latter effect was demonstrated at doses of CCK ranging from 4 to 500 ng in ovariectomized female rats. Inhibition also was seen after peripheral CCK administration but only at very high doses that also produced marked elevations of serum prolactin; therefore, this was felt to be the result of nonspecific stress effects. Similar results were obtained by subsequent investigators (Fuxe *et al.*, 1985), and *in vitro* studies demonstrating that CCK inhibited KCl depolarization-induced release of gonadotropin releasing hormone from isolated rat hypothalami also support the possibility that central CCK may inhibit gonadotropin secretion (Micevych *et al.*, 1986).

However, other studies have suggested a stimulatory effect of CCK on gonadotropin secretion. Direct injection of CCK-8 into the medial preoptic area of estrogen-primed ovariectomized rats caused an enhancement of the circadian rise in LH in the afternoon following injection (Hashimoto and Kimura, 1986). Interestingly, this effect of CCK could be completely blocked by the dopamine receptor antagonist pimozide, suggesting an absolute requirement for dopamine. Given the co-localization of CCK with dopamine in at least some groups of midbrain neurons, it is interesting to speculate that these findings may represent a postsynaptic action of CCK to enhance primary dopaminergic effects, similar to the recent demonstration that injection of CCK into the nucleus accumbens can enhance dopamine-induced hyperlocomotion at some doses (Crawley, 1985b).

Although these studies appear to reach contradictory conclusions, this is likely the result of complications arising from the ubiquitous distribution of CCK in the brain. By activating more widely distributed neuronal systems, icv injections may produce different effects than those produced by more selective injections into discrete brain areas. These findings therefore emphasize the desirability of using site-specific injections whenever possible in attempting to define more clearly the potential neurophysiological actions of CCK.

Oxytocin and Vasopressin

Given the extensive co-localization of CCK in oxytocinergic and vasopressinergic neurons, many of the initial studies of CCK's effects on neuroendocrine secretion involved neurohypophyseal hormones. As a result, probably more is known today about the effects of CCK on secretion of OT and AVP than any other hormone. However, analogous to the previously described effects of CCK on gonadotropin secretion, which illustrate the importance of differentiating the effects of site-specific injections from more generalized icv injections, in the case of OT and AVP it is also essential to differentiate the effects produced by peripherally injected CCK from those resulting from central injections. Accordingly, this section will consider separately the effects produced by these differing routes of CCK administration.

Peripheral administration. In 1986, CCK was first reported to be a potent stimulant of pituitary OT secretion when administered peripherally in doses ranging from 1 to 100 µg/kg (Verbalis *et al.*, 1986a). These effects were present whether CCK was administered intravenously or intraperitoneally, and were remarkable in that CCK had no significant stimulatory effect on AVP secretion (Fig. 1). CCK therefore represents a pharmacological stimulus in rats analogous to the physiological stimulus of suckling, which also causes a selective activation of oxytocinergic but not vasopressinergic neurons. These findings also were remarkable because of the relative sensitivity of this response, demonstrating a threshold for stimulation of OT secretion (1 µg/kg) corresponding to the threshold for inhibition of food intake by CCK.

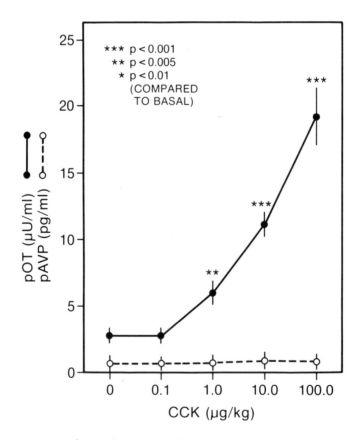

Fig. 1. Plasma oxytocin and vasopressin levels measured 5 min after injection of various doses of sulfated CCK-8 intraperitoneally (data from Verbalis *et al.*, 1986a).

Various chemical agents (LiCl, $CuSO_4$ and apomorphine) that inhibit food intake when injected into rats produce an analogous pattern of stimulating relatively selective OT secretion (Verbalis et al., 1986b). Because these agents are known to cause nausea in other species as well as the related phenomenon of learned taste aversions in rats (Garcia and Ervin, 1968), the findings suggest the possibility that CCK at these doses is similarly causing nausea-like symptoms in rats. Although OT secretion is known to occur in rats in response to a variety of stressful stimuli, including immobilization and swimming (Lang et al., 1983), it seems unlikely that CCK-induced OT secretion merely reflects a generalized stress response because (1) the magnitude of the induced OT secretion using CCK or toxic chemical agents generally exceeds the lesser magnitudes observed in response to a heterogeneous group of stressors (Verbalis et al., 1986b); (2) doses of dexamethasone that block hypothalamic CRH release did not significantly inhibit the OT response to nausea-producing agents (Verbalis et al., 1986b); and (3) the OT responses to CCK were nearly completely abolished in animals given a total abdominal or capsaicin-induced vagotomy (McCann et al., 1988; Verbalis et al., 1986a).

These and other results (Smith et al., 1985) suggest that whatever the subjective sensations produced by injections of CCK, they clearly originate in the periphery, specifically in the visceral organs innervated by the gastric vagus nerve. Consequently, if CCK-induced OT secretion results from the presence of stress, it is a form of stress related to abdominal discomfort. For lack of better terminology, we have used the term "visceral malaise" to describe the potential origin of the CCK-induced effects on OT secretion. In this regard, food intake itself caused a lesser though still significant stimulation of pituitary OT secretion (Verbalis et al., 1986a). However, the volumes consumed to produce such increases were quite large for rats (20-30 ml of liquid diet in 30 min), and consequently the OT secretion probably reflects a response to extreme gastric distention rather than to normal amounts of food intake. Subsequent studies have since verified that not only CCK but also acute mechanical gastric distention can cause firing of magnocellular oxytocinergic neurons in the SON as well as pituitary OT secretion (Renaud et al., 1987). These electrophysiological studies also confirmed that AVP secretion was not stimulated by CCK and demonstrated a transient inhibition of firing of vasopressinergic neurons following CCK administration. That both CCK or gastric distention can result in pituitary OT secretion suggests the possibility that these two treatments may be activating similar neural pathways. This hypothesis is supported by observations that the response to both is eliminated by capsaicin-induced vagotomy (McCann et al.,

1988), and that electrophysiological recordings from brainstem neurons in the nucleus tractus solitarius have shown that many neurons responding to gastric distention respond similarly to CCK administration (Raybould *et al.*, 1985). Furthermore, vagally transmitted responses to these two stimuli appear to have a synergistic effect on OT secretion when applied concurrently (Verbalis *et al.*, 1986a; Fig. 2). The potential physiological significance of these observations will be discussed in later sections.

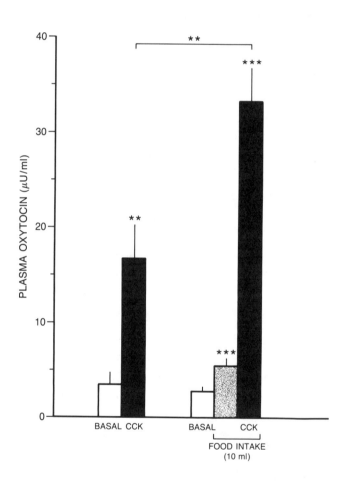

Fig. 2. Plasma oxytocin levels before (open bars) and 5 min after (solid bars) injection of sulfated CCK-8 (10 μg/kg) intraperitoneally. The bars on the left show the response to CCK in fasted rats, the bars on the right show the response to the same dose following a 10 ml gastric preload of liquid diet ingested over 10 min. The shaded bar on the right indicates the increment in plasma OT produced by the gastric distention from the preload prior to CCK injection (data from Verbalis *et al.*, 1986a).

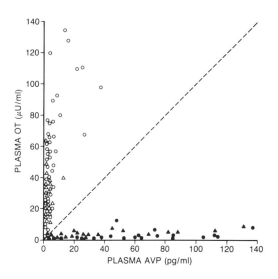

Fig. 3. Comparison of plasma oxytocin and vasopressin levels in monkeys (solid symbols) and rats (open symbols) in response to administration of CCK-8 (triangles) or nausea-producing toxins (LiCl or CuSO$_4$, circles). The dashed line indicates equivalency of the measured responses (reproduced with permission from Verbalis *et al.*, 1987).

Although both physiological and electrophysiological data have clearly demonstrated the specificity of peripheral CCK injections to activate oxytocinergic but not vasopressinergic pathways, there is a reversal of this phenomenon in other species. Specifically, in both monkeys (Verbalis *et al.*, 1987) and humans (Miaskiewicz *et al.*, 1989) CCK causes marked pituitary AVP secretion without any measurable OT secretion. This is true not only of CCK but also of various other chemical agents that inhibit food intake (Verbalis *et al.*, 1987; Fig. 3). Because CCK and these other agents can all produce nausea in humans, and because nausea has long been known to be a very potent stimulant of pituitary AVP secretion (Rowe *et al.*, 1979), this suggests that the OT secretion induced by these treatments in rats may be produced by activation of central neural pathways that are equivalent to those activated by symptomatic nausea in primates. Whatever subjective sensations are associated with the injection of CCK and similar agents in rats, the intriguing reversal of the neurohypophyseal response in primates illustrates that there are basic differences across species, in the afferent pathways from vagal brainstem centers to the hypothalamus, which must be taken into account in any physiological studies involving peripherally injected CCK.

Central administration. The effects produced by CCK administered peripherally must be differentiated from the effects produced by CCK administered centrally. CCK receptors have been found both on magnocellular neurons in the SON and PVN (Day *et al.*, 1987) as well as in the posterior pituitary (Bondy *et al.*, 1989). Moreover, direct icv injections of CCK stimulated AVP secretion in rats (Marley *et al.*, 1984) and caused activation of both AVP and OT neurons measured electrophysiologically (Jarvis *et al.*, 1988). Similarly, incubation of neural lobes with CCK *in vitro* led to secretion of both AVP and OT (Bondy *et al.*, 1989). However, such direct central CCK-mediated neurohypophyseal secretion differs significantly from the release seen in response to peripheral CCK administration in that *both* neurohypophyseal peptides, AVP and OT, are released with the former treatment in contrast to the selective release of OT seen with the latter. Furthermore, the selective OT release following peripheral CCK injection is vagally dependent, which is not the case with central injections. Consequently, it should be clear that the effect of peripherally administered CCK on OT secretion results from an action on vagal receptors in the gastrointestinal tract, rather than directly at the neural lobe or deeper in the hypothalamus, in which case the injected CCK would be likely to activate receptors responsible for centrally mediated release of both AVP and OT.

Although the general significance of centrally stimulated neurohypophyseal secretion in response to CCK is unclear, these findings may be important with regard to the small amounts of CCK that are co-localized within magnocellular oxytocinergic neurons. The recent work of Bondy *et al.* (1989), using isolated neural lobes, demonstrated several very interesting characteristics of the CCK-induced secretion of AVP and OT seen *in vitro*. First, this response was of relatively long latency, peaking at 30-40 min after incubation of the neural lobes with CCK (in marked contrast to the vagally mediated OT secretion following peripheral injections of CCK, which peaks in only 2 min and then dissipates rapidly). Second, exocytosis from the neural lobe appears to occur via a calcium-independent mechanism that can be blocked by the protein kinase-C inhibitor, staurosporine, suggesting activation of a second messenger system that mobilizes intracellular calcium stores mediating the CCK action at the neural lobe. Third, AVP and OT release following incubation with CCK occurred in the nanomolar dose range, far above physiological concentrations of plasma CCK which are now known to be in the picomolar range (Liddle *et al.*, 1986; Reidelberger *et al.*, 1989). Finally, desulfated CCK-8 was equipotent with sulphated CCK-8, indicating that the neural lobe receptors are of the B receptor type (again, quite different from the neurohypophyseal OT secretion produced by peripheral injections of CCK, which requires injection of the sulphated peptide). As a result of these observations, Bondy *et al.* (1989) hypothesized that CCK co-localized within magnocellular oxytocinergic neurons is released with OT at the nerve terminals in the pituitary where it may have a local paracrine action presynaptically to potentiate additional OT secretion.

These data suggest, therefore, that central administration of CCK produces a different spectrum of actions on neurohypophyseal secretion than the vagally mediated effects stimulated by peripheral injections. That this has proven to be the case serves to emphasize the importance of differentiating the effects produced by centrally administered peptides from the effects produced when the same peptides are administered peripherally; entirely different responses may be possible in each case depending upon the mechanisms and pathways by which the peripherally injected peptides ascend to activate the hypothalamus.

Prolactin

Several studies have examined prolactin secretion in response to CCK administration. As mentioned previously, high doses of peripherally injected CCK stimulated prolactin secretion in rats (Vijayan *et al.*, 1979), and a more recent study in pigs (Ebenezer *et al.*, 1989) has also demonstrated prolactin secretion in response to peripheral injections of CCK in this species. However, Vijayan *et al.* (1979) found that central injections of CCK did not consistently stimulate prolactin secretion, and subsequent studies have actually demonstrated a dose-related inhibition of prolactin secretion following icv administration of CCK (Fuxe *et al.*, 1985). These results again emphasize the importance of differentiating effects produced by central versus peripheral injections of peptides.

Given that prolactin secretion occurs in response to stress in many species (Neill, 1970; Noel *et al.*, 1972; Parrott *et al.*, 1988), this response to CCK, like OT secretion in rats, may simply be a manifestation of a generalized stress state. Although this issue remains unanswered at present, it is interesting to consider the pathways by which prolactin secretion might be stimulated in response to CCK. In this regard, it seems noteworthy that central OT pathways have been implicated in prolactin secretion produced by icv injection of vasoactive intestinal peptide (Samson *et al.*, 1989). Because CCK is such a potent and reliable activator of hypothalamic OT secretion in rats, it seems reasonable that the prolactin response to peripheral administration of CCK may be secondary to activation of a central component of OT secretion occurring in parallel with the pituitary secretion.   In this case, the prolactin response should similarly be abolished by vagotomy, and in addition it might be significantly attenuated by pretreatment centrally with an OT receptor antagonist.   These possibilities remain to be evaluated.

CRH and ACTH

If peripheral administration of CCK produces a viscerally mediated stress response, as suggested by the OT and prolactin responses, then one would expect activation of CRH-mediated pituitary ACTH secretion as well.   There is now increasing data that this response indeed occurs.   Early studies indicated that CCK-33 administered intraperitoneally in rats caused significant increases in plasma corticosterone levels peaking at 30 min postinjection (Sander and Porter, 1988).   Because CCK-33 had no effects on isolated adrenal cells, stimulation of pituitary ACTH secretion appears to account for these results.   More recent studies by Kamilaris *et al.* (1989) have since confirmed a sensitive response of plasma ACTH as well as corticosterone to doses of CCK administered peripherally in the range of 0.1 to 10 μg/kg.   Interestingly, this response could be blocked by capsaicin-induced afferent vagotomy similar to the OT response. These data, therefore, support the likelihood that CCK injected peripherally produces aversive visceral symptoms which ascend to the brainstem via vagal pathways and from there result in activation of a variety of hypothalamic hormonal "stress" responses.

Because CCK in parvocellular neurons of the PVN appears to be co-localized within AVP- and CRH-containing neurons projecting to the median eminence, it seems possible that this CCK might be directly involved in regulating ACTH secretion.   Studies utilizing incubation of anterior pituitary cells in primary culture with CCK have supported this possibility by demonstrating potentiation of ACTH release by AVP (Mezey *et al.*, 1986), although the physiological significance of these effects *in vivo* remains to be determined.

Physiological Functions of CCK

The preceding section mentioned two possible neurophysiological functions for CCK which is co-localized within hypothalamic neurons: (1) CCK in magnocellular oxytocinergic neurons may be acting in a paracrine manner within the neural lobe to further enhance OT secretion, and (2) CCK in parvocellular vasopressinergic neurons projecting to the median eminence may be acting to potentiate the CRF-like activity of AVP on corticotrophs in the anterior pituitary. This section will discuss potential actions of CCK and CCK-activated pathways in the broader context of coordinated physiological and behavioral functions. Although our knowledge in this area is less certain, sufficient data already exist in support of likely involvement of central and peripheral CCK and CCK-activated pathways in several selected behavioral areas.

Grooming Behavior

Although many peptides injected peripherally or centrally cause grooming, there is reason to believe that CCK may be more specifically involved in mediating grooming behavior. Studies by Kaltwasser and Crawley (1987) have demonstrated that direct injection of CCK into the VTA, where CCK is co-localized within dopaminergic neurons, produces significant increases in grooming behavior in rats. Interestingly, injection of OT into the same area similarly stimulates grooming behavior. Based on the previously mentioned co-localization of CCK in some oxytocinergic neurons, these authors have postulated that a projection of such neurons from the PVN to the VTA may influence the activity of dopaminergic neurons projecting from the VTA to forebrain areas involved with activation of grooming behavior. Although additional studies must be done to assess this hypothesis, this represents one example of a potential integrative function of hypothalamic CCK acting in concert with other neuropeptides and neural pathways.

Reproductive Behavior

There is substantial evidence supporting the possible involvement of brain CCK in reproductive behavior. Specifically, intraperitoneal injection of CCK-8 has been shown to potentiate lordosis behavior in estrogen-primed rats (Bloch *et al.*, 1987), although induction of this behavior is crucially dependent upon the state of receptivity of the rat (Mendelson and Gorzalka, 1984; Bloch *et al.*, 1987). In addition, analogous to the contradictory effects of CCK which have been observed on LH secretion, site-specific intracerebral injections into different areas have produced quite different results. Direct injection of CCK into the VMH has been shown to inhibit lordosis behavior in female rats (Babcock *et al.*, 1988a), while injections into the medial preoptic area (MPOA) facilitated lordosis behavior in both female and male rats (Priest *et al.*, 1988; Bloch *et al.*, 1988). Because both the VMH and MPOA are areas where sexual dimorphism of CCK-positive neurons and terminal fields has been demonstrated (Micevych *et al.*,

1987), and because estrogen appears to decrease CCK receptor density in the VMH (Akesson *et al.*, 1987), it has been suggested that hypothalamic CCK may modify the reproductive behavior of estrogen-primed animals via reciprocal actions at the MPOA and VMH. Although substantial indirect support for this hypothesis exists, recent evidence that CCK-immunoreactive neurons in the hypothalamus do not appear to concentrate estrogen suggests the absence of any direct effects of ovarian steroids to modulate CCK neuronal activity in female rats (Akesson and Micevych, 1988).

Clearly more work will be needed to unravel the exact role of central CCK in modulating lordosis behavior. However, as in the case of interactions within the neurohypophyseal system, interpretation of the effects of CCK on lordosis behavior must take into account the differing neural pathways activated by peripheral and central injections. In this regard, it is interesting to consider that peripheral CCK injections activate central oxytocinergic pathways via a vagally dependent route. Because central injection of OT can induce lordosis behavior in rats as well (Gorzalka and Lester, 1987), peripheral CCK may facilitate lordosis in part by activating brain oxytocinergic pathways. The recent demonstration that vagotomy blocks the stimulation of lordosis behavior by peripheral injections of CCK (Babcock *et al.*, 1988b) is consistent with this possibility. It will therefore be of interest in future studies to ascertain the effects of centrally active OT antagonists on CCK-induced lordosis behavior, with the expectation that this treatment may block the effects of CCK injected peripherally but not centrally.

Another potential mechanism that should be considered involves interactions of CCK with dopaminergic systems. Lordosis behavior is increased in rats following either dopamine-depleting brain lesions (Caggiula *et al.*, 1979a) or treatment with drugs blocking central dopamine receptors (Everitt *et al.*, 1975; Caggiula *et al.*, 1979b). Because some studies have suggested that CCK-8 can depress potassium-stimulated release of dopamine from caudate nucleus slices *in vitro* (Hökfelt *et al.*, 1985), these data suggest the possibility that some of the effects of CCK on lordosis behavior could result from presynaptic inhibition of central dopaminergic activity. This again emphasizes the necessity to consider potential interactive effects of a peptide as widely distributed in the brain as CCK.

Ingestive Behavior

Perhaps the largest volume of data addressing potential physiological and behavioral effects of central CCK and CCK-activated pathways has been generated with regard to the control of food intake. Peripheral injection of CCK has long been noted to inhibit food intake in rats (Gibbs *et al.*, 1973; Gibbs and Smith, 1984) and other species including man (Kissileff *et al.*, 1981). However, despite numerous studies during the last decade, the potential physiological role of CCK in controlling food intake still remains much debated. Many investigators have suggested that endogenously secreted CCK from the gastrointestinal tract causes satiety, or to some degree contributes to it, while others have argued that injections of CCK decrease food intake primarily by inducing visceral malaise. This contention is supported by the observed CCK-induced production of learned taste aversions in rats (Deutsch and Hardy, 1977) and subjective symptoms of nausea or gastric discomfort in humans (Miaskiewicz *et al.*, 1989). In view of this ongoing controversy, it is instructive to attempt to relate the stimulatory effects of CCK on hypothalamic neuroendocrine secretion to its inhibitory effects on food intake in order to evaluate the central pathways by which CCK may be mediating these effects.

The first neuropeptide to be studied in relation to CCK's effects on food intake was OT. Not only was OT secretion potently induced by CCK in rats, but there was a close temporal relation between the onset of OT secretion and the inhibition of food intake. In addition, there was a striking relation between the degree of OT secretion induced and the magnitude of inhibition of food intake (Verbalis *et al.*, 1986a), and this same relation also was obtained using a more diverse group of treatments that similarly inhibit food intake, including chemical agents such as LiCl as well as hyperosmolality (McCann *et al.*, 1989; Flanagan *et al.*, 1989a). The inverse correlation that has been described between stimulation of pituitary OT secretion and food intake in response to CCK and other treatments, suggests the possible involvement of hypothalamic oxytocinergic pathways in mediating some part of the CCK inhibition of food ingestion. However, it is important to note that the pituitary OT secretion itself is not causally related to the decrease in food intake (Verbalis *et al.*, 1986b), but simply reflects activation of neural pathways that are associated with the decrease in food intake.

In considering what central pathways might mediate inhibition of food intake following peripheral injections of CCK, two potential candidates stand out. The first of these includes the spectrum of neural pathways activated by stress responses. Observations that peripheral injection of CCK in rats increases secretion of OT, prolactin, and ACTH is consistent with the concept that this represents a neuroendocrine stress response to a viscerally aversive agent. However, this may be too simplistic an explanation to account for the observed

OT secretion. Specifically, it is well known that all stressors do not stimulate secretion of OT in rats. For example, hypoglycemia and hypothermia do not stimulate OT secretion (Gibbs, 1986), and it is noteworthy that both of these stressors are known to increase rather than decrease food intake. Thus, central oxytocinergic pathways seem to be activated by stressors that decrease food intake, but not by stressors that do not change or even increase food intake. In this regard, it is interesting to note that intracerebroventricular CRH also has been shown to be a potent inhibitor of food intake (Morley and Levine, 1982), that stress-induced hypothalamic CRH secretion has been shown to mediate the anorexia accompanying various stress states (Lenz *et al.*, 1988), and that intracerebroventricular injection of CRH has been shown to cause an increase in pituitary OT secretion (Bruhn *et al.*, 1986). Thus, these observations may represent another example of activation of the oxytocinergic component of the stress response that may be more specifically associated with inhibition of food intake.

The second likely possibility for central pathways mediating inhibition of food intake following peripheral injections of CCK are the efferent projections from the PVN to the brainstem. The PVN has long been known to exert a tonic inhibitory influence on food intake because PVN lesions have been associated with hyperphagia in rats (Leibowitz *et al.*, 1981), an effect also produced by parasagittal knife cuts that block caudal projections from the PVN, which would include oxytocinergic fibers (Kirchgessner and Sclafani, 1988). Although it is not yet clear exactly how such pathways might inhibit food intake, there is compelling evidence that these pathways provide significant inhibitory effects on complementary aspects of gastric function. Rogers and Hermann (1987) have demonstrated involvement of an oxytocinergic projection from the PVN to the dorsal motor nucleus (DMN) of the vagus that modulates gastric motility. Direct microinjection of OT (4 pM) into the DMN decreased gastric motility, while injection of an OT antagonist into the DMN increased spontaneous gastric motility and abolished the inhibitory component resulting from electrical stimulation of the PVN. These data strongly suggest that a descending parvocellular oxytocinergic pathway originating in the PVN acts to inhibit gastric motility. In view of this, it is noteworthy that not only CCK, but many other agents that both decrease food intake and stimulate pituitary secretion of OT in rats similarly inhibit gastric motility (Flanagan *et al.*, 1989b) and gastric emptying (McCann *et al.*, 1989). These findings suggest that stimuli such as CCK, which clearly activate magnocellular oxytocinergic neurons in the PVN, also likely co-activate a subset of parvocellular projections projecting to the

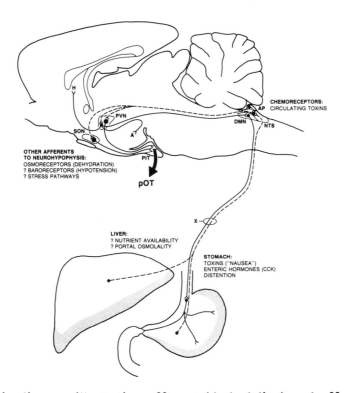

Fig. 4. Schematic diagram illustrating afferent (dashed line) and efferent (solid line) connections of the paraventricular nucleus potentially influencing ingestive behavior (PIT = pituitary; SON = supraoptic nucleus; PVN = paraventricular nucleus; H = hippocampus; A = amygdala; DMN = dorsal motor nucleus of the vagus nerve; NTS = nucleus tractus solitarius; AP = area postrema; X = vagus nerve).

brainstem, and that at least a portion of these parvocellular projections are oxytocinergic and act to inhibit gastric motility (Stricker *et al.*, 1988). This hypothesis is schematically illustrated in Fig. 4, demonstrating how a variety of afferent inputs into the PVN, including CCK, might simultaneously co-activate both magnocellular pituitary-projecting and parvocellular centrally projecting pathways, the latter influencing gastric function and possibly food intake as well.

As with all other proposed central actions of CCK, this hypothesis remains to be evaluated more fully. Several important qualifying points should be stressed, however, concerning the activation of central oxytocinergic pathways by CCK: (1) CCK and other agents that stimulate pituitary OT secretion in rats appear to be functioning as non-caloric modulators of food intake. To date, there is no convincing data that activation of oxytocinergic pathways is in any way involved with a caloric homeostatic influence on food intake and body weight maintenance. (2) Although nausea-producing chemical agents, as well as CCK, activate pituitary OT secretion in rats, this does not imply that pituitary OT secretion is always associated with nausea; obviously many physiological stimuli to OT secretion, such as increases in plasma osmolality or suckling, are not accompanied by subjective sensations of nausea. This fact serves to emphasize that multiple stimuli can cause secretion of individual neuropeptides from the hypothalamus, and it is inappropriate to generalize about equivalency of different stimuli simply because they cause secretion of the same neuropeptide. (3) As noted previously, although OT is secreted in response to CCK and nausea-producing agents in rats, in other species including man these agents cause secretion of AVP rather than an OT response. Consequently, the relevant peptide to consider in these species with regard to potential effects on gastric motility and food intake is AVP and not OT, and it would not be surprising if a correlation between OT secretion and food intake were absent in these species in contrast to rats. Undoubtedly, other qualifications will be discovered in the future. Nonetheless, it seems likely that central oxytocinergic pathways are in some way associated with many of the effects produced by peripheral injections of CCK, and the data presented here would suggest that they may play a prominent role in at least some of these effects, particularly gastric motility.

In addition to peripheral injections of CCK, icv administration of CCK also has been shown to be inhibitory to food intake (Della-Fera and Baile, 1979). These effects have been demonstrated most conclusively in sheep, which appear to be exquisitely sensitive to icv injection of CCK. Furthermore, pretreatment with antibodies to CCK has been shown to cause stimulation of food intake in this species (Della-Fera *et al.*, 1981). Although the efferent pathways involved in this response have not been as well studied as those stimulated by peripheral injections of CCK, from previous examples it might be anticipated that the two treatments are likely activating an entirely different set of central pathways from one another.

In summary, evidence to date suggests that CCK is likely to be involved in some aspects of ingestive behavior in mammals. It is intriguing to consider the additional possibility that there may be a coordinated action of endogenously secreted CCK acting peripherally on vagal receptors with centrally secreted CCK acting at receptors within the brain to produce a negative modulation of food intake. However, it remains to be ascertained to what extent such effects are physiological and mediated by endogenous CCK secretion, versus pharmacologically produced in response to administration of exogenous CCK. Hopefully, CCK antagonist studies will aid in this determination.

Future Directions

The preceding sections have summarized what is presently known about the effects of peripherally and centrally administered CCK on neuroendocrine secretion and associated behaviors. Throughout the discussion, speculation has been raised as to whether analogous effects are caused by endogenous CCK secretion and, if so, what physiologically relevant functions this secretion subserves. As yet no physiological function for either peripherally or centrally secreted CCK has been proven. This final section will, therefore, discuss two types of future studies which may provide some of the answers needed to define the physiological functions of endogenously secreted CCK, especially those related to neuroendocrine functions.

Receptor Antagonists

The use of receptor antagonists is perhaps the most powerful tool available to understand physiological functions of endogenously secreted peptides. Past studies of CCK have been hampered by the lack of sensitive and specific receptor antagonists. Although numerous studies were done with the CCK receptor antagonist, proglumide (see review, Schneider *et al.*, 1988), very large amounts of this compound are necessary to block CCK binding (IC_{50} of 250 ± 60 μM in rat pancreatic tissue and 800 ± 26 μM in guinea pig brain; Chang and Lotti, 1986). Injected amounts to achieve such concentrations are not feasible in the central nervous system, and even in the periphery the specter of producing simultaneous effects from nonspecific actions of the antagonist often precludes a clear interpretation of the results.

Fortunately, more sensitive and specific compounds than proglumide have recently become available. The most promising among these is MK 329 (initially named L-364,718), which displays marked selectivity for CCK A receptors (IC_{50} of 0.08 ± 0.02 nM for rat pancreatic tissue versus 245 ± 97 nM for guinea pig brain tissue; Chang and Lotti, 1986). Although not all studies using peripheral injection of MK 329 have shown increases in food intake (*e.g.*, Khosla and Crawley, 1988; Schneider *et al.*, 1988), several now have (Hewson *et al.*, 1988; Reidelberger and O'Rourke, 1989). However, several qualifying points should be considered concerning such studies: (1) increments in food intake have generally been modest, ranging from 5 to 20% in most positive studies; (2) protocols employing highly palatable diets generally have been more successful in showing larger increments in food intake (Hewson *et al.*, 1988); (3) effects have not been sustained in chronic studies, suggesting some degree of tolerance to the effects of the antagonist (Watson *et al.*, 1988); (4) because MK 329 can cross the blood-brain barrier, effects on food intake cannot be definitely assigned to peripheral CCK receptor blockade because of significant binding of this compound to CCK B receptors; moreover, the fact that much larger doses of MK 329 are required to increase food intake than to block the effects of peripherally administered CCK (see review, Schneider *et al.*, 1988) can be viewed as consistent with the possibility of a central rather than a peripheral effect of the antagonist; and (5) CCK antagonists have been shown to have significant antinociceptive properties (Dourish *et al.*, 1988), suggesting that at high doses such agents could increase food intake by eliminating or reducing the discomfort associated with gastric distention rather than by blocking peripheral CCK effects. Despite these and other uncertainties, further studies using MK 329 and similar compounds hold out the best promise of understanding the function of endogenously secreted CCK. Future development of a more selective CCK B receptor antagonist will allow even better evaluation of central CCK actions, and specifically of its effects on neuroendocrine secretion.

Not only can CCK receptor antagonists aid in ascertaining physiological functions of CCK, the use of receptor antagonists to peptides known to be stimulated by CCK can help to identify the efferent pathways activated by CCK. For example, recent studies using an OT receptor antagonist $(d(CH_2)_5$-Tyr(Me)-Orn^8-vasotocin) demonstrated a spontaneous increase in food intake of a magnitude quite similar to that shown for CCK antagonists (Arletti *et al.*, 1989). This finding is consistent with the hypothesis proposed in Fig. 4, namely that CCK-activated oxytocinergic neurons may mediate some of the inhibitory effects of CCK on gastric motility and food intake. In this case not only should central administration of an OT antagonist increase spontaneous food intake to whatever degree endogenously secreted CCK contributes to satiety, but it also should block some of the effects of peripherally administered CCK on food intake, as well as other proposed OT-mediated effects on prolactin secretion and lordosis behavior. Answers to these issues should be soon forthcoming.

Functional Neuroanatomical Mapping

Neuroanatomical studies employing immunohistochemical staining techniques in combination with retrograde tracing methods have provided neuroscientists with detailed maps showing where peptidergic neurons are localized and to what areas they project. However, such mapping is static in nature and cannot identify the subsets of neurons activated by specific physiological or pharmacological stimuli. Functional mapping of pathways activated by specific treatments would clearly represent a major advance in understanding the mechanisms responsible for the effects produced by administration of peptides such as CCK, and would be particularly useful for studying the CCK connectivity with the PVN, given the structural complexity of this region (Swanson and Sawchenko, 1983).

One method available for functional neuroanatomical mapping is *in situ* hybridization with cDNA probes complementary to mRNA of a particular peptide. However, this method is best suited to more chronic stimuli because of the need for sufficient stimulus intensity and duration to produce measurable increases in cytoplasmic mRNA. Given the short half-life of CCK, this would not appear to be the best method for studying neural pathways activated by peripheral and/or central administration of this peptide.

A potential alternative to *in situ* hybridization is to use the expression of products of early gene activation, such as c-fos antigen, to identify neurons activated by a specific stimulus (Sagar *et al.*, 1988). Recent studies have confirmed that activated neurons can be identified even after relatively brief exposure to acute stimuli such as CCK (Verbalis *et al.*, 1991). Future use of such techniques in combination with immunohistochemical staining and retrograde tracing should, therefore, allow determination of the neuronal pathways activated by CCK administration, as well as comparisons between pathways activated by CCK and by physiological stimuli such as gastric distention.

Summary

CCK is likely to be a significant modulator of multiple neuronal systems, including several systems involving neuroendocrine secretion. Studies to date have shown that peripheral administration of CCK in rats causes pituitary secretion of OT, ACTH, and prolactin, the first two via activation of vagally mediated pathways. In contrast, central CCK administration appears to stimulate both AVP and OT secretion (though with very different temporal characteristics than the vagally mediated OT secretion following peripheral CCK administration) with more variable effects on LH, ACTH, and prolactin secretion, depending upon the site injected, the dose of CCK administered, and possibly the gonadal steroid hormonal environment. The neuroendocrine secretion observed in response to peripheral CCK administration is consistent with activation of central stress pathways likely resulting from visceral discomfort, but there appears to be no obvious pattern to account for the spectrum of actions produced by central administration of CCK. Evidence has been presented implicating the CCK co-localized within magnocellular oxytocinergic cells in mediating enhanced pituitary OT secretion, and the CCK co-localized in vasopressinergic parvocellular neurons in potentiating the effects of AVP on pituitary ACTH secretion. Although many behavioral studies have suggested participation of CCK in grooming, reproductive, and ingestive behaviors, most of the evidence to date for such involvement is based upon responses to peripheral and central injections of CCK. No definite proof for involvement of endogenously secreted CCK in neurophysiological or behavioral functions exists at this time. However, the recent availability of potent and selective CCK receptor antagonists should aid in the elucidation of such neurophysiological and behavioral functions of CCK. Nonetheless, investigators attempting to interpret studies of potential CCK actions, including those involving receptor antagonists, must be aware of the caveat on excluding simultaneous effects on other neuronal systems in which CCK acts as a neurotransmitter or neuromodulator, and particularly central dopaminergic pathways involved with initiation of motivated behaviors. Although obstacles to a clear understanding of central functions of CCK on neuroendocrine secretion and other CNS functions are substantial, it is anticipated that studies during the next decade should provide important insights into the physiological importance of this peptide both in the brain and in the periphery.

REFERENCES

Akesson, T.R., P.W. Mantyh, C.R. Mantyh, D. Matt, and P.E. Micevych. 1987. Estroud cyclicity of ^{125}I-CCK octapeptide binding in the ventromedial hypothalamic nucleus: Evidence for down modulation by estrogen. *Neuroendocrinology* 45: 257-262.

Akesson, T.R. and P.E. Micevych. 1988. Evidence for an absence of estrogen-concentration by CCK-immunoreactive neurons in the hypothalamus of the female rat. *J. Neurobiol.* 19: 3-16.

Anhut, H., D.K. Meyer and W. Knepel. 1983. Cholecystokinin-like immunoreactivity of rat medial basal hypothalamus: Investigations on a possible hypophysiotropic function. *Neuroendocrinology* 36: 119-124.

Arletti, R., A. Benelli and A. Bertolini. 1989. Influence of oxytocin on feeding behavior in the rat. *Peptides* 10: 89-93.

Babcock, A.M., G.J. Bloch and P.E. Micevych. 1988a. Injections of cholecystokinin into the ventromedial hypothalamic nucleus inhibit lordosis behavior in the rat. *Physiol. Behav.* 43: 195-199.

Babcock, A.M., J.C. Barton, G.J. Bloch and P.E. Micevych. 1988b. Vagotomy blocks the peripheral effects of CCK-8 on lordosis. *Soc. Neurosci. Abstr.* 14: 290.

Beinfeld, M.C., D.K. Meyer and M.J. Brownstein. 1980. Cholecystokinin octapeptide in the rat hypothalamo-neurohypophysial system. *Nature* 288: 376-378.

Bloch, G.J., A.M. Babcock, R.A. Gorski and P.E. Micevych. 1987. Cholecystokinin stimulates and inhibits lordosis behavior in female rats. *Physiol. Behav.* 39: 217-224.

Bloch, G.J., W.A. Dornan, A.M. Babcock, R.A. Gorski and P.E. Micevych. 1988. Lordosis response to site-specific cholecystokinin in the male. *Soc. Neurosci. Abstr.* 14: 290.

Bondy, C.A., H. Gainer, R.T. Jensen and L.S. Brady. 1989. Cholecystokinin evokes secretion of oxytocin and vasopressin from rat neural lobe independent of external calcium. *Proc. Natl. Acad. Sci. U.S.A.* 86: 5198-5201.

Brownstein, M.J. and J.F. Rehfeld. 1985. Molecular forms of cholecystokinin in the nervous system. *Ann. N.Y. Acad. Sci.* 488: 9-10.

Bruhn, T.O., S.W. Sutton, P.M. Plotsky and W.W. Vale. 1986. Central administration of corticotropin-releasing factor modulates oxytocin secretion in the rat. *Endocrinology* 119: 1558-1563.

Caggiula, A.R., J.G. Herndon Jr., R. Scanlon, D. Greenstone, W. Bradshaw and D. Sharp. 1979a. Dissociation of active from immobility components of sexual behavior in female rats by central 6-hydroxydopamine: Implications for CA involvement in sexual behavior and sensorimotor responsiveness. *Brain Res.* 172: 505-520.

Caggiula, A.R., S.M. Antelman, L.A. Chiodo and C.G. Lineberry. 1979b. Brain dopamine and sexual behavior: Psychopharmacological and electrophysiological evidence for an antagonism between active and passive components. In *Catecholamines: Basic and Clinical Frontiers*, eds. E. Usdin, I. Kopin and J. Barchas, 1765-1767. Pergamon Press, New York.

Chang, R.S.L. and V.J. Lotti. 1986. Biochemical and pharmacological characterization of an extremely potent and selective nonpeptide cholecystokinin antagonist. *Proc. Natl. Acad. Sci. U.S.A.* 83: 4923-4926.

Crawley, J.N. 1985a. Comparative distribution of cholecystokinin and other neuropeptides. *Ann. N.Y. Acad. Sci.* 488: 1-8.

Crawley, J.N. 1985b. Cholecystokinin potentiation of dopamine-mediated behaviors in the nucleus accumbens. *Ann. N.Y. Acad. Sci.* 488: 283-292.

Day, N.C., M.D. Hall, R.G. Hill and J. Hughes. 1987. Modulation of hypothalamic cholecystokinin receptors with changes in magnocellular activity. *Soc. Neurosci. Abstr.* 13: 1524.

Della-Fera, M.A. and C.A. Baile. 1979. Cholecystokinin octapeptide: Continuous picomole injections into the cerebral ventricles of sheep suppress feeding. *Science* 206: 471-473.

Della-Fera, M.A., C.A. Baile, B.S. Schneider and J.A. Grinker. 1981. Cholecystokinin antibody injected in cerebral ventricles stimulates feeding in sheep. *Science* 212: 687-689.

Deschepper, C., F. Lotstra, F. Vandesande and J.J. Vanderhaeghen. 1983. Cholecystokinin varies in the posterior pituitary and external median eminence of the rat according to factors affecting vasopressin and oxytocin. *Life Sci.* 32: 2571-2577.

Deutsch, J.A. and W.T. Hardy. 1977. Cholecystokinin produces bait shyness in rats. *Nature* 266: 196.

Dockray, G.J. 1976. Immunochemical evidence of cholecystokinin-like peptides in brain. *Nature* 264: 568-570.

Dockray, G.J., H. Desmond, R.J. Gayton, A-C. Jonsson, H. Raybould, K.A. Sharkey, A. Varro and R.G. Williams. 1985. Cholecystokinin and gastrin forms in the nervous system. *Ann. N.Y. Acad. Sci.* 488: 32-43.

Dourish, C.T., M.L. Clark and S.D. Iversen. 1988. Analgesia induced by restraint stress is attenuated by CCK and enhanced by the CCK antagonists MK-329, L-365,031 and CR 1409. *Soc. Neurosci. Abstr.* 14: 290.

Ebenezer, I.S., S.N. Thornton and R.F. Parrott. 1989. Anterior and posterior pituitary hormone release induced in sheep by cholecystokinin. *Am. J. Physiol.* 256: R1355-R1357.

Everitt, B.J., K. Fuxe, T. Hökfelt and G. Jansson. 1975. Role of monoamines in the control by hormones of sexual receptivity in the female rat. *J. Comp. Physiol. Psychol.* 89: 556-572.

Fallon, J.H. and K.B. Seroogy. 1985. The distribution and some connections of cholecystokinin neurons in the rat brain. *Ann. N.Y. Acad. Sci.* 488: 121-132.

Flanagan, L.M., J.G. Verbalis and E.M. Stricker. 1989a. Effects of osmotic dehydration on food intake, gastric motility and oxytocin secretion in rats. *Appetite* 12: 209.

Flanagan, L.M., J.G. Verbalis and E.M. Stricker. 1989b. Effects of anorexigenic treatments on gastric motility in rats. *Am. J. Physiol.* 256: R955-R961.

Fuxe, K., L.F. Agnati, J-J. Vanderhaeghen, K. Tatemoto, K. Andersson, P. Eneroth, A. Härfstrand, G. Von Euler, R. Toni, M. Goldstein and V. Mutt. 1985. Cholecystokinin neuron systems and their interactions with the presynaptic features of the dopamine neuron systems. *Ann. N.Y. Acad. Sci.* 488: 231-254.

Garcia, J. and F.R. Ervin. 1968. Gustatory-visceral and telereceptor-cutaneous conditioning: Adaptation in external and internal milieus. *Commun. Behav. Biol.* 1: 389-415.

Gibbs, D.M. 1986. Vasopressin and oxytocin: Hypothalamic modulators of the stress response: A review. *Psychoneuroendocrinology* 11: 131-140.

Gibbs, J. and G.P. Smith. 1984. The neuroendocrinology of postprandial satiety. In *Frontiers in Neuroendocrinology*, Vol. 8, eds. L. Martini and W.F. Ganong, 223-245. Raven Press, New York.

Gibbs, J., R.C. Young and G.P. Smith. 1973. Cholecystokinin decreases food intake in rats. *J. Comp. Physiol. Psychol.* 84: 488-495.

Gorzalka, B.B. and G.L. Lester. 1987. Oxytocin-induced facilitation of lordosis behavior in rats is progesterone-dependent. *Neuropeptides* 10: 55-65.

Hashimoto, R. and F. Kimura. 1986. Inhibition of gonadotropin secretion induced by cholecystokinin implants in the medial preoptic area by the dopamine receptor blocker, pimozide, in the rat. *Neuroendocrinology* 42: 32-37.

Hewson, G., G.E. Leighton, R.G. Hill and J. Hughes. 1988. The cholecystokinin receptor antagonist L364,718 increases food intake in the rat by attenuation of the action of endogenous cholecystokinin. *Br. J. Pharmacol.* 93: 79-84.

Hökfelt, T., L. Skirboll, B. Everitt, B. Meister, M. Brownstein, T. Jacobs, A. Faden, S. Kuga, M. Goldstein, R. Markstein, G. Dockray and J. Rehfeld. 1985. Distribution of cholecystokinin-like immunoreactivity in the nervous system. *Ann. N.Y. Acad. Sci.* 488: 255-274.

Innis, R.B. and S.H. Snyder. 1980. Distinct cholecystokinin receptors in brain and pancreas. *Proc. Natl. Acad. Sci. U.S.A.* 77: 6917-6921.

Jarvis, C.R., C.W. Borque and L.P. Renaud. 1988. Cholecystokinin depolarizes rat supraoptic nucleus neurosecretory neurons. *Soc. Neurosci. Abstr.* 14: 145.

Kaltwasser, M.-T. and J.N. Crawley. 1987. Oxytocin and cholecystokinin induce grooming behavior in the ventral tegmentum of the rat. *Brain Res.* 426: 1-7.

Kamilaris, T.C., A.E. Calogero, R. Bernardini, E.O. Johnson, T. Geracioti, P.W. Gold and G.P. Chrousos. 1989. Effect of cholecystokinin octapeptide on pituitary-adrenal function *in vivo* in the rat. *Clin. Res.* 37: 452A.

Khosla, S. and J. Crawley. 1988. Potency of L-364,718 as an antagonist of the behavioral effects of peripherally administered cholecystokinin. *Life Sci.* 42: 153-160.

Kirchgessner, A.L. and A. Sclafani. 1988. PVN-hindbrain pathway involved in the hypothalamic hyperphagia-obesity syndrome. *Physiol. Behav.* 42: 517-528.

Kiss, J.Z., T.H. Williams and M. Palkovits. 1984. Distribution and projections of cholecystokinin-immunoreactive neurons in the hypothalamic paraventricular nucleus of rat. *J. Comp. Neurol.* 227: 173-181.

Kissileff, H.R., F.X. Pi-Sunyer, J. Thornton and G.P. Smith. 1981. Cholecystokinin-octapeptide (CCK-8) decreases food intake. *Am. J. Clin. Nutr.* 34: 154-60.

Lang, R.E., J.W.E. Heil, D. Ganten, K. Herman, T. Unger and W. Rascher. 1983. Oxytocin unlike vasopressin is a stress hormone in the rat. *Neuroendocrinology* 37: 314-316.

Leibowitz, S.F., N.J. Hammer and K. Chang. 1981. Hypothalamic paraventricular nucleus lesions produce overeating and obesity in the rat. *Physiol. Behav.* 27: 1031-1040.

Lenz, H.J., A. Raedler, H. Greten, W.W. Vale and J.E. Rivier. 1988. Stress-induced gastrointestinal secretory and motor responses in rats are mediated by endogenous corticotropin-releasing factor. *Gastroenterology* 95: 1510-1517.

Liddle, R.A., E.T. Morita, C.K. Conrad and J.A. Williams. 1986. Regulation of gastric emptying in humans by cholecystokinin. *J. Clin. Invest.* 77: 992-996.

Lorén, I., J. Alumets, R. Hakanson and F. Sundler. 1979. Distribution of gastrin and CCK-like peptides in rat brain. An immunocytochemical study. *Histochemistry* 59: 249-257.

Marley, P.D., S.L. Lightman, M.L. Forsling, K. Todd, M. Goedert, J.F. Rehfeld and P.C. Emson. 1984. Localization and actions of cholecystokinin in the rat pituitary neurointermediate lobe. *Endocrinology* 114: 1902-1911.

McCann, M.J., J.G. Verbalis and E.M. Stricker. 1988. Capsaicin pretreatment attenuates multiple responses to cholecystokinin in rats. *J. Auton. Nerv. Syst.* 23: 265-272.

McCann, M.J., J.G. Verbalis and E.M. Stricker. 1989. LiCl and CCK inhibit gastric emptying and feeding and stimulate OT secretion in rats. *Am. J. Physiol.* 256: R463-R468.

Mendelson, S.D. and B.B. Gorzalka. 1984. Cholecystokinin octapeptide produces inhibition of lordosis in the female rat. *Pharmacol. Biochem. Behav.* 21: 755-759.

Mezey, E., T.D. Reisine, L. Skirboll, M. Beinfeld and J.Z. Kiss. 1986. Role of cholecystokinin in corticotropin release: Coexistence with vasopressin and corticotropin-releasing factor in cells of the rat hypothalamic paraventricular nucleus. *Proc. Natl. Acad. Sci. U.S.A.* 83: 3510-3512.

Miaskiewicz, S.L., E.M. Stricker and J.G. Verbalis. 1989. Neurohypophyseal secretion in response to cholecystokinin but not meal-induced gastric distention in humans. *J. Clin. Endocrinol. Metab.* 68: 837-843.

Micevych, P.E., S.S. Park, T.R. Akesson and R. Elde. 1987. Distribution of cholecystokinin-immunoreactive cell bodies in the male and female rat: I. Hypothalamus. *J. Comp. Neurol.* 255: 124-136.

Micevych, P.E., A.M. Babcock, D.W. Mott, and J.K.H. Lui. 1986. Cholecystokinin inhibits K^+ stimulated LHRH release from male and female rat hypothalami in vitro. *Soc. Neurosci. Abstr.* 12: 1415.

Moran, T.H., P.H. Robinson, M.S. Goldrich and P.R. McHugh. 1986. Two brain cholecystokinin receptors: Implications for behavioral actions. Brain Res. 362: 175-179.

Morley, J.E. and A.S. Levine. 1982. Corticotropin releasing factor, grooming and ingestive behavior. Life Sci. 31: 1459-1464.

Neill, J.D. 1970. Effect of stress on serum prolactin and luteinizing hormone levels during the estrous cycle of the rat. Endocrinology 87: 1192-1197.

Noel, G.L., H.K. Suh, J.G. Stone and A.G. Frantz. 1972. Human prolactin and growth hormone release during surgery and other conditions of stress. *J. Clin. Endocrinol. Metab.* 35: 840-851.

Palkovits, M., J.Z. Kiss, M.C. Beinfeld and M.J. Brownstein. 1984. Cholecystokinin in the hypothalamo-hypophyseal system. *Brain Res.* 299: 186-189.

Parrott, R.F., S.N. Thornton and J.E. Robinson. 1988. Endocrine responses to acute stress in castrated rams: No increase in oxytocin but evidence for an inverse relationship between cortisol and vasopressin. *Acta Endocrinol.* 177: 381-386.

Priest, C.A., W.A. Dorman, G. Bloch and P.E. Micevych. 1988. Facilitation of lordosis by injection of CCK-8 into the medial preoptic area in the female rat. *Soc. Neurosci. Abstr.* 14: 290.

Raybould, H.E., R.J. Gayton and G.J. Dockray. 1985. CNS effects of circulating CCK8: Involvement of brainstem neurones responding to gastric distention. *Brain Res.* 342: 187-190.

Reidelberger, R.D. and M.F. O'Rourke. 1989. Potent cholecystokinin antagonist L 364718 stimulated food intake in rats. *Am. J. Physiol.* 257: R1512-R1518.

Reidelberger, R.D., T.J. Kalogeris and T.E. Solomon. 1989. Plasma CCK levels after food intake and infusion of CCK analogues that inhibit feeding in dogs. *Am. J. Physiol.* 256: R1148-R1154.

Renaud, L.P., M. Tang, M.J. McCann, E.M. Stricker and J.G. Verbalis. 1987. Cholecystokinin and gastric distension activate oxytocinergic cells in rat hypothalamus. Am. J. Physiol. 253: R661-R665.

Rogers, R.C. and G.E. Hermann. 1987. Oxytocin, oxytocin antagonist, TRH, and hypothalamic paraventricular nucleus stimulation effects on gastric motility. *Peptides* 8: 505-513.

Rowe, J.W., R.L. Shelton, J.H. Helderman, R.E. Vestal and G.L. Robertson. 1979. Influence of the emetic reflex on vasopressin release in man. *Kidney Int.* 16: 729-735.

Sagar, S.M., F.R. Sharp and T. Curran. 1988. Expression of c-fos protein in brain: Metabolic mapping at the cellular level. *Science* 240: 1328-1331.

Samson, W.K., R. Bianchi, R.J. Mogg, J. Rivier, W. Vale and P. Melin. 1989. Oxytocin mediates the hypothalamic action of vasoactive intestinal peptide to stimulate prolactin secretion. *Endocrinology* 124: 812-819.

Sander, L.D. and J.R. Porter. 1988. Influence of bombesin, CCK, secretion and CRF on corticosterone concentration in the rat. *Peptides* 9: 113-117.

Schneider, L.H., R.B. Murphy, J. Gibbs and G.P. Smith. 1988. Comparative potencies of CCK antagonists for the reversal of the satiating effect of cholecystokinin. In *Cholecystokinin Antagonists*, Vol. 47, *Neurology and Neurobiology*, eds. R.Y. Wang and R. Schoenfeld, 263-305. Alan R. Liss, New York.

Smith, G.P., C. Jerome and R. Norgren. 1985. Afferent axons in the abdominal vagus mediate satiety effect of cholecystokinin in rats. *Am. J. Physiol.* 249: R638-R641.

Stricker, E.M., M.J. McCann, L.M. Flanagan and J.G. Verbalis. 1988. Neurohypophyseal secretion and gastric function: Biological correlates of nausea. In *Biowarning System in the Brain*, eds. H. Takagi, Y. Oomura, M. Ito and M. Otsuka, 295-307. University of Tokyo Press, Tokyo.

Swanson, L.W. and P.E. Sawchenko. 1983. Hypothalamic integration: Organization of the paraventricular and supraoptic nuclei. *Annu. Rev. Neurosci.* 6: 269-324.

Vanderhaeghen, J.J., F. Lotstra, J. De Mey and C. Gilles. 1980. Immunohistochemical localization of cholecystokinin- and gastrin-like peptides in the brain and hypophysis of the rat. *Proc. Natl. Acad. Sci. U.S.A.* 77: 1190-1194.

Vanderhaeghen, J.J., F. Lotstra, F. Vandesande and K. Dierickx. 1981. Coexistence of cholecystokinin and oxytocin-neurophysin in some magnocellular hypothalamo-hypophyseal neurons. *Cell Tissue Res.* 221: 227-231.

Verbalis, J.G., M.J. McCann, C.M. McHale and E.M. Stricker. 1986a. Oxytocin secretion in response to cholecystokinin and food intake: Differentiation of nausea from satiety. *Science* 232: 1417-1419.

Verbalis, J.G., C.M. McHale, T.W. Gardiner and E.M. Stricker. 1986b. Oxytocin and vasopressin secretion in response to stimuli producing learned taste aversions in rats. *Behav. Neurosci.* 100: 466-475.

Verbalis, J.G., D.W. Richardson and E.M. Stricker. 1987. Vasopressin release in response to nausea-producing agents and cholecystokinin in monkeys. *Am. J. Physiol.* 252: R749-R753.

Verbalis, J.G., G.E. Hoffman, E.M. Stricker and A.G. Robinson. 1989. Cholecystokinin activates c-fos expression in hypothalamic oxytocin and corticotropin-releasing hormone neurons. *J. Neuroendocrinol.* In press.

Vijayan, E., W.K. Samson and S.M. McCann. 1979. *In vivo* and *in vitro* effects of cholecystokinin on gonadotropin, prolactin, growth hormone and thyrotropin release in the rat. *Brain Res.* 172: 295-302.

Watson, C.A., L.H. Schneider, E.S. Corp, S.C. Weatherford, R. Shindledecker, R.B. Murphy, G.P. Smith and J. Gibbs. 1988. The effects of chronic and acute treatment with the potent peripheral cholecystokinin antagonist L-364,718 on food and water intake in the rat. *Soc. Neurosci. Abstr.* 14: 1196.

Williams, J.A., D.J. McChesney, M.C. Calayag, V.R. Lingappa and C.D. Logsdon. 1988. Expression of receptors for cholecystokinin and other Ca^{2+}-mobilizing hormones in Xenopus oocytes. *Proc. Natl. Acad. Sci. U.S.A.* 85: 4939-4943.

MOTILIN AND GASTRIC INHIBITORY POLYPEPTIDE (GIP)

C.H.S. McIntosh
Regulatory Peptide Group
Medical Research Council of Canada
Department of Physiology
University of British Columbia
Vancouver, B.C., Canada

I.   Introduction

II.   Cellular Localization of Gastrointestinal Motilin and its Existence in Other Species

III.   Motilin in the Brain

IV.   Motilin Secretion

V.   Actions of Motilin

VI.   GIP: Isolation and Characterization

VII.   Cellular Localization of GIP and the Existence of GIP in Other Species

VIII.   GIP Secretion

IX.   GIP as a Releasing Hormone in the Gastrointestinal Tract and Pancreas

X.   Metabolic Effects of GIP

XI.   Other Actions of GIP

XII.   Effects of GIP on Reproductive Hormone Secretion

Introduction: Motilin, Isolation and Structure

Porcine Motilin

Studies by Brown *et al.* (1966, 1967) have shown that alkalinization of the duodenum increased motor activity in denervated and in transplanted gastric pouches; it was suggested that a hormone was responsible. The active compound was isolated from a side-fraction obtained during the purification of secretin from porcine intestinal extracts. Amino acid analyses indicated that it was unlike any known gastrointestinal peptide; it was named "motilin" because of its stimulatory action on gastric motor activity (Brown *et al.*, 1972; Brown and McIntosh, 1989). The pure peptide (porcine I) had 22 amino acids, a calculated molecular weight of 2700, and the sequence shown in figure 1 (Brown *et al.*, 1973). With a new preparation of porcine motilin (II), it was later shown that residue 14 was glutamine (Schubert and Brown, 1974) (Fig. 1). Evidence in support of the existence of both forms was obtained when a radioimmunoassay (RIA) was used to follow the purification of motilin: Two regions of immunoreactive motilin (IRM) were detected in fractions from CM-cellulose chromatography which could represent the amidated and deamidated peptides (Brown and Dryburgh, 1978). Both species are biologically active. Cloned

Porcine I

Phe-Val-Pro-Ile-Phe-Thr-Tyr-Gly-Glu-Leu-Gln-Arg-Met-Glu-Glu-Lys-Glu-Arg-Asn-Lys-Gly-Gln

1 2 3 4 5 6 7 8 9 10 11 12 13 14 15 16 17 18 19 20 21 22

Porcine II and Human

Phe-Val-Pro-Ile-Phe-Thr-Tyr-Gly-Glu-Leu-Gln-Arg-Met-Gln-Glu-Lys-Glu-Arg-Asn-Lys-Gly-Gln

1 2 3 4 5 6 7 8 9 10 11 12 13 14 15 16 17 18 19 20 21 22

Canine

Phe-Val-Pro-Ile-Phe-Thr-His-Ser-Glu-Leu-Gln-Lys-Ile-Arg-Glu-Lys-Glu-Arg-Asn-Lys-Gly-Gln

1 2 3 4 5 6 7 8 9 10 11 12 13 14 15 16 17 18 19 20 21 22

Fig. 1. Amino Acid Sequences of Porcine, Human and Canine Motilin.

cDNAs encoding the porcine motilin precursor have recently been isolated from an intestinal library using synthetic oligonucleotide probes (Bond *et al.*, 1988). The predicted amino acid sequence indicated that the precursor consists of 119 amino acids. This is made up of a 25-amino acid N-terminal signal peptide followed by motilin and a 70-residue C-terminal peptide. Heterogeneity of porcine IRM in tissue extracts and plasma has been demonstrated by RIA (Shin *et al.*, 1980) and larger forms may represent differentially processed products of the precursor.

Canine and Human Motilins

An antiserum raised against porcine motilin was used to follow the purification of canine motilin, from intestinal mucosal extracts, by RIA (Poitras *et al.*, 1983; Reeve *et al.*, 1985). Canine motilin differs from the porcine peptide(s) at positions 7, 8, 12, 13 and 14 (Fig. 1).

The presence of IRM in human gastrointestinal tissues has been demonstrated by immunocytochemistry (ICC) (Pearse *et al.*, 1974; Polak *et al.*, 1975a) and RIA (Bloom *et al.*, 1976, Mitznegg *et al.*, 1976). Although no full description of the purification of human motilin has yet been published, a sequence obtained by Reeve (quoted in Beinfeld and Bailey, 1985) showed identity between porcine and human motilins. Studies similar to those used to isolate porcine cDNAs encoding the motilin precursor have confirmed that human motilin has the same sequence as porcine motilin (Seino *et al.*, 1987). The precursor is identical in size to the porcine peptide at the N-terminus but the C-terminal peptide consists of 68-residues and differs considerably in amino acid sequence. As in the pig, both canine and human tissue extracts and plasma contain heterogeneous forms of IRM which, in the case of man, appear to be extended at the C-terminus (Christofides *et al.*, 1981).

Cellular Localization of Gastrointestinal Motilin and its Existence in Other Species

A number of studies have provided evidence for the existence of gastrointestinal motilin in other species (McIntosh and Brown, 1989, 1990). Based on immunological techniques using antisera raised against porcine motilin, they have resulted in both confusion and contradictions in the literature. This is due to problems inherent to these techniques. Even in man and pig, where the existence of motilin is certain, there is still controversy as to which cell type of the gastrointestinal tract produces the peptide. In early studies a major population of enterochromaffin (EC) cells and a minor (15%) population of argyrophil and nonargentaffin cells were found to contain IRM (Pearse *et al.*, 1974; Polak *et al.*, 1975a). The majority of motilin-containing endocrine cells were in the duodenum and upper jejunum, with lesser numbers in the ileum. From differential staining studies it was concluded that motilin was localized to EC_2 cells, distinct from substance P-containing EC_1 cells (Heitz *et al.*, 1978; Polak *et al.*, 1976). With other antisera the majority of IRM was found in a nonargentaffin sub-group of D_1 cells, termed M-cells (Forssmann *et al.*, 1976; Usellini *et al.*, 1984a). These cells are characterized by relatively small granules (180 nm in man; 200 nm in dog) and they may correspond to the minor population of cells reported by Pearse *et al.* (1974). The disparate results obtained with different antisera have not been explained but there is evidence for the existence of more than one motilin or motilin-like molecule. Further, the ability to detect IRM in some species depends on the regional specificity of the antiserum used.

Species for which there is evidence for gastrointestinal IRM include: goldfish, frog, lizard, chick, quail, platypus, rat, hamster, rabbit, opossum, cow, baboon and tupaia (McIntosh and Brown, 1990a). In view of the problems mentioned above, it is possible that all that is seen as motilin may not be motilin; one can only take the immunological data as possible evidence until confirmation of the identity has been obtained. An example of such uncertainty is the difficulty many investigators have had in demonstrating the presence of motilin in the rat gastrointestinal tract. Although O'Donahue *et al.* (1981) were clearly able to demonstrate the presence of IRM in rat intestinal extracts with some antisera, several polyclonal antisera and monoclonal antibodies failed to detect motilin cells in the rat small intestine by ICC (Smith *et al.*, 1981; Vogel, 1987), and RIA of tissue extracts has often shown an extremely low content of IRM (Shin *et al.*, 1980; Yanaihara *et al.*, 1978; Vogel, 1987) which was only detectable with C-terminal-specific antibodies. Therefore, if motilin exists in the rat intestine it may differ considerably from the porcine peptide.

Motilin in the Brain

There have been a number of studies on the presence of motilin in the brain. Either RIA or ICC has demonstrated IRM in the brain of such diverse species as the colorado beetle, Atlantic hagfish, scorpion, quail, hamster, rat, guinea pig, opossum, pig and cow (Reviewed in McIntosh *et al.*, 1990a). An extensive RIA study by O'Donahue *et al.* (1981) showed that IRM was widely distributed in the rat brain. The highest content (pg/region) in grossly dissected areas of brain tissue was found in the cerebral cortex and cerebellum, with a moderate content in the midbrain and forebrain subcortical areas. The lowest amounts were in the pons, medulla and spinal cord. The pituitary and pineal gland also contained significant amounts of IRM, in agreement with earlier studies (Yanaihara *et al.*, 1978). In microdissected regions of the brain, IRM content varied over a 7 to 8-fold range, with the organum vasculosum of the lamina terminalis (OVLT) having the highest content and the caudate, putamen and cingulate cortex the least. Relatively high levels of IRM were also found in the medial preoptic and arcuate nuclei and in the basal hypothalamus. A number of subsequent studies have extended and confirmed these findings, although differences in the absolute and relative amounts of IRM have been found, depending on the antiserum used (Beinfeld and Korchak, 1985). Immunocytochemical studies have, to a large extent, supported the RIA data. Cell processes were mainly observed within the hypothalamus, preoptic area and mammillary body (Jacobowitz *et al.*, 1981). There was a rich innervation of varicose fibers in the median eminence (ME) and OVLT region (Jacobowitz *et al.*, 1981; Nilaver *et al.*, 1982). Following colchicine treatment, cell bodies were revealed in the hypothalamus. Jacobowitz *et al.* (1981) pointed out that the high concentration of fibres in the ME resembled the distribution of LHRH fibres. In the forebrain IRM was detected in clusters of cells in the pars distalis in the rat and mouse and also in the pars intermedia of the rat (Nilaver *et al.*, 1982). In adult rat pituitary about 80% of total motilin content was concentrated in the anterior lobe, and in rat, guinea pig and man it was found in many, but not all, somatotrophs (Loftus *et al.*, 1986) where it was often co-localized with growth hormone. An extensive innervation occurred in the cerebellum with most Purkinje cells containing IRM, and some neurons demonstrated co-existence of IRM with glutamic acid decarboxylase, the synthesizing enzyme for GABA (Chan-Palay, 1982; Chan-Palay *et al.*, 1981). Korchak *et al.* (1984), in a developmental study of the cerebellum and pituitary, showed that IRM is present before the onset of synaptogenesis and suggested that it may be involved in development, possibly acting as a trophic factor.

Chromatographic studies of brain IRM have shown significant differences from porcine intestinal motilin. Gel filtration studies showed the presence of both high and low molecular size rat brain IRM with the larger form predominating (O'Donahue *et al.*, 1981). The smaller form could not be distinguished from porcine motilin on gel permeation HPLC but demonstrated a different retention time on reverse phase HPLC. Both high and low molecular size forms have also been detected in extracts of monkey, pig, cow and dog brains (Beinfeld and Korchak, 1985; Poitras *et al.*, 1987b). Interestingly,

Korchak *et al.* (1985) found that on subcellular fractionation of brain the majority of the IRM was present in the nuclear fraction and not synaptosomes. The authors interpreted this as possibly being due to motilin binding to nuclear proteins. This would support a role in synaptogenesis. An alternative explanation, however, is that the antisera detect part of a nuclear protein. Such a possibility cannot be ruled out in view of the ambiguities mentioned and the fact that other groups have been unable to detect IRM in rat brain using specific monoclonal antibodies (Vincent and Brown, unpublished). The elution characteristics on reverse phase HPLC of low molecular size porcine, bovine, guinea pig and rhesus monkey brain IRMs (Beinfeld and Bailey, 1985; Beinfeld and Korchak, 1985) have also been shown to be different from porcine intestinal motilin. In addition, Bond *et al.* (1988), using Northern blot analysis, were unable to detect pre-promotilin mRNA in several brain regions tested, further substantiating important differences between the peptides. In relation to this, it is of interest that there is homology within the motilin 9-17 region and the 142-150 region of rabbit skeletal tropomyosin (Beinfeld and Korchak, 1985).

Motilin Secretion

The factors regulating gastrointestinal motilin secretion are still unclear and a number of stimulatory and inhibitory pathways may be involved (Fig. 2). A complete description of these various factors is beyond the scope of this review and the reader is referred to previous reviews for a more comprehensive outline and the relevant references (McIntosh and Brown, 1988, 1990b).
Basal circulating levels of IRM in infants are higher than in adults (Christofides *et al.*, 1978), but they gradually fall until the age of 20. While there is a postnatal surge in basal IRM to levels higher than those in adults (Lucas *et al.*, 1982), there is no evidence for sex-related differences. The basal levels of motilin exhibit a cyclic pattern which is associated with the inter-digestive migrating motor complex (MMC) (Itoh *et al.*, 1978a,b; Itoh, 1981; Peters *et al.*, 1980). The relationship between these two phenomena will be discussed later.

In response to feeding a mixed meal, either an increase in IRM (Mitznegg *et al.*, 1976; Jenssen *et al.*, 1984) or a sustained drop (Christofides *et al.*, 1978; Itoh *et al.*, 1978a,b) in IRM levels may occur depending on the dietary composition of the meal. Although there is some controversy in the literature over the responses to individual nutrients (McIntosh and Brown, 1988, 1990b), the majority of investigators have found a decrease in response to glucose, amino acids and water but an increase in response to fat or protein (Mitznegg *et al.*, 1976; Christofides

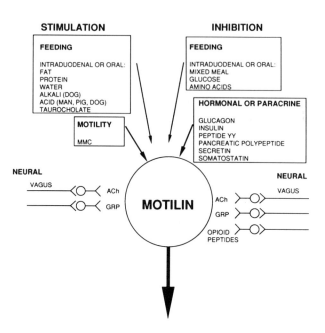

Fig. 2. Proposed Regulatory Pathways for Motilin Secretion. There is evidence for all of the pathways outlined, but in many cases there are also contradictory studies. The roles of the various hormones, paracrines and non-adrenergic non-cholinergic transmitters which can influence motilin secretion are also not clear.

et al., 1978; Itoh *et al.*, 1978a; Jenssen *et al.*, 1984). It has not been determined whether the nutrients themselves directly affect IRM secretion or whether responses are secondary to changes in intraduodenal pH, secreted bile, or via release of hormones, paracrine substances or neurotransmitters. Both alkalinization and acidification of the duodenum have been reported to stimulate IRM secretion in different species (Dryburgh and Brown, 1975; Mitznegg *et al.*, 1976; Jenssen *et al.*, 1984). A number of hormones and paracrines have also been shown to modify nutrient-induced IRM secretion (Fig. 2) (McIntosh and Brown, 1988, 1990b), the most completely studied being inhibition of IRM secretion by somatostatin (Jenssen *et al.*, 1986a). This peptide may act locally to modulate motilin secretion.

The autonomic nervous system is also probably involved in the regulation of IRM secretion but the pathways are complicated and not completely elucidated. In the dog, studies using nerve stimulation and various pharmacological manipulations have indicated that there is a vagal cholinergic stimulatory pathway (Lee *et al.*, 1981). Other studies in man and dog, including responses to insulin-induced hypoglycemia, sham feeding and vagal cooling at the cervical level (Hall *et al.*, 1983; Lemoyne *et al.*, 1984), have provided evidence for a vagal inhibitory pathway involving cholinergic and non-adrenergic non-cholinergic (NANC) nerves. The neuropeptides which have been implicated in the regulation of motilin secretion are summarized in figure 2.

Actions of Motilin

Effects on Motility

Motilin has been shown to exert effects on motility of the esophagus, stomach, small intestine and gallbladder (McIntosh and Brown, 1988, 1990b). In particular, there is considerable evidence for a physiological role in the regulation of contractions during the interdigestive period: the interdigestive MMC (Itoh, 1981; Itoh *et al.*, 1976, 1978a,b, 1982; Itoh and Sekiguchi, 1983; McIntosh and Brown, 1990b).

Spontaneous contractions of the lower esophageal sphincter (LES) during the interdigestive state represent the most orad component of the MMC. Motilin was shown to induce premature contractions of the LES during the fasted state, but not in the digestive state, and peak circulating IRM levels coincided with increased LES motor activity in the dog (Itoh *et al.*, 1978b, 1982). The relationship between the release of IRM and changes in LES pressure in man was not so convincing as in the dog.

Motilin was shown to be a powerful stimulator of gastric motor activity and tachyphylaxis to the effect of motilin occurred upon continued infusion (Dryburgh and Brown, 1975; McIntosh and Brown, 1988). Effects of motilin on gastric emptying depend on the type of meal, but have generally been found to be stimulatory (McIntosh and Brown, 1990b). Since the major actions of motilin appear to be mediated during the interdigestive period, these effects may be pharmacological rather than physiological. Infusion of porcine motilin in fasted dogs evoked increased myoelectrical activity and strong contractions, similar to naturally occurring phase III contractions of the MMC, in both the stomach and duodenum. Motilin-induced activity fronts in the stomach were propagated along the small intestine. A major debate has been whether motilin initiates MMCs or MMCs result in IRM secretion. The arguments for and against motilin being the physiological initiator of the MMC have been reviewed (McIntosh and Brown, 1989, 1990) and cannot be dealt with in depth here, but in the author's opinion the majority of evidence favors a role for motilin in the initiation of the MMC in the proximal gut.

A number of studies have demonstrated that motilin can act both directly on smooth muscle and via interaction with intrinsic gastrointestinal nerves. In addition, there is species specificity in smooth muscle responses to motilin with rat tissues, in particular, being refractory to porcine motilin (McIntosh and Brown, 1989, 1990). The canine duodenum contracted in response to both porcine and canine motilins *in vivo* (Itoh *et al.*, 1990) but only to canine motilin *in vitro* (Poitras *et al.*, 1987a).

Effects on Exocrine Secretion

Motilin also has a number of actions on exocrine secretion (McIntosh and Brown, 1989, 1990). Duodenal phase III activity is preceded by an increase in hydrochloric acid, pepsinogen, and bile acid output and followed by secretion of bicarbonate and amylase. Motilin administration stimulated basal pepsin secretion, and peaks of endogenous IRM coincided with maximal pepsin levels in the interdigestive state. Basal acid secretion was also stimulated by motilin, but stimulated levels of acid and pepsin were inhibited.

In dogs, basal secretion of bicarbonate and fluid secretion were weakly stimulated by motilin whereas secretin-stimulated fluid and bicarbonate secretion were inhibited. In the interdigestive state the major effect of motilin infusion was an advanced onset of a normal peak of secretion. Peaks of pancreatic exocrine and IRM secretion were inhibited by atropine, and anti-motilin serum blocked the pancreatic secretory cycles. These studies, therefore, support a role for motilin in the regulation of peak cyclic pancreatic secretion. Evidence has also been presented for motilin involvement in the regulation of biliary flow into the duodenum in the interdigestive state, whereas other studies have suggested that bile secretion regulates the secretion of motilin.

Central Nervous System Effects

The broad distribution of IRM in the brain suggests that it may be involved in neural function. Among the effects of motilin on the central nervous system are: increased food intake in rats (Rosenfeld and Grathwaite, 1987), excitation of neurons in the rat cerebellum and toad spinal cord (Phillis and Kirkpatrick, 1979), inhibition of neurons in the lateral vestibular nucleus (Chan-Palay *et al.*, 1982), and either stimulation or inhibition of different populations of neurons in the rat lower brain stem (Okada and Miura, 1985).

Motilin and Reproductive Hormones

There is limited information suggesting that motilin can influence hypothalamic and/or pituitary function, although the radioimmunological data suggesting that it is present in both regions would support such a role. No effect of motilin was found on LH or prolactin secretion from pituitary cells *in vitro* or *in vivo* following intravenous administration (Samson *et al.*, 1984); however, administration of motilin systemically, or its addition to pituitary cells *in vitro* resulted in a stimulation of growth hormone secretion, whereas following intracerebroventricular administration it suppressed release (McCann *et al.*, 1984; Samson *et al.*, 1984). It has been proposed that motilin released into the portal blood may act as a physiological releasing factor of growth hormone. Two proposals were made to explain the inhibitory effect of intracerebroventricular motilin: Motilin acting via a short loop feedback pathway inhibits its own secretion, or motilin releases somatostatin (Samson *et al.*, 1984). Passive immunization with motilin antiserum lowered basal growth hormone levels (Samson *et al.*, 1984), suggesting that circulating motilin could regulate growth hormone secretion. However, it is unlikely that the source of the motilin is the pituitary since circulating IRM levels were found to be normal in hypophysectomized human patients (Peeters *et al.*, 1988). These studies favor a role for motilin in the regulation of growth hormone secretion rather than the pituitary reproductive hormones.

It is possible that either there is interaction between motilin and sex hormone changes during pregnancy or that motilin plays a role during delivery. Basal IRM levels were found to be lower in the mother during late pregnancy than those postpartum, and meal-stimulated release was also reduced (Jenssen *et al.*, 1988). Jenssen *et al.* (1986b) also observed a transient rise in IRM during delivery. Although this may represent a compensatory mechanism leading to increased gastrointestinal motility following birth (Jenssen *et al.*, 1986b) motilin may have an alternative role, for example, in regulation of uterine motility. The effect of motilin on uterine smooth muscle has not, to my knowledge, been investigated. Breast feeding was not found to influence IRM levels in the mother, so it is unlikely to play a role in this reflex (Holst *et al.*, 1986).

At present, there is no evidence for motilin being involved in the regulation of anterior pituitary reproductive hormone secretion, although it is possible that the correct experimental protocols have not as yet been established. In particular, since rat motilin may differ considerably from the porcine peptide, the biological activities may also be species specific. As discussed later for GIP, access of motilin to the circumventricular organs is a possible route by which peripheral motilin could influence the brain. There have been no studies reported on the effect of motilin on steroid hormone synthesis or secretion by the testes or ovaries.

GIP: Isolation and Characterization

Porcine GIP

The isolation of GIP resulted from studies which demonstrated that crude preparations of cholecystokinin (CCK) contained gastric acid secretion inhibitor (Brown and Pederson, 1970). Purification was achieved from a side fraction produced during the isolation of CCK (Brown *et al.*, 1969, 1970). Determination of the amino acid sequence was performed by the dansyl Edman technique on fragments obtained by cleavage with cyanogen bromide, trypsin, chymotrypsin and thermolysin (Brown 1971; Brown and Dryburgh, 1971). A corrected sequence was later reported by Jörnvall *et al.* (1981). Porcine GIP is a straight-chain 42-amino acid polypeptide with a molecular weight of 4976 and the sequence shown in figure 3. A minor component, GIP 3-42, exists in GIP preparations obtained by conventional chromatography. GIP belongs to the glucagon family of peptides and shares regional homology with glucagon, glicentin, glucagon-like peptide (GLP) I, GLP II, secretin, vasoactive intestinal peptide (VIP), peptide histidine isoleucine (PHI), and growth hormone releasing hormone. Heterogeneity of GIP has been demonstrated in porcine intestinal extracts (Brown *et al.*, 1989). The size of the major higher molecular weight form is approximately 8 kD, but it has not as yet been characterized.

Human and Bovine GIP

Human GIP was isolated by Moody *et al.* (1984) from postmortem intestinal tissue. Sequence analysis showed that the arginine at position 18 in porcine GIP was substituted by histidine and serine at position 34 by asparagine (Fig. 3). This sequence has been confirmed from cDNAs encoding the human GIP precursor (Takeda *et al.*, 1987). The predicted amino acid sequence showed that the GIP precursor consists of 153 amino acids. The GIP sequence is flanked by 51- and 60-amino acid peptides at the N- and C-termini, respectively. The N-terminus consists of a 21-amino acid signal peptide and a propeptide of 30 amino acids.

STRUCTURES OF GASTRIC INHIBITORY POLYPEPTIDE

PORCINE

Tyr-Ala-Glu-Gly-Thr-Phe-Ile-Ser-Asp-Tyr-Ser-Ile-Ala-Met-Asp-Lys-Ile-Arg-Gln-Gln-Asp-Phe-Val-Asn-Trp-Leu-Leu-Ala-Gln-Lys-Gly-Lys-

1 2 3 4 5 6 7 8 9 10 11 12 13 14 15 16 17 18 19 20 21 22 23 24 25 26 27 28 29 30 31 32

Lys-Ser-Asp-Trp-Lys-His-Asn-Ile-Thr-Gln

33 34 35 36 37 38 39 40 41 42

HUMAN

Tyr-Ala-Glu-Gly-Thr-Phe-Ile-Ser-Asp-Tyr-Ser-Ile-Ala-Met-Asp-Lys-Ile-**His**-Gln-Gln-Asp-Phe-Val-Asn-Trp-Leu-Leu-Ala-Gln-Lys-Gly-Lys-

Lys-**Asn**-Asp-Trp-Lys-His-Asn-Ile-Thr-Gln

BOVINE

Tyr-Ala-Glu-Gly-Thr-Phe-Ile-Ser-Asp-Tyr-Ser-Ile-Ala-Met-Asp-Lys-Ile-Arg-Gln-Gln-Asp-Phe-Val-Asn-Trp-Leu-Leu-Ala-Gln-Lys-Gly-Lys-

Lys-Ser-Asp-Trp-**Ile**-His-Asn-Ile-Thr-Gln

Fig. 3. Amino Acid Sequences of Porcine and Human GIP.

The isolation of bovine GIP from intestine, by ion-exchange and reverse phase HPLC, was achieved by Carlquist *et al.* (1984) using a radioreceptor assay to follow the purification. This peptide differs from porcine GIP at only one position: lysine at position 37 in porcine GIP being substituted by an isoleucine (Fig. 3).

Cellular Localization of GIP and the Existence of GIP in Other Species

In man, dog, rat, and baboon, GIP-containing cells have been demonstrated by ICC (using antisera or monoclonal antibodies) to be confined to epithelial cells in the mucosa of the small intestine and not the large intestine (Polak *et al.*, 1973). This corresponds with measurements of extracted material measured by RIA (Bryant and Bloom, 1979). GIP has not been found in either peripheral or central neurons in mammals although immunoreactive GIP (IR-GIP) has been detected by ICC in the brain of a species of lepidoptera (El-Salhy *et al.*, 1983).

Using a monoclonal antibody, the GIP cells have been ultrastructurally characterized in man as being the K-cells (Buchan *et al.*, 1982) which are characterized by round, fairly osmiophilic granules with a mean size of 188 ± 34 nm. In the dog, the GIP-containing cells differed from those in man since they resembled cells classified as I-cells in this species (Usellini *et al.*, 1984b). Among the species for which there is evidence from ICC for IR-GIP in the gastrointestinal tract are teleost fish (El-Salhy, 1984; Elbal *et al.*, 1988) and the toad (El-Salhy *et al.*, 1981).

The localization of GIP in the endocrine pancreas is controversial, yet it is important since if it were present it would imply an intra-islet function. Using conventional rabbit antisera, IR-GIP was initially reported to be present in the glucagon-containing A-cells (Alumets *et al.*, 1978; Smith *et al.*, 1977). However it was later demonstrated that staining with the antisera used could be blocked with glicentin (Larsson and Moody, 1980) or glucagon (Buchan *et al.*, 1982). Other studies indicated that there was a further GIP-like peptide in the pancreas (Smith, 1983; Smith *et al.*, 1977; Sjölund *et al.*, 1983) of the rat, dog and man. GIP immunoreactive cells have also been reported to be present in the lizard pancreas (El-Salhy and Grimelius, 1981), but no evidence for such material was obtained in a survey of several reptile species (Buchan *et al.*, 1982). At the time of the earlier studies on pancreatic GIP, the structures of the mammalian proglucagon-derived peptides were unknown. In view of the sequence homology with GLP-1 and GLP-2 it is likely that the controversy over GIP's existence in the pancreas is due to cross-reactivity with these peptides.

GIP Secretion

Numerous studies have been performed on the secretion of IR-GIP (Fig. 4). Glucose ingestion induced a rapid increase in circulating IR-GIP levels (Cataland *et al.*, 1974; Pederson *et al.*, 1975a; Morgan *et al.*, 1979); galactose and sucrose also stimulated release but fructose was without effect (Morgan *et al.*, 1979). It has been suggested that secretion of GIP is dependent upon active transport of monosaccharides (Sykes *et al.*, 1980), and there is strong support for such a mechanism: Sugars which are not transported (mannose) or not metabolized (2-deoxyglucose) do not stimulate GIP secretion, and when glucose transport is blocked by phloridzin the release of IR-GIP is abolished (Creutzfeldt *et al.*, 1983; Ebert and Creutzfeldt, 1980).

Oral administration of fat is also a potent stimulant for GIP release. The responses were shown to be slower than with glucose, and circulating levels can remain elevated for periods of over 4 h (Cleator and Gourlay, 1975; Falko *et al.*,

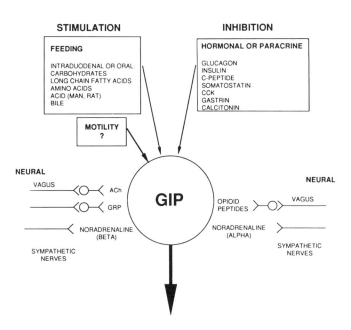

Fig. 4. Proposed Regulatory Pathways for GIP Secretion. The responses to nutrients are well-characterized, but the involvement of the hormonal, paracrine and neural pathways remains to be clarified.

1975; Pederson *et al.*, 1975a). These differences are probably due to the slower gastric emptying and absorption of fat. Such elevations of GIP are not accompanied by increased insulin release under euglycemic conditions but are in agreement with a possible role of GIP as an inhibitor of acid secretion (enterogastrone). Metabolites of triacylglycerol digestion, the free fatty acids, and not triacylglycerol itself, are the stimuli for release (O'Dorisio *et al.*, 1976; Ross and Shaffer, 1981; Ohneda *et al.*, 1984). Intraduodenal long-chain fatty acids increased GIP secretion whereas medium chain triglycerides produced only small increases or had no effect on IR-GIP secretion (O'Dorisio *et al.*, 1976; Ross and Shaffer, 1981). The ability of long-chain fatty acids to stimulate insulin release may be associated with chylomicron formation since these fatty acids are carried by chylomicrons, and treatment of rats with Pluronic L-81, which blocks chylomicron formation, inhibited GIP responses to long-chain fatty acids (Ebert and Creutzfeldt, 1987).

Protein ingestion had no effect on GIP secretion (Cleator and Gourlay, 1975), but intraduodenal administration of a mixture of amino acids (arginine, histidine, isoleucine, lysine and tyrosine) produced an increase in both GIP and insulin secretion (Thomas *et al.*, 1978). Much smaller responses were obtained with a mixture containing methionine, phenylalanine, tryptophan and valine.

Intraduodenal administration of acid in the rat and in man also stimulated release of GIP (Ebert *et al.*, 1979b), but there was no such response in the dog (Brown *et al.*, 1975). The amount of acid required to produce relatively small increases in IR-GIP in man were very high and it therefore seems unlikely that it is a physiological stimulus. Intraduodenal bile also stimulated release (Burhol *et al.*, 1980; Flaten *et al.*, 1981).

Although nutrients act directly on the GIP cell to induce secretion, there is evidence that hormones, paracrine substances and neurotransmitters can influence GIP release. Insulin (Brown *et al.*, 1975; Sirinek *et al.*, 1978), C-peptide (Dryburgh *et al.*, 1980), glucagon (Ebert *et al.*, 1977), somatostatin (Pederson *et al.*, 1975b; Jorde *et al.*, 1981a), calcitonin (Stevenson *et al.*, 1985), gastrin and CCK (Sirinek *et al.*, 1977) have all been shown to inhibit GIP secretion. Although it is not certain whether all of these are physiological effects, in the case of insulin and C-peptide there is probably a negative feedback loop between the B-cell and GIP secretion which protects against hypoglycemia following a fatty meal. The evidence for this has come from studies showing that fat-induced GIP secretion is attenuated by intravenous infusion of glucose (Falko *et al.*, 1975; Andersen *et al.*, 1978; Ebert *et al.*, 1979a), insulin (Brown *et al.*, 1975; Sirinek *et al.*, 1978; Talaulicar *et al.*, 1981) and C-peptide (Dryburgh *et al.*, 1980), whereas glucose-induced secretion was not affected by endogenous (Andersen *et al.*, 1978) or exogenous insulin (Talaulicar *et al.*, 1981).

Possible neural pathways involved in the regulation of GIP secretion are unclear and there is controversy as to whether or not they play a role. Some studies have suggested that the GIP cell is under vagal control, since glucose-induced secretion was attenuated by atropine (Larrimer *et al.*, 1978), vagotomy resulted in increased basal and glucose-stimulated levels of IR-GIP (Yoshiya *et al.*, 1985), and insulin-induced hypoglycemia increased GIP secretion in an atropine-sensitive manner (Jorde *et al.*, 1981b). However, neither sham feeding nor direct vagal stimulation affected GIP secretion in the rat (Berthoud *et al.*, 1982) and sham feeding had no effect in man (Taylor and Feldman, 1982). Bombesin stimulated (McDonald *et al.*, 1981) and an opioid peptide inhibited (Sullivan *et al.*, 1983) secretion. These are potential NANC transmitters although evidence for such a pathway is lacking.

A role has also been proposed for the sympathetic system with an α-adrenergic-inhibitory (Salera *et al.*, 1982) and a β-adrenergic-stimulatory (Flaten *et al.*, 1982) component, but physiological increases in circulating catecholamines had no effect on GIP secretion (Graffner *et al.*, 1987). Nelson *et al.* (1986) found no effect of cholinergic or adrenergic blockade on glucose-stimulated GIP secretion in man.

GIP as a Releasing Hormone in the Gastrointestinal Tract and Pancreas

Effects of GIP on Gastric Secretion

The acid inhibitory properties of porcine GIP, and the fact that it was released in response to fat, suggested that it could be a physiological enterogastrone. The model used in the first studies on its acid inhibitory activity was the vagally and sympathetically denervated pouch of the body of the dog stomach, and in this model GIP was an extremely potent inhibitor of pentagastrin-stimulated acid secretion (Pederson and Brown, 1972). Pepsin secretion was also inhibited. These studies and a number of subsequent investigations, both in man and dog, showed that the innervated stomach was relatively refractory to GIP (Soon-Shiong *et al.*, 1979; Yamagishi and Debas, 1980). Nevertheless, the demonstration by Wolfe *et al.* (1983), that following administration of anti-GIP antibodies to dogs both gastric acid output and integrated gastrin release were increased, suggested that meal-released GIP did inhibit acid secretion physiologically. A possible explanation for the lack of effect of GIP on the innervated stomach was proposed by McIntosh *et al.* (1979), namely that GIP was acting as a releasing hormone for an intragastric mediator of its actions. The observation of reduced effectiveness in the innervated stomach and the blockade of GIP's action by cholinomimetics (Soon-Shiong *et al.*, 1979, 1984) could then be explained by an antagonistic parasympathetic input on the release of this intermediate. A major candidate was somatostatin since there is a high gastric somatostatin content and somatostatin was shown to be a potent inhibitor of acid and gastrin secretion (McIntosh, 1985). An isolated perfused stomach preparation has been used to study the suggestion that GIP releases gastric somatostatin (McIntosh *et al.*, 1979, 1981). Infusion of GIP in this preparation increased secretion of somatostatin-like immunoreactivity (SLI) and the threshold concentration of GIP was within the physiological range found following a mixed meal (Brown, 1982; McIntosh *et al.*, 1981). The secretion of GIP-stimulated SLI was shown to be inhibited by electrical stimulation of the vagus nerves (McIntosh *et al.*, 1979, 1981, 1983, 1985) or vascular perfusion of parasympathomimetics (McIntosh *et al.*, 1979, 1981, 1983, 1985). The parasympathetic inhibition of SLI release is probably mediated via both cholinergic and NANC nerves. The NANC transmitter(s) has not as yet been determined with certainty, but evidence is accumulating that opioid peptides and substance P are potential regulators (Kwok *et al.*, 1985; McIntosh *et al.*, 1983, 1985).

It has been proposed that one possible way by which GIP could be acting as a physiological enterogastrone is that fat in the small intestine activates a sympathetic nervous component at the same time as releasing GIP. Inhibition of parasympathetic activity and direct stimulation of D-cell secretion by catecholamines may then allow GIP to also stimulate somatostatin release. An alternative, or additional, possibility is that the inhibitory effect of GIP will be most profound when the parasympathetic activity on the stomach is at a minimum. Since GIP is elevated throughout the day, in man, it is possible that GIP inhibition is more important in the interdigestive period. It is also possible that GIP is not the only enterogastrone and that it normally acts in concert with other peptides such as secretin. There is also evidence that GIP influences gastrin secretion (Villar *et al.*, 1976; Wolfe *et al.*, 1983; McIntosh et al., 1984; reviewed in Brown *et al.*, 1989).

Effects of GIP on Insulin and Glucagon Secretion

The major physiological role for GIP is probably as a component of the enteroinsular axis. This incretory function was first demonstrated by Dupré *et al.* (1973) who infused GIP into normal human volunteers together with intravenous glucose and compared the response to glucose infusion alone. GIP improved glucose tolerance and increased insulin to levels greater than those observed with glucose. Later studies demonstrated it was also insulinotropic in the rat (Pederson and Brown, 1976) and dog (Pederson *et al.*, 1975a). In the perfused rat pancreas (Pederson and Brown, 1976) and in man, using the *in vivo* glucose clamp (Elahi *et al.*, 1979), it was shown that the insulinotropic action of GIP was glucose dependent and that below a certain glucose threshold GIP was not insulinotropic. The threshold in the perfused rat pancreas was 100 mg/dl similar to that in man where it is approximately 25 mg/dl above basal levels. The maximal potentiating effect of GIP occurred at a glucose concentration of around 290 mg/dl. This concentration of glucose produced a maximal insulin release in the absence of GIP, but in its presence the levels of insulin released were several-fold higher. When GIP was administered to the perfused pancreas as a square wave pulse, the insulin response was biphasic in nature, whereas in the presence of a glucose gradient the response was monophasic. It has been suggested that the transient decrease in insulin secretion during the square wave stimulus was due to somatostatin secretion (Brown *et al.*, 1980), but this proposal has not as yet been substantiated. *In vitro* studies have shown that GIP is also insulinotropic in fresh or monolayer cultures of rat pancreatic islets (Schauder *et al.*, 1975; Fujimoto *et al.*, 1978; Siegel and Creutzfeldt, 1985). The presence of high-affinity receptors for GIP on membranes from hamster B-cell tumors (Amiranoff *et al.*, 1986; Maletti *et al.*, 1987) and the demonstration of GIP-stimulated release of insulin from rat insulinoma cells (Swanson-Flatt and Flatt, 1988) or isolated B-cells (Verchere and Brown, unpublished) have shown that GIP acts directly on the B-cell. Second messengers, etc. could be thought of as intermediates.

The effect of GIP on insulin secretion can be modulated by various substances. CCK (Zawalich, 1988) and cholinomimetics (McCullough *et al.*, 1985) act synergistically with GIP while somatostatin (Pederson et al., 1975), galanin (Schuerer *et al.*, 1987; Miralles *et al.*, 1988) and calcitonin gene-related peptide (CGRP) (Ishizuka *et al.*, 1988) inhibit GIP-stimulated insulin secretion.

Studies have also been performed on the effect of GIP on glucagon secretion. In the isolated perfused rat pancreas, GIP stimulated glucagon secretion in a glucose-dependent manner and the response was completely suppressed at a glucose concentration of 10 mM (Pederson and Brown, 1978). In addition, arginine stimulation of glucagon secretion was potentiated by GIP and this response was also reversed by elevating the glucose level. GIP did not stimulate glucagon secretion in the dog (Sirinek *et al.*, 1980).

Metabolic Effects of GIP

Since GIP is released by long-chain fatty acids, the question has been raised as to whether GIP also has an effect on fat metabolism. Although there are a number of contradictory reports, there is evidence that GIP may play such a role. GIP was shown to promote the clearance of chylomicrons in dogs (Wasada *et al.*, 1981) and stimulate lipoprotein lipase activity in cultured preadipocytes (Eckel *et al.*, 1979). From the latter study it was concluded that GIP may be physiologically important in the clearance of chylomicron triglyceride after feeding. GIP also inhibited glucagon-stimulated lipolysis (Dupré *et al.*, 1976) and enhanced insulin-stimulated fatty acid incorporation into adipose tissue (Beck and Max, 1983).

There is also evidence that GIP affects hepatic glucose production. GIP inhibited insulin-dependent, glucagon-stimulated glycogenolysis in the perfused liver (Hartmann *et al.*, 1986), and this is consistent with studies of Andersson *et al.* (1980) in conscious dogs.

Other Actions of GIP

Both *in vivo* and *in vitro* studies have demonstrated a number of other effects of GIP (reviewed in Brown *et al.*, 1989), although the physiological significance of some of these is unclear since in these studies the concentrations used were above those normally found following a meal--or the blood levels of GIP were not measured, making it difficult to assess whether the circulating levels achieved were in the physiological range.

Among these actions are a reduction in lower esophageal sphincter pressure, inhibition of pentagastrin- and acetylcholine-stimulated gastric motor activity and relaxation of the stomach, inhibition of intestinal electrical and motor activity, decreased intraluminal pressure of the gallbladder, increased superior mesenteric blood flow, and effects on salivary and intestinal secretion.

Effects of GIP on Reproductive Hormone Secretion

At present there is no data indicating that GIP has a direct effect on gonadal steroid metabolism and only limited information suggesting that it can influence reproductive hormone secretion from the pituitary. Since there is no evidence for the existence of GIP in the mammalian brain, the only way that it could have a function in the regulation of pituitary secretion is via the bloodstream. There are a number of possible pathways by which this could occur: a direct action on the anterior pituitary, a direct effect on the secretion of releasing hormones and release inhibiting hormones, or via neurons regulating the latter. In addition, there is now evidence that insulin can be considered to be a reproductive hormone, and it is possible that GIP is involved in this function.

Effects of GIP on the Release of Anterior Pituitary Hormones

Ottlecz *et al.* (1985) showed that intraventricular injection of GIP in rats produced a decrease in plasma FSH. No change in LH occurred. Following administration of estradiol benzoate, FSH and LH levels were decreased, and GIP had no further effect. With *in vitro* studies, high GIP concentrations (10^{-7}-10^{-6} M) increased both FSH and LH release from overnight-cultured, dispersed anterior pituitary cells in a concentration-dependent manner. GIP was less potent than LHRH and the effect of the two peptides was additive. The inhibitory action observed on FSH secretion *in vivo* was clearly not due to a direct action on the pituitary, since the opposite effect was observed with *in vitro* pituitary cells. It seems likely, therefore, that GIP has a selective inhibitory action on FSH release via interaction with neuronal systems in the brain and a direct effect on the pituitary to release both FSH and LH. Intraventricular injection of GIP also stimulated growth hormone secretion which again suggests that GIP can influence central neurons involved in the regulation of anterior pituitary secretion. In contrast to the responses following intraventricular administration, Murphy *et al.* (1983) showed that GIP inhibited growth hormone secretion in sodium pentobarbital-anesthetized rats when administered subcutaneously. Whether this was mediated by an action at the pituitary or indirectly was not clear.

If GIP can interact with neurons involved in the regulation of pituitary secretion, what are the possible pathways? There is increasing evidence that although peripherally derived peptides cannot cross the blood-brain barrier a number of them can influence brain function via interaction with the circumventricular organs (CVOs) (Van Houten and Posner, 1981; Phillips, 1988). These organs, including the subfornical organ (SFO), the OVLT and the ME, Since GIP can induce peptide release in the periphery, it could have a similar effect on releasing and release-inhibiting hormones in the ME. The effect on FSH secretion could be explained on the basis of an inhibitory action on release of the hypothetical FSH-releasing factor or by influencing inhibin's actions/release. The effects on growth hormone secretion may also be explained in several ways. If there is a central counterpart of GIP's potent stimulatory action on somatostatin secretion, one would expect an inhibition of growth hormone secretion, as observed by Murphy *et al.* (1983). However, the data of Ottlecz *et al.* (1985) suggest that following ICV administration GIP activates stimulatory pathways, possibly via release of GHRH; if GIP acts as a partial agonist at the pituitary, competitively antagonizing the full agonist activity of GHRH at the level of the pituitary, inhibition of growth hormone secretion could occur following peripheral administration (Murphy *et al.*, 1983). Alternatively, a similar stimulation could result from an action of GIP to inhibit somatostatin secretion. The possible pathways involved in the action of GIP on the anterior pituitary are summarized in figure 5.

Possible Involvement of GIP in the Reproductive Functions of Insulin

Insulin can be considered to be a "reproductive hormone" since direct effects of insulin on the gonads have been demonstrated and gonadal function is often altered in situations in which there are pathological changes in insulin levels (Poretsky and Kalin, 1987). It is also possible, although not yet examined, that GIP stimulation of insulin secretion is important for the normal reproductive

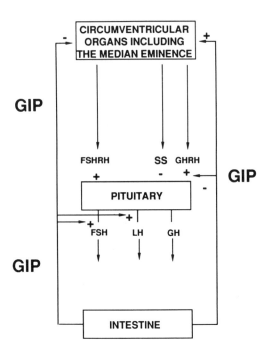

Fig. 5. Pathways by which GIP may influence FSH, LH and growth hormone secretion. See text for details.

hormone actions of insulin and that pathological changes in GIP may be involved in gonadal changes associated with hyper- or hypoinsulinemia.

Insulin is found both in the pancreas and the brain (Havrankova *et al.*, 1978; Bernstein *et al.*, 1984), but the major reproductive hormone-related actions of insulin are probably exerted peripherally. Both insulin and insulin-like growth factors are thought to play a role in the regulation of ovarian function. Receptors for both peptides have been found in ovarian tissues of animals and man, and these are present in stromal and follicular tissue and isolated granulosa cells (Adashi *et al.*, 1985; Poretsky and Kalin, 1987). There have been numerous studies demonstrating an effect of insulin on steroidogenesis, and secretion of progesterone, estrogen, androstenedione, and testosterone have all been reported to be increased (Poretsky and Kalin, 1987). There appear to be a number of pathways which are influenced by insulin. These include effects on uptake and metabolism of lipoproteins (Veldhuis *et al.*, 1986) and on levels of cytochrome P_{450} side-chain cleavage enzyme (Veldhuis *et al.*, 1983), 3β-hydroxysteroid dehydrogenase (Blanco *et al.*, 1981) and aromatase (Garzo and Dorrington, 1984).

Other actions include modulation of LH and FSH receptor number, synergistic activity with FSH and/or LH, and enhancement of cell viability (Poretsky and Kalin, 1987). It is not clear whether insulin's action on ovarian tissue is mediated via interaction with insulin or insulin-like growth factor receptors, or both (Adashi *et al.*, 1985).

Direct evidence of a stimulatory effect of insulin on *in vitro* biosynthesis of androgens by ovarian tissue has also been demonstrated in tissue from patients with hyperandrogenism (Barbieri *et al.*, 1986). Both testosterone and Δ^4-androstenedione, but not estradiol or progesterone, were increased by insulin in thecal cultures. In stromal cultures, testosterone, Δ^4-androstenedione, and dihydrotestosterone synthesis were all stimulated. Interaction between insulin and LH or FSH on steroid synthesis also occurred, but the responses obtained were complex.

Other effects of insulin on reproductive hormones, or their associated carrier proteins include inhibition of inhibin production by seminiferous tubules (Gonzales *et al.*, 1989) and of sex-hormone binding globulin production (Plymate *et al.*, 1988).

GIP is probably important for the regulation of circulating insulin at the levels required for normal gonadal function. Could it also be involved in the events which result in the pathological changes in insulin found in different disease states associated with changes in gonadal function? A number of studies have shown a relationship between hyperinsulinemia and hyperandrogenism. This association could be a result of hyperinsulinemia causing hyperandrogenism, hyperandrogenism causing hyperinsulinemia (or both) resulting from some independent factors (Barbieri and Hornstein, 1988). This question has not been resolved, but for this review the concept that hyperinsulinemia can result in hyperandrogenism and the possible involvement of GIP will be considered.

Hyperinsulinemia has three major origins: peripheral insulin resistance, increased pancreatic sensitivity to secretagogues and decreased insulin clearance. A number of clinical states are commonly associated with insulin resistance; these include obesity, Kahn type A and type B diabetes, lipoatrophic diabetes and leprichaunism (Barbieri and Hornstein, 1988). In addition, many women with ovarian hyperandrogenism are insulin resistant and obesity alone does not account for this resistance since weight-matched nonhyperandrogenic controls are less insulin resistant. There is also evidence that hyperandrogenic women demonstrate altered pancreatic sensitivity to secretagogues: Chang and Geffner (1985) found an increased sensitivity of the pancreas to glucose stimulation of insulin secretion.

The evidence supporting the hypothesis that hyperinsulinemia may result in hyperandrogenism has been summarized by Barbieri and Hornstein (1988). This evidence includes: 1) the close association between severe hyperinsulinemic states and hyperandrogenism; 2) a positive correlation between insulin and androgen levels in women with hyperandrogenism; 3) glucose administration to hyperinsulinemic hyperandrogenic women increases insulin and androgen levels; 4) insulin administration to women with hyperandrogenism increases circulating androstenedione; 5) weight loss, fasting or a hypocaloric diet results in decreased androgen levels in women; 6) *in vitro* insulin stimulates human ovarian stromal androgen production.

If hyperinsulinemia is a cause of hyperandrogenism, which clinical syndromes could involve GIP? Two major categories of ovarian hyperandrogenic diseases have been proposed by Barbieri and Hornstein (1988). The first of these have hyperandrogenism and insulin resistance (HA-IR group). These patients have elevated fasting and greatly exaggerated insulin responses to a glucose load, normal LH and prolactin, and often have stromal hyperthecosis rather than polycystic ovaries. The second category have hyperandrogenism but are not insulin resistant, and are characterized by polycystic ovaries, slightly elevated fasting insulin and nearly normal insulin responses to a glucose load, and elevated LH and prolactin levels. The origin of the marked hyperinsulinemia in the HA-IR group is not clear; however, Barbieri and Hornstein (1988) reported that in response to an oral glucose load HA-IR patients had large increases in insulin and increases in testosterone, androstenedione, and dihydrotestosterone. Such a response could result from an overactive enteroinsular axis, with an increased secretion of GIP producing an excessive insulin secretion. Such a possibility has not as yet been studied, but one clinical situation for which evidence exists for an exaggerated GIP response is in obesity, where insulin resistance is often present. Some studies have demonstrated an elevated basal IR-GIP level and an exaggerated GIP and insulin response to ingestion of a high-caloric liquid test meal or a large oral fat load (Creutzfeldt *et al.*, 1978; Elahi *et al.*, 1984). Two factors may play a role: a more rapid gastric emptying in obese patients (Creutzfeldt, 1987) and insensitivity of GIP-releasing cells to feedback inhibition by insulin. In obese subjects with glucose intolerance, i.v. glucose failed to lower the exaggerated secretion of GIP in response to oral fat, whereas in normals there was a suppression (Ebert *et al.*, 1979a). The exaggerated rise of GIP was reversed by reducing the body weight, and the lowering effect of i.v. glucose was restored. Steroid levels were not examined in any of these studies so it is not known whether there was any relationship between androgen levels, insulin and

GIP. It is possible, as summarized in figure 6, that prolonged and excessive nutrient ingestion leads to both hyperGIPemia and hyperinsulinemia. The associated insulin resistance contributes to the hyperinsulinemia. The GIP cell may also become insulin resistant, thus explaining the lack of feedback regulation. Stimulation of the ovaries by insulin results in hyperandrogenism, as proposed by Poretsky and Karlin (1987).

The enteroinsular axis has also been found to be hyperactive in animal models of obesity. In some of these there are increased numbers of GIP-containing cells and altered sensitivity of the B-cell to GIP. The obese hyperglycemic mouse (ob/ob) and diabetic-obese mouse (db/db) both exhibit hyperphagia, hyperglycemia and hyperinsulinemia. The ob/ob strain exhibits hyperplasia of GIP cells in the intestinal mucosa (Polak *et al.*, 1975b; Bailey *et al.*, 1986) and intestinal content of GIP, and elevated plasma IR-GIP levels have been demonstrated in both types of animals (Flatt *et al.*, 1983). Further studies on ob/ob mice have shown that increasing the dietary fat stimulates the production and secretion of GIP and enhances the K cell density (Bailey *et al.*,

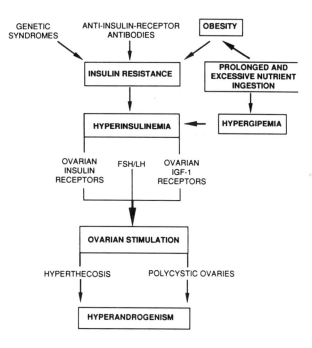

Fig. 6. Possible Involvement of GIP in the Pathogenesis of Hyperandrogenism in Insulin-resistant States (modified from Poretsky and Kalin, 1987). Chronically elevated GIP levels may result from excessive and prolonged nutrient ingestion. The resulting hyperinsulinemia produces ovarian hyperstimulation.

1986). This was not due to hyperalimentation and it has been proposed that the hyperGIPemia might result in hyperglucagonemia, resulting in hyperglycemia and increased lipogenesis leading to obesity. It is also possible that these animals exhibit insulin resistance and a deficient feedback from insulin (Flatt *et al.*, 1983). In another animal model, the obese Zucker rat, the fat (fa/fa) rat is hyperphagic, hyperinsulinemic, hyperlipidemic and has increased adiposity. The obese rats respond to GIP under low conditions of glycemia, unlike their lean litter mates (Chan *et al.*, 1984). The normal glucose threshold is, therefore, missing in these animals. Whether this is due to a change in receptors or post-receptor events is unclear. There do not appear to have been any systematic studies examining whether there is an association between the elevated GIP levels, hyperinsulinemia and gonad function in such animals. However, recently it was shown that the obese (fa/fa) rat had reduced LH surges during estrus, and the maximal concentrations of FSH and prolactin declined more slowly (Whitaker and Robinson, 1989). Progesterone concentrations were higher for most of the estrous cycle and the concentration of hypothalamic estrogen receptors was reduced. Whether hyperinsulinemia and hyperGIPemia are involved is not known, but such changes may well contribute to the infertility of obese female Zucker rats.

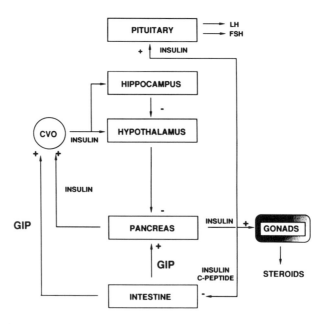

Fig. 7.   Possible Involvement of GIP in Insulin Feedback to the Brain and Insulin's Action on Gonadal Steroidogenesis (modified from Phillips, 1988). See text for details.

One final, and equally speculative, role for GIP is in the regulation of brain insulin secretion. As previously mentioned, peptides may influence brain function via interaction with the CVOs. Phillips (1988) has suggested that insulinacts in a feedback loop involving release of brain insulin. The evidence for such a role for insulin has come from several studies. Van Houten *et al.* (1979) showed that the CVOs specifically sequester peptides such as insulin from plasma, and glial cells may limit access of these peptides to other parts of the brain. Insulin (Havrankova *et al.*, 1978; Bernstein *et al.*, 1984) and insulin mRNA (Young, 1986) are present in the brain, and release of immunoreactive insulin from neurons in culture has been demonstrated. Insulin receptors are present in various regions of the brain including the CVOs (Van Houten *et al.*, 1979), and insulin has been shown to inhibit spontaneous firing of hippocampal pyramidal cells in isolated brain slices (Phillips, 1988). In the model proposed by Phillips (1988), feedback of insulin from the pancreas to the CVOs influences hippocampal and hypothalamic function which, in turn, inhibits insulin secretion by the pancreas, possibly via sympathetic pathways. There is at present no evidence for a role for GIP in such feedback pathways. However, if GIP receptors are present in the CVOs, then stimulation of central insulin and feedback to the pancreas would reinforce the effect of insulin itself (Fig. 7). This could have several functions: It could guard against hypoglycemia, alter the levels of insulin reaching the gonads, and reduce the stimulatory effect of insulin on LH and FSH secretion from the pituitary (Adashi *et al.*, 1981).

REFERENCES

Adashi, E.Y., A.J.W. Hsueh and D. Yen. 1981. Insulin enhancement of luteinizing hormone and follicle stimulating hormone release by cultured pituitary cells. *Endocrinology* 108: 1441.

Adashi, E.Y., C.E. Resnick, J. D'Ercole, M.E. Svoboda and J.J. Van Wyck. 1985. Insulin-like growth factors as intraovarian regulators of granulosa cell growth and function. *Endocrine Reviews* 6: 400-420.

Alumets, J., R. Håkanson, T. O'Dorisio, K. Sjölund and F. Sundler. 1978. Is GIP a glucagon cell constituent? *Histochem.* 58: 253-257.

Amiranoff, B., A. Couvineau, N. Vauclin-Jacques and M. Laburthe. 1986. Gastric inhibitory polypeptide receptor in hamster pancreatic β cells. Direct cross-linking, solubilization and characterization as a glycoprotein. *Eur. J. Biochem.* 159: 353-358.

Andersen, D.K., D. Elahi, J.C. Brown, J.D. Tobin and R. Andres. 1978. Oral glucose augmentation of insulin secretion. Interactions of gastric inhibitory polypeptide with ambient glucose and insulin levels. *J. Clin. Invest.* 62: 152-161.

Andersen, D.K., W.S. Putnam, J.B. Hanks, J.E. Wise, H.E. Lebovitz and R.S. Jones. 1980. Gastric inhibitory polypeptide (GIP) suppression of hepatic glucose production. *Reg. Peptides* 1 (Suppl.2): 4A.

Bailey, C.J., P.R. Flatt, P. Kwasowski, C.J. Powell and V. Marks. 1986. Immunoreactive gastric inhibitory polypeptide and K cell hyperplasia in obese hyperglycaemic (ob/ob) mice fed high fat and high carbohydrate cafeteria diets. *Acta-Endocrinol. (Copenh).* 112: 224-229.

Barbieri, R.L. and H.J. Hornstein. 1988. Hyperinsulinemia and ovarian hyperandrogenism. Cause and effect. Endocrinol. Metab. Clin. (N. America). 17: 685-703.

Barbieri, R.L., A. Makris, R.W. Randall, G. Daniels, R.W. Kistner and K.J. Ryan. 1986. Insulin stimulates androgen accumulation in incubations of ovarian stroma obtained from women with hyperandrogenism. *J. Clin. Invest.* 46: 904-910.

Beck, B. and J.P. Max. 1983. Gastric inhibitory polypeptide enhancement of the insulin effect on fatty acid incorporation into adipose tissue in the rat. *Reg. Peptides* 7: 3-8.

Beinfeld, M.C. and G.J. Bailey. 1985. The distribution of motilin-like peptides in rhesus monkey brain as determined by radioimmunoassay. *Neurosci. Lett.* 54: 345-350.

Beinfeld, M.C. and D.M. Korchak. 1985. The regional distribution and the chemical, chromatographic, and immunologic characterization of motilin brain peptides: The evidence for a difference between brain and intestinal motilin-immunoreactive peptides. *J. Neurosci.* 5: 2502-2509.

Bernstein, H.G., A. Dorn, M. Reiser and M. Ziegler. 1984. Cerebral insulin-like immunoreactivity in rats and mice. *Acta Histochem.* 74: 33-36.

Berthoud, H.R., E.R. Trimble and A.J. Moody. 1982. Lack of gastric inhibitory polypeptide (GIP) response to vagal stimulation in the rat. *Peptides (Fayetteville)* 3: 907-912.

Blanco, F.L., L.F. Fanjul and C.M. Ruiz de Galareta. 1981. The effect of insulin and luteinizing hormone treatment on serum concentrations of testosterone and dihydrotestosterone and testicular 3β-hydroxysteroid dehydrogenase activity in intact and hypophysectomized diabetic rats. *Endocrinology* 109: 1248-1253.

Bloom, S.R., P. Mitznegg and M.G. Bryant. 1976. Measurement of human plasma motilin. *Scand. J. Gastroenterol.* 11 (Suppl.39): 47-52.

Bond, C.T., G. Nilaver, B. Godfrey, E.A. Zimmerman and J.P. Adelman. 1988. Characterization of complementary deoxyribonucleic acid for precursor of porcine motilin. *Mol. Endocrinology* 2: 175-180.

Brown, J.C. 1971. A gastric inhibitory polypeptide. I. The amino acid composition and the tryptic peptides. *Can. J. Biochem.* 49: 255-261.

Brown, J.C. 1982. Gastric Inhibitory Polypeptide (Monogr.) *Endocrinology* 24: 1-88.

Brown, J.C. and J.R. Dryburgh. 1971. A gastric inhibitory polypeptide. II. The complete amino acid sequence. *Can. J. Biochem.* 49: 867-872.

Brown, J.C. and J.R. Dryburgh. 1978. Isolation of motilin. In *Gut Hormones*, 1st Edition, ed. S.R. Bloom, 327-331. Churchill Livingstone, Edinburgh, London, New York.

Brown, J.C. and C.H.S. McIntosh. 1990. The discovery of motilin. In *Motilin*, ed. Z. Itoh, 5-12. Academic Press, New York.

Brown, J.C. and R.A. Pederson. 1970. A multiparameter study on the action of preparations containing cholecystokinin-pancreozymin. *Scand. J. Gastroenterol.* 5: 537-541.

Brown, J.C., L.P. Johnson and D.F. Magee. 1966. Effect of duodenal alkalinization on gastric motility. *Gastroenterol.* 50: 333-339.

Brown, J.C., L.P. Johnson and D.F. Magee. 1967. The inhibition of fundic pouch activity in transplanted fundic pouches. *J. Physiol.* 188: 45-52.

Brown, J.C., R.A. Pederson, J.E. Jorpes and V. Mutt. 1969. Preparation of a highly active enterogastrone. *Can. J. Physiol. Pharmacol.* 47: 113-114.

Brown, J.C., V. Mutt and R.A. Pederson. 1970. Further purification of a polypeptide demonstrating enterogastrone activity. *J. Physiol.* 209: 57-64.

Brown, J.C., M.A. Cook and J.R. Dryburgh. 1972. Motilin, a gastric motor activity-stimulating polypeptide: Final purification, amino acid composition and C-terminal residues. *Gastroenterol.* 62: 401-404.

Brown, J.C., M.A. Cook and J.R. Dryburgh. 1973. Motilin, a gastric motor activity stimulating polypeptide: The complete amino acid sequence. *Can. J. Biochem.* 51: 533-537.

Brown, J.C., J.R. Dryburgh, S.A. Ross and J. Dupré. 1975. Identification and actions of gastric inhibitory polypeptide. *Recent. Prog. Horm. Res.* 31: 487-532.

Brown, J.C., H. Koop, C.H.S. McIntosh, S.C. Otte and R.A. Pederson. 1980. Physiology of gastric inhibitory polypeptide. *Front. Horm. Res.* (Karger Basel) 7: 132-144.

Brown, J.C., A.M.J. Buchan, C.H.S. McIntosh and R.A. Pederson. 1989. Gastric Inhibitory Polypeptide. In *Handbook of Physiology*, Section 6, Vol. II. ed. G.M. Makhlouf, 403-430. Bethesda.

Bryant, M.G. and S.R. Bloom. 1979. Distribution of the gut hormones in the primate intestinal tract. *Gut* 20: 653-659.

Buchan, A.M.J., J. Ingman-Baker, J. Levy and J.C. Brown. 1982. A comparison of the ability of serum and monoclonal antibodies to gastric inhibitory polypeptide to detect immunoreactive cells in the gastroenteropancreatic system of mammals and reptiles. *Histochem.* 76: 341-349.

Burhol, P.G., R. Jorde and H.L. Waldum. 1980. Radioimmunoassay of plasma gastric inhibitory polypeptide (GIP), release of GIP after a test meal and duodenal infusion of bile, and immunoreactive plasma GIP components in man. *Digestion* 20: 336-345.

Carlquist, M., M. Maletti, H. Jürnvall and V. Mutt. 1984. A novel form of gastric inhibitory polypeptide (GIP) isolated from bovine intestine using a radioreceptor assay. Fragmentation with staphylococcal protease results in GIP1-3 and GIP4-42, fragmentation with enterokinase in GIP1-16 and GIP17-42. *Eur. J. Biochem.* 145 (3): 573-577.

Cataland, S., S.E. Crockett, J.C. Brown and E.L. Mazzaferri. 1974. Gastric inhibitory polypeptide (GIP) stimulation by oral glucose in man. J. Clin. Endocrinol. Metab. 39: 223-228.

Chan, C.B., R.A. Pederson, A.M.J. Buchan, K.B. Tubesing and J.C. Brown. 1984. Gastric inhibitory polypeptide (GIP) and insulin release in the obese Zucker rat. Diabetes 33: 536-542.

Chan-Palay, V. 1982. Coexistence of traditional neurotransmitters with peptides in the mammalian brain: 5-hydroxytryptamine and substance P in the raphé and γ-aminobutyric acid and motilin in the cerebellum. In *Co-transmission*, ed. E.C. Cuello, 1-24. Macmillan, London.

Chan-Palay, V., G. Nilaver, S.L. Palay, M.C. Beinfeld, E.A. Zimmerman, J.Y. Wu and T.L. O'Donohue. 1981. Chemical heterogeneity in cerebellar Purkinje cells: Existence and coexistence of glutamic acid decarboxylase-like and motilin-like immunoreactivities. Proc. Natl. Acad. Sci. U.S.A. 78: 7787-7791.

Chan-Palay, V., M. Ito, P. Tongroach, M. Sakurai and S. Palay. 1982. Inhibitory effects of motilin, somatostatin, [Leu]enkephalin, [Met]enkephalin, and taurine on neurons of the lateral vestibular nucleus: Interactions with γ-aminobutyric acid. *Proc. Natl. Acad. Sci. U.S.A. 9: 3355-3359.*

Chang, R.J. and M.E. Geffner. 1985. Associated non-ovarian problems of polycystic ovarian disease: Insulin resistance. *Clin. Obstet. Gynecol.* 12: 675-685.

Christofides, N.D., S.R. Bloom and H.S. Besterman. 1978. Physiology of motilin II. In *Gut Hormones,* 1st Edition, ed. S.R. Bloom, 343-350. Churchill Livingstone, Edinburgh, London, New York.

Christofides, N.D., T.M.G. Bryan, M.A. Ghatei, S. Kishimoto, A.M.J. Buchan, J.M. Polak and S.R. Bloom. 1981. Molecular forms of motilin in the mammalian and human gut and human plasma. *Gastroenterol.* 80: 292-300.

Cleator, I.G. and R.H. Gourlay. 1975. Release of immunoreactive gastric inhibitory polypeptide (IR-GIP) by oral ingestion of food substances. *Am. J. Surg.* 130: 128-135.

Creutzfeldt, W. 1987. The changing concept of the enteroinsular axis. In: *Frontiers of Hormone Research* 16, Gut Regulatory Peptides: Their Role in Health and Disease, ed. E. Blazques, 1-14. Karger, Basel, Switzerland.

Creutzfeldt, W., R. Ebert, B. Willms, H. Frerichs and J.C. Brown. 1978. Gastric inhibitory polypeptide (GIP) and insulin in obesity: Increased response to stimulation and defective feedback control of serum levels. *Diabetologia* 14: 15-24.

Creutzfeldt, W., R. Ebert, M. Nauk and F. Stockman. 1983. Disturbances of the enteroinsular axis. *Scand. J. Gastroenterol.* 18 (Suppl.83): 111-119.

Dryburgh, J.R. and J.C. Brown. 1975. Radioimmunoassay for motilin. *Gastroenterol.* 68: 1169-1176.

Dryburgh, J.R., S.M. Hampton and V. Marks. 1980. Endocrine pancreatic control of the release of gastric inhibitory polypeptide. A possible physiological role for C-peptide. *Diabetologia* 19: 397-401.

Dupré, J., S.A. Ross, D. Watson and J.C. Brown. 1973. Stimulation of insulin secretion by gastric inhibitory polypeptide in man. *J. Clin. Endocrinol. Metab.* 37: 826-828.

Dupré, J., N. Greenidge, T.J. McDonald, S.A. Ross and D. Rubinstein. 1976. Inhibition of actions of glucagon in adipocytes by gastric inhibitory polypeptide. *Metabolism* 25: 1197-1199.

Ebert, R. and W. Creutzfeldt. 1980. Decreased GIP secretion through impairment of absorption. *Front. Horm. Res.* (Karger Basel) 7: 192-201.

Ebert, R. and W. Creutzfeldt. 1987. Metabolic effects of gastric inhibitory polypeptide. *Front. Horm. Res.* (Karger Basel) 16: 175-185.

Ebert, R., R. Arnold and W. Creutzfeldt. 1977. Lowering of fasting and food stimulated serum immunoreactive gastric inhibitory polypeptide (GIP) by glucagon. *Gut* 18: 121-127.

Ebert, R., H. Frerichs and W. Creutzfeldt. 1979a. Impaired feedback control of fat induced gastric inhibitory polypeptide (GIP) secretion by insulin in obesity and glucose intolerance. *Eur. J. Clin. Invest.* 9: 129-135.

Ebert, R., K. Illmer and W. Creutzfeldt. 1979b. Release of gastric inhibitory polypeptide (GIP) by intraduodenal acidification in rats and humans and abolishment of the incretin effect of acid by GIP-antiserum in rats. *Gastroenterol.* 76: 515-523.

Eckel, R.H., W.Y. Fujimoto and J.D. Brunzell. 1979. Gastric inhibitory polypeptide enhanced lipoprotein lipase activity in cultured preadipocytes. *Diabetes* 28: 1141-1142.

Elahi, D., D.K. Andersen, J.C. Brown, H.T. Debas, R.-J. Hershcopf, G.S. Raizes, J.D. Tobin and R. Andres. 1979. Pancreatic α- and β-cell responses to GIP infusion in normal man. *Am. J. Physiol.* 237: E185-E191.

Elahi, D., D.K. Andersen, D.C. Muller, J.D. Tobin, J.C. Brown and R. Andres. 1984. The enteric enhancement of glucose-stimulated insulin release. The role of GIP in aging, obesity, and non-insulin-dependent diabetes mellitus. *Diabetes* 33: 950-957.

Elbal, M.T., M.T. Lozano and B. Agulleiro. 1988. The endocrine cells in the gut of Mugil saliens Risso, 181 (Teleostei): An immunocytochemical and ultrastructural study. *Gen. Comp. Endocrinol.* 70: 231-246.

El-Salhy, M. 1984. Immunocytochemical investigation of the gastro-entero-pancreatic (GEP) neurohormonal peptides in the pancreas and gastrointestinal tract of the dogfish Squalus acanthias. *Histochem.* 80: 193-205.

El-Salhy, M. and L. Grimelius. 1981. Immunohistochemical localization of gastrin C-terminus, gastric inhibitory peptide (GIP) and endorphin in the pancreas of lizards with special reference to the hibernation period. *Reg. Peptides* 2: 97-111.

El-Salhy, M., L. Grimelius, E. Wilander, G. Abu-Sinna and G. Lundqvist. 1981. Histological and immunohistochemical studies of the endocrine cells of the gastrointestinal mucosa of the toad (Bufo regularis). *Histochemistry* 71: 53-65.

El-Salhy, M., S. Falkmer, K.J. Kramer and R.D. Speirs. 1983. Immunohistochemical investigations of neuropeptides in the brain, corpora cardiaca, and corpora allata of an adult lepidopteran insect, Manduca sexta (L). *Cell Tissue Res.* 232: 295-317.

Falko, J.M., S.E. Crockett, S. Cataland and E.L. Mazzaferri. 1975. Gastric inhibitory polypeptide (GIP) stimulated by fat ingestion in man. *J. Clin. Endocrinol. Metab.* 41: 260-265.

Flaten, O., L.E. Hanssen, M. Osnes and J. Myren. 1981. Plasma concentrations of gastric inhibitory polypeptide after intraduodenal infusion of cattle bile and synthetic bile salts in man. *Scand. J. Gastroenterol.* 16: 1073-1075.

Flaten, O., T. Sand and J. Myren. 1982. β-adrenergic stimulation and blockade of the release of gastric inhibitory polypeptide and insulin in man. *Scand. J. Gastroenterol.* 17: 283-288.

Flatt, P.R., C.J. Bailey, P. Kwasowski, S.K. Swanston-Flatt and V. Marks. 1983. Abnormalities of GIP in spontaneous syndromes of obesity and diabetes in mice. *Diabetes* 32: 433-435.

Forssmann, W.G., N. Yanaihara, V. Helmstaedter and D. Grube. 1976. Differential demonstration of the motilin-cell and the enterochromaffin-cell. *Scand. J. Gastroenterol.* 11 (Suppl.39): 43-45.

Fujimoto, W.Y., J.W. Ensinck, F.W. Merchant, R.H. Williams, P.H. Smith and D.G. Johnson. 1978. Stimulation by gastric inhibitory polypeptide of insulin and glucagon secretion by rat islet cultures. *Proc. Soc. Exp. Biol. Med.* 157: 89-93.

Garzo, G. and J.H. Dorrington. 1984. Aromatase activity in human granulosa cells during follicular development and the modulation by follicle stimulating hormone and insulin. *Am. J. Obstet.* 148: 657-662.

Gonzales, G.F., G.P. Risbridger and D.M. De Kretser. 1989. The effect of insulin on inhibition production in isolated seminiferous tubule segments from adult rats cultured *in vitro*. *Mol. Cell. Endocrinol.* 61: 209-216.

Graffner, H., S.R. Bloom, L.O. Farnebo and J. Jarhult. 1987. Effects of physiological increases of plasma noradrenaline on gastric acid secretion and gastrointestinal hormones. *Dig. Dis. Sci.* 32: 715-719.

Hall, K.E., G.R. Greenberg, T.Y. El-Sharkawy and N.E. Diamant. 1983. Vagal control of migrating motor complex-related peaks in canine plasma motilin, pancreatic polypeptide, and gastrin. *Can. J. Physiol. Pharmacol.* 61: 1289-1298.

Hartmann, H., R. Ebert and W. Creutzfeldt. 1986. Insulin-dependent inhibition of hepatic glycogenolysis by gastric inhibitory polypeptide (GIP) in perfused rat liver. *Diabetologia* 29: 112-114.

Havrankova, J., D. Schmechel, J. Roth and M.J. Brownstein. 1978. Identification of insulin in rat brain. *Proc. Natl. Acad. Sci. U.S.A.* 75: 5737-5741.

Heitz, P.H., J.M. Polak and A.G.E. Pearse. 1978. Cellular origin of motilin. In *Gut Hormones*, 1st Edition, ed. S.R. Bloom, 332-324. Churchill Livingstone, Edinburgh, London, New York.

Holst, N., T.G. Jenssen, P.G. Burhol, R. Jorde, J.M. Maltau and E. Haug. 1986. Gut peptides in lactation. *Br. J. Obstet. Gynaecol.* 93: 188-193.

Ishizuka, J., G.H. Greeley Jr., C.W. Cooper and J.C. Thompson. 1988. Effect of calcitonin gene-related peptide on glucose and gastric inhibitory polypeptide-stimulated insulin release from cultured newborn and adult rat islet cells. *Reg. Peptides* 20: 73-82.

Itoh, Z. 1981. Effect of motilin on gastrointestinal tract motility. In *Gut Hormones*, 2nd Edition, ed. S.R. Bloom, J.R. Polak, 280-289. Churchill Livingstone, Edinburgh, London, New York.

Itoh, Z. 1990. Effect of motilin on gastrointestinal motor activity in the dog. In *Motilin*, ed. Z. Itoh, 133-153. Academic Press, New York.

Itoh, Z. and T. Sekiguchi. 1983. Interdigestive motor activity in health and disease. *Scand. J. Gastroenterol.* 82 [Suppl]: 121-134.

Itoh, Z., R. Honda, K. Hiwatashi, S. Takeuchi, I. Aizawa, R. Takayanagi and E.F. Couch. 1976. Motilin-induced mechanical activity in the canine alimentary tract. *Scand. J. Gastroenterol.* 11 [Suppl]: 93-110.

Itoh, Z., S. Takeuchi, I. Aizawa, K. Mori, T. Taminato, Y. Seino, H. Imura and N. Yanaihara. 1978a. Changes in plasma motilin concentration and gastrointestinal contractile activity in conscious dogs. *Am. J. Dig. Dis.* 23: 929-935.

Itoh, Z., S. Takeuchi, I. Aizawa, R. Takayanagi, K. Mori, T. Taminato, Y. Seino, H. Imura and N. Yanaihara. 1978b. Recent advances in motilin research: Its physiological and clinical significance. In *Advances in Experimental Medicine. Gastrointestinal Hormones and Pathology of the Digestive System*, ed. M. Grossman, V. Speranza, N. Basso, E. Lezoche, 241-257. Plenum Press, New York, London.

Itoh, Z., I. Aizawa and T. Sekiguchi. 1982. The interdigestive migrating complex and its significance in man. *Clin. Gastroenterol.* 11: 497-521.

Jacobowitz, D.M., T.L. O'Donahue, W.Y. Chey and T.M. Chang. 1981. Mapping of motilin-immunoreactive neurons of the rat brain. *Peptides* 2: 479-487.

Jenssen, T.G., P.G. Burhol, R. Jorde, J. Florholmen and I. Lygren. 1984. Radioimmuno-assayable plasma motilin in man. Secretagogues, insulin-induced suppression, renal removal, and plasma components. *Scand. J. Gastroenterol.* 19: 717-723.

Jenssen, T.G., H.H. Haukland, J. Florholmen, R. Jorde and P.G. Burhol. 1986a. Evidence of somatostatin as a humoral modulator of motilin release in man. A study of plasma motilin and somatostatin during intravenous infusion of somatostatin, secretin, cholecystokinin, and gastric inhibitory polypeptide. *Scand. J. Gastroenterol.* 21: 273-280.

Jenssen, T.G., N. Holst, P.G. Burhol, R. Jorde, J.M. Maltau and B. Vonen. 1986b. Plasma concentrations of motilin, somatostatin and pancreatic polypeptide before, during and after parturition. *Acta Obstet. Gynecol. Scand.* 65: 153-156.

Jenssen, T.G., H.H. Haukland, B. Vonen, J. Florholmen, P.G. Burhol and J.M. Maltau. 1988. Changes in postprandial release patterns of gastrointestinal hormones in late pregnancy and the early postpartum period. *Br. J. Obstet. Gynaecol.* 95: 565-570.

Jorde, R., H.L. Waldum, P.G. Burhol, I. Lygren, T.B. Schulz, J. Florholmen and T.G. Jenssen. 1981a. The effect of somatostatin on fasting and postprandial plasma GIP, serum insulin, and blood glucose in man. *Scand. J. Gastroenterol.* 16: 113-119.

Jorde, R., H.L. Waldum and P.G. Burhol. 1981b. The effect of insulin-induced hypoglycemia with and without atropine on plasma GIP in man. *Scand. J. Gastroenterol.* 16(2): 219-223.

Jörnvall, H., M. Carlquist, S. Kwauk, S.C. Otte, C.H. McIntosh, J.C. Brown and V. Mutt. 1981. Amino acid sequence and heterogeneity of gastric inhibitory polypeptide (GIP). *FEBS Lett.* 123: 205-210.

Korchak, D.M., G. Nilaver and M.C. Beinfeld. 1984. The development of motilin-like immunoreactivity in the rat cerebellum and pituitary as determined by radioimmunoassay. *Neurosci. Lett.* 48: 267-272.

Korchak, D.M., M. Laskowski and M.C. Beinfeld. 1985. Brain motilin-like peptides are localized within the cell nucleus. *Peptides* 6: 1119-1123.

Kwok, Y.N., C.H.S. McIntosh, R.A. Pederson and J.C. Brown. 1985. Effect of substance P on somatostatin release from the isolated perfused rat stomach. *Gastroenterol.* 88: 90-95.

Larrimer, J.N., E.L. Mazzaferri, S. Cataland and H.S. Mekhjian. 1978. Effect of atropine on glucose-stimulated gastric inhibitory polypeptide. *Diabetes* 27: 638-642.

Larsson, L.I. and A.J. Moody. 1980. Glicentin and gastric inhibitory polypeptide immunoreactivity in endocrine cells of the gut and pancreas. *J. Histochem. Cytochem.* 28: 925-933.

Lee, K.Y., T.M. Chang and W.Y. Chey. 1981. Effect of electrical stimulation of the vagus on plasma motilin concentration in dog. *Life Sci.* 29: 1093-1097.

Lemoyne, M., R. Wassef, D. Tasse, L. Trudel and P. Poitras. 1984. Motilin and the vagus in dogs. *Can. J. Physiol. Pharmacol.* 62: 1092-1096.

Loftus, C.M., G. Nilaver, M.C. Beinfeld and K.D. Post. 1986. Motilin-related immunoreactivity in mammalian adenohypophysis. *Neurosurgery* 19: 201-204.

Lucas, A., S.R. Bloom and A. Aynsley-Green. 1982. Postnatal surges in plasma gut hormones in term and preterm infants. *Biol. Neonate* 41: 63-67.

McCann, S.M., M.D. Lumpkin, H. Mizunuma, O. Khorram, W.K. Samson. 1984. Recent studies on the role of brain peptides in control of anterior pituitary hormone secretion. *Peptides* 5 (Suppl.1): 3-7.

McCullough, A.J., J.B. Marshall, C.P. Bingham, B.L. Rice, L.D. Manning and S.C. Kalhan. 1985. Carbachol modulates GIP-mediated insulin release from rat pancreatic lobules *in vitro*. *Am. J. Physiol.* 248: E299-E303.

McDonald, T.J., M.A. Ghatei, S.R. Bloom, N.S. Track, J. Radziuk, J. Dupré and V. Mutt. 1981. A qualitative comparison of canine plasma gastroenteropancreatic hormone response to bombesin and the porcine gastrin-releasing peptide (GRP). *Reg. Peptides* 2: 293-304.

McIntosh, C.H.S. 1985. Minireview. Gastrointestinal somatostatin: Distribution, secretion and physiological significance. *Life Sci.* 37: 2043-2058.

McIntosh, C.H.S. and J.C. Brown. 1988. Motilin. *Adv. Metab. Disorders* 11: 439-455.

McIntosh, C.H.S. and J.C. Brown. 1990a. Purification and chemical structure of porcine and canine motilins and evidence for the existence of motilin in other species. In *Motilin*, ed. Z. Itoh, 13-30. Academic Press, New York.

McIntosh, C.H.S. and J.C. Brown. 1990b. Motilin: Isolation, secretion, actions and pathophysiology. *Front. Gastrointestinal Res.* (Karger Basel), 17: 307-352.

McIntosh, C.H.S., R.A. Pederson, H. Koop and J.C. Brown. 1979. Inhibition of GIP stimulated somatostatin-like immunoreactivity (SLI) by acetylcholine and vagal stimulation. In *Gut Peptides: Secretion, Function and Clinical Aspects*, ed. A. Miyoshi, 100-104. Kodansha, Japan.

McIntosh, C.H.S., Y.N. Kwok, T. Mordhorst, E. Nishimura, R.A. Pederson and J.C. Brown. 1983. Enkephalinergic control of somatostatin secretion from the perfused rat stomach. *Can. J. Physiol. Pharmacol.* 61: 657-663.

McIntosh, C.H.S., R.A. Pederson, H. Koop and J.C. Brown. 1981. Gastric inhibitory polypeptide stimulated secretion of somatostatinlike immunoreactivity from the stomach: Inhibition by acetylcholine or vagal stimulation. *Can. J. Physiol. Pharmacol.* 59: 468-472.

McIntosh, C., V. Bakich, T. Trotter, Y.N. Kwok, E. Nishimura, R. Pederson and J. Brown. 1984. Effect of cysteamine on secretion of gastrin and somatostatin from the rat stomach. *Gastroenterol.* 86: 834-838.

McIntosh, C.H.S., T. Mordhorst, R.A. Pederson, M.K. Mueller and J.C. Brown. 1985. The control of rat gastric somatostatin secretion by the autonomic nervous system. In *2nd International Somatostatin Symposium*, 258-262. Attempto Verlag, Tübingen.

Maletti, M., J.J. Altman, D.H. Hoa, M. Carlquist and G. Rosselin. 1987. Evidence of functional gastric inhibitory polypeptide (GIP) receptors in human insulinoma. Binding of synthetic human GIP 1-31 and activation of adenylate cyclase. *Diabetes* 36: 1336-1340.

Miralles, P., E. Peiro, R.A. Silvestre, M.L. Villanueva and J. Marco. 1988. Effects of galanin on islet cell secretory responses to VIP, GIP, 8-CCK, and glucagon by the perfused rat pancreas. *Metabolism* 37: 766-770.

Mitznegg, P., S.R. Bloom, N. Christofides, H. Besterman, W. Domschke, S. Domschke, E. Wunsch and L. Demling. 1976. Release of motilin in man. *Scand. J. Gastroenterol.* 11 (Suppl. 39): 53-56.

Moody, A.J., L. Thim and I. Valverde. 1984. The isolation and sequencing of human gastric inhibitory peptide (GIP). *FEBS Lett.* 172: 42-48.

Morgan, L.M., J.W. Wright and V. Marks. 1979. The effect of oral galactose on GIP and insulin secretion in man. *Diabetologia* 16: 235-239.

Murphy, W.A., V.A. Lance, J. Sueiras-Diaz and D.H. Coy. 1983. Effects of secretin and gastric inhibitory polypeptide on human pancreatic growth hormone-releasing factor (1-40)-stimulated growth hormone levels in the rat. *Biochem. Biophys. Res. Commun.* 112: 469-474.

Nelson, R.L., V.L. Go, A.J. McCullough, D.M. Ilstrup and F.J. Service. 1986. Lack of a direct effect of the autonomic nervous system on glucose-stimulated gastric inhibitory polypeptide (GIP) secretion in man. *Dig. Dis. Sci.* 31: 929-935.

Nilaver, G., R. Defendi, E.A. Zimmerman, M.C. Beinfeld and T.L. O'Donahue. 1982. Motilin in the Purkinje cell of the cerebellum. *Nature* 295: 597-599.

O'Donahue, T.L., M.C. Beinfeld, W.Y. Chey, T.M. Chang, G. Nilaver, E.A. Zimmerman, H. Yajima, H. Adachi, M. Poth, R.P. McDevitt and D.M. Jacobowitz. 1981. Identification, characterization and distribution of motilin immunoreactivity in the rat nervous system. *Peptides* 2: 467-477.

O'Dorisio, T.M., S. Cataland, M. Stevenson and E.L. Mazzaferri. 1976. Gastric inhibitory polypeptide (GIP). Intestinal distribution and stimulation by amino acids and medium-chain triglycerides. *Am. J. Dig. Dis.* 21: 761-765.

Ohneda, A., T. Kobayashi and J. Nihei. 1984. Response of gastric inhibitory polypeptide to fat ingestion in normal dogs. *Reg. Peptides* 8: 123-130.

Okada, J. and M. Miura. 1985. Sites of action of brain-gut peptides in cultured neurons of rat brainstem. *Brain Res.* 348: 175-179.

Ottlecz, A., W.K. Samson and S.M. McCann. 1985. The effects of gastric inhibitory polypeptide (GIP) on the release of anterior pituitary hormones. *Peptides* (Fayetteville) 6: 115-119.

Pearse, A.G.E., J.M. Polak, S.R. Bloom, C. Adams, J.R. Dryburgh and J.C. Brown. 1974. Enterochromaffin cells of the mammalian small intestine as the source of motilin. *Virch. Arch. B. Cell. Path.* 16: 111-120.

Pederson, R.A. and J.C. Brown. 1972. Inhibition of histamine-, pentagastrin-, and insulin-stimulated canine gastric secretion by pure "gastric inhibitory polypeptide". *Gastroenterol.* 62: 393-400.

Pederson, R.A. and J.C. Brown. 1976. The insulinotropic action of gastric inhibitory polypeptide in the perfused isolated rat pancreas. *Endocrinology* 99: 780-785.

Pederson, R.A. and J.C. Brown. 1978. Interaction of gastric inhibitory polypeptide, glucose, and arginine on insulin and glucagon secretion from the perfused rat pancreas. *Endocrinology* 103: 610-615.

Pederson, R.A., H.E. Schubert and J.C. Brown. 1975a. Gastric inhibitory polypeptide. Its physiologic release and insulinotropic action in the dog. *Diabetes* 24: 1050-1056.

Pederson, R.A., J.R. Dryburgh and J.C. Brown. 1975b. The effect of somatostatin on release and insulinotropic action of gastric inhibitory polypeptide. *Can. J. Physiol. Pharmacol.* 53: 1200-1205.

Peeters, T., G. Vantrappen and J. Janssens. 1980. Fasting plasma motilin levels are related to the interdigestive motility complex. *Gastroenterol.* 79: 716-719.

Peeters, T.L., J. Janssens, C. Plets and G. Vantrappen. 1988. Interdigestive motility and motilin in hypophysectomized patients. *Scand. J. Gastroenterol.* 23: 71-74.

Phillips, M.I. 1988. Insulin in the brain: A feedback loop involving brain insulin and circumventricular organs. In *Insulin, insulin-like growth factors, and their receptors in the central nervous system*, ed. M.K. Raizada, M.I. Phillips and D. LeRoith, 163-175. Plenum Press, New York, London.

Phillis, J.W. and J.R. Kirkpatrick. 1979. Motilin excites neurons in the cerebral cortex and spinal cord. *Eur. J. Pharmacol.* 58: 469-472.

Plymate, S.R., L.A. Matej, R.E. Jones and K.E. Friedl. 1988. Inhibition of sex hormone-binding globulin production in the human hepatoma (Hep G2) cell line by insulin and prolactin. *J. Clin. Endocrinol. Metab.* 67: 460-464.

Poitras, P., J.R. Reeve Jr., M.W. Hunkapillar, L.E. Hood and J.H. Walsh. 1983. Purification and characterization of canine intestinal motilin. *Reg. Peptides* 5: 197-208.

Poitras, P., R.G. Lahaie, S. St-Pierre and L. Trudel. 1987a. Comparative stimulation of motilin duodenal receptor by porcine or canine motilin. *Gastroenterol.* 92: 658-662.

Poitras, P., L. Trudel, R.G. Lahaie and G. Pomier-Layrargue. 1987b. Motilin-like immunoreactivity in intestine and brain of dog. *Life Sci.* 40: 1391-1395.

Polak, J.M., S.R. Bloom, M. Kuzio, J.C. Brown and A.G. Pearse. 1973. Cellular localization of gastric inhibitory polypeptide in the duodenum and jejunum. *Gut* 14: 284-288.

Polak, J.M., A.G.E. Pearse and C.M. Heath. 1975a. Complete identification of endocrine cells in the gastrointestinal tract using semi-thin sections to identify motilin cells in human and animal intestine. *Gut* 16: 225-229.

Polak, J.M., A.G. Pearse, L. Grimelius and V. Marks. 1975b. Gastrointestinal apudosis in obese hyperglycaemic mice. *Virchows-Archiv. [Cell Pathol].* 19: 135-150.

Polak, J.M., P. Heitz and A.G.E. Pearse. 1976. Differential localization of substance P and motilin. *Scand. J. Gastroenterol.* 11 (Suppl.39): 39-42.

Poretsky, L. and M.A. Karlin. 1987. The gonadotropic function of insulin. *Endocrine Reviews* 8: 132-141.

Reeve, J.R. Jr., F.J. Ho, J.H. Walsh, C.M. Ben-Avram and J.E. Shively. 1985. Rapid high-yield purification of canine intestinal motilin and its complete sequence determination. *J. Chromatog.* 321: 421-432.

Rosenfeld, D.J. and T.L. Garthwaite. 1987. Central administration of motilin stimulates feeding in rats. *Physiol. Behav.* 39: 753-756.

Ross, S.A. and E.A. Shaffer. 1981. The importance of triglyceride hydrolysis for the release of gastric inhibitory polypeptide. *Gastroenterol.* 80: 108-111.

Salera, M., R. Ebert, P. Giacomoni, L. Pironi, S. Venturi, R. Corinaldesi, M. Miglioli and L. Barbara. 1982. Adrenergic modulation of gastric inhibitory polypeptide secretion in man. *Dig. Dis. Sci.* 27: 794-800.

Samson, W.K., M.D. Lumpkin, G. Nilaver and S.M. McCann. 1984. Motilin: A novel growth hormone releasing agent. *Brain Res. Bull.* 12: 57-62.

Schauder, P., J.C. Brown, H. Frerichs and W. Creutzfeldt. 1975. Gastric inhibitory polypeptide: Effect on glucose-induced insulin release from isolated rat pancreatic islets *in vitro*. *Diabetologia* 11: 483-484.

Schnuerer, E.M., T.J. McDonald and J. Dupré. 1987. Inhibition of insulin release by galanin and gastrin-releasing peptide in the anaesthetized rat. *Reg. Peptides* 18: 307-320.

Schubert, H. and J.C. Brown. 1974. Correction to the amino acid sequence of porcine motilin. *Can. J. Biochem.* 52: 7-8.

Seino, Y., K. Tanaka, J. Takeda, H. Takahashi, T. Mitani, M. Kurono, T. Kayano, G. Koh, H. Fukumoto, H. Yano, J. Fujita, N. Inagaki, Y. Yamada and H. Imura. 1987. Sequence of an intestinal cDNA encoding human motilin precursor. *FEBS Lett.* 223: 74-76.

Shin, S., K. Imagawa, F. Shimizu, E. Hashimura, K. Nagai, C. Yanaihara and N. Yanaihara. 1980. Heterogeneity of immunoreactive motilin. *Endocrinol. Japon* 27 (Suppl.1): 141-149.

Siegel, E.G. and W. Creutzfeldt. 1985. Stimulation of insulin release in isolated rat islets by GIP in physiological concentrations and its relation to islet cyclic AMP content. *Diabetologia* 28: 857-861.

Sirinek, K.R., S. Cataland, T.M. O'Dorisio, E.L. Mazzaferri, S.E. Crockett and W.G. Pace. 1977. Augmented gastric inhibitory polypeptide response to intraduodenal glucose by exogenous gastrin and cholecystokinin. *Surgery* 82: 438-42.

Sirinek, K.R., W.G. Pace, S.E. Crockett, T.M. O'Dorisio, E.L. Mazzaferri and S. Cataland. 1978. Insulin-induced attenuation of glucose-stimulated gastric inhibitory polypeptide secretion. *Am. J. Surg.* 135: 151-155.

Sirinek, K.R., B.A. Levine, S.E. Crockett and S. Cataland. 1980. Cholecystokinin and gastric inhibitory polypeptide not glucagonotropic in dogs. *J. Surg. Res.* 28: 367-372.

Sjölund, K., M. Ekelund, R. Håkanson, A.J. Moody and F. Sundler. 1983. Gastric inhibitory peptide-like immunoreactivity in glucagon and glicentin cells: Properties and origin. An immunocytochemical study using several antisera. *J. Histochem. Cytochem.* 31: 811-817.

Smith, P.H. 1983. Immunocytochemical localization of glucagonlike and gastric inhibitory polypeptidelike peptides in the pancreatic islets and gastrointestinal tract. *Am. J. Anat.* 168: 109-118.

Smith, P.H., F.W. Merchant, D.G. Johnson, W.Y. Fujimoto and R.H. Williams. 1977. Immunocytochemical localization of a gastric inhibitory polypeptide-like material within A-cells of the endocrine pancreas. *Am. J. Anat.* 149: 585-590.

Smith, P.H., B.J. Davis, Y. Seino and N. Yanaihara. 1981. Localization of motilin-containing cells in the intestinal tract of mammals: A further comparison using region-specific motilin antisera. *Gen. Comp. Endocrinol.* 44: 288-291.

Soon-Shiong, P., H.T. Debas and J.C. Brown. 1979. The evaluation of gastric inhibitory polypeptide (GIP) as the enterogastrone. *J. Surg. Res.* 26: 681-686.

Soon-Shiong, P., H.T. Debas and J.C. Brown. 1984. Bethanechol prevents inhibition of gastric acid secretion by gastric inhibitory polypeptide. *Am. J. Physiol.* 247: G171-G175.

Stevenson, J.C., T.E. Adrian, N.D. Christofides and S.R. Bloom. 1985 Effect of calcitonin on gastrointestinal regulatory peptides in man. *Clin. Endocrinol. (Oxf).* 22: 655-660.

Sullivan, S.N., S.R. Bloom, T.E. Adrian, L. Lamki, T. McDonald and P. Corcoran. 1983. The effect of a synthetic enkephalin analogue on postprandial gastrointestinal and pancreatic hormone secretion. *Am. J. Gastroenterol.* 78: 287-290.

Swanston-Flatt, S.K. and P.R. Flatt. 1988. Effects of amino acids, hormones and drugs on insulin release and ^{45}Ca uptake by transplantable rat insulinoma cells maintained in tissue culture. *Gen. Pharmacol.* 19: 239-242.

Sykes, S., L.M. Morgan, J. English and V. Marks. 1980. Evidence for preferential stimulation of gastric inhibitory polypeptide secretion in the rat by actively transported carbohydrates and their analogues. *J. Endocrinol.* 85: 201-207.

Takeda, J., Y. Seino, K. Tanaka, H. Fukumoto, T. Kayano, H. Takahashi, T. Mitani, M. Kurono, T. Suzuki, T. Tobe, et al. 1987. Sequence of an intestinal cDNA encoding human gastric inhibitory polypeptide precursor. *Proc. Natl. Acad. Sci. U.S.A.* 84: 7005-7008.

Talaulicar, M., R. Ebert, B. Willms and W. Creutzfeldt. 1981. Effect of exogenous insulin on fasting serum levels of gastric inhibitory polypeptide (GIP) in juvenile diabetes. *Clin. Endocrinol. (Oxf).* 14: 175-180.

Taylor, I.L. and M. Feldman. 1982. Effect of cephalic-vagal stimulation on insulin, gastric inhibitory polypeptide, and pancreatic polypeptide release in humans. *J. Clin. Endocrinol. Metab.* 55: 1114-1117.

Thomas, F.B., D. Sinar, E.L. Mazzaferri, S. Cataland, H.S. Mekhjian, J.H. Caldwell and J.J. Fromkes. 1978. Selective release of gastric inhibitory polypeptide by intraduodenal amino acid perfusion in man. *Gastroenterol.* 74: 1261-1265.

Usellini, L., A.M.J. Buchan, J.M. Polak, C. Capella, M. Cornaggia and E. Solcia. 1984a. Ultrastructural localization of motilin in endocrine cells of human and dog intestine by the immunogold technique. *Histochem.* 81: 363-368.

Usellini, L., C. Capella, E. Solcia, A.M. Buchan and J.C. Brown. 1984b. Ultrastructural localization of gastric inhibitory polypeptide (GIP) in a well characterized endocrine cell of canine duodenal mucosa. *Histochem.* 80: 85-89.

Van Houten, M., B.I. Posner, B.M. Kopriwa and J.R. Brawer. 1979. Insulin binding sites in the rat brain: *In vivo* localization to the circumventricular organs by quantitative autoradiography. *Endocrinology* 105: 666-673.

Van Houten, M. and B.I. Posner. 1981. Cellular basis of direct insulin action in the central nervous system. *Diabetologia* 20: 255-267.

Veldhuis, J.D., L.A. Kolp, M.E. Toaff, J.F. Strauss III and L.M. Demers. 1983. Mechanisms subserving the trophic actions of insulin on ovarian cells: *In vitro* studies using swine granulosa cells. *J. Clin. Invest.* 72: 1046.

Veldhuis, J.D., J.E. Nestler and J.T. Strauss III-Gwynne. 1986. Insulin regulates low-density lipoprotein metabolism by swine granulosa cells. *Endocrinology* 118: 2242-2253.

Villar, H.V., H.R. Fender, P.L. Rayford, S.R. Bloom, N.I. Ramus and J.C. Thompson. 1976. Suppression of gastrin release and gastric secretion by gastric inhibitory polypeptide (GIP) and vasoactive intestinal polypeptide (VIP). *Ann. Surg.* 184: 97-102.

Vogel, L. 1987. *Characterization of rat intestinal immunoreactive motilin.* M.Sc. Thesis, University of British Columbia.

Wasada, T., K. McCorkle, V. Harris, K. Kawai, B. Howard and R.H. Unger. 1981. Effect of gastric inhibitory polypeptide on plasma levels of chylomicron triglycerides in dogs. *J. Clin. Invest.* 68: 1106-1107.

Weindl, A. 1973. Neuroendocrine aspects of circumventricular organs. In *Frontiers in Neuroendocrinology*, ed. W.F. Ganong and L. Martini, 3-32. Oxford University Press, New York.

Whitaker, E.M. and A.C. Robinson. 1989. Circulating reproductive hormones and hypothalamic oestradiol and progestin receptors in infertile Zucker rats. *J. Endocrinol.* 120: 331-336.

Wolfe, M.M., M.P. Hocking, D.G. Maico and J.E. McGuigan. 1983. Effects of antibodies to gastric inhibitory peptide on gastric acid secretion and gastrin release in the dog. *Gastroenterol.* 84: 941-948.

Yamagishi, T. and H.T. Debas. 1980. Gastric inhibitory polypeptide (GIP) is not the primary mediator of the enterogastrone action of fat in the dog. *Gastroenterol.* 78: 931-936.

Yanaihara, C., H. Sato, N. Yanaihara, S. Naruse, W.G. Forssmann, V. Helmstaedter, T. Fujita, K. Yamaguchi and K. Abe. 1978. Motilin-, substance P- and somatostatin-like immunoreactivities in extracts from dog, tupaia and monkey brain and GI tract. *Adv. Exp. Med. Biol.* 106: 269-283.

Young, W.S. 1986. Periventricular hypothalamic cells in the rat brain contain insulin mRNA. *Neuropeptides* 8: 93-97.

Yoshiya, K., T. Yamamura, Y. Ishikawa, J. Utsunomiya, J. Takemura, J. Takeda, Y. Seino and H. Imura. 1985. Effect of truncal vagotomy on GIP release induced by intraduodenal glucose or fat in dogs. *Digestion* 31: 41-46.

Zawalich, W.S. 1988. Synergistic impact of cholecystokinin and gastric inhibitory polypeptide on the regulation of insulin secretion. *Metabolism* 37: 778-781.

INDEX

A

A-cells, 252
Acetylcholine, 87, 258
Adipose tissue, 264
Adrenergic neurons
  angiotensin II and
    ductus deferens, 55
    uterus, 30
  and GIP, 254–255
  neurotensin regulation of prolactin release, 182–183
  NPY in, 116, 121
  opioid-LHRH neuron functional links, 164–165
Adrenergic receptors, and angiotensin II effects on LH release, 7
Adrenergic turnover, and LH surge, 170
Adrenocorticotropin
  angiotensin II and, 10–11
  CCK and, 222, 225
  NPY and, 112, 122, 133–134
Aldosterone, 122
Amino acids, and GIP, 253–254
Amino acid sequences
  GIP, 251
  motilin, 241
  neurotensin, 181
  NPY, 107–110
Amniotic fluid, 35–36
Amygdala, 87, 113
Androgens, insulin and, 260, 261, see also Testosterone
Angiogenesis, angiotensin II and, 27, 30
Angiotensin converting enzyme (ACE)
  in females, 23, 25, 37
  in males, 45
    ductus deferens, 54–55
    epididymis, 49
    prostate, 56
    seminal fluid, 56–57
    seminal vessels, 56
    testis, 46
  ovary, 25
Angiotensin I, 23, 25
  in females, 30, 37
  in males, 48

and sperm velocity, 57
testis, 46
Angiotensin II
  in females, 37, see also Renin-angiotensin system in female reproductive structures
    cervix, 34
    ovarian responsiveness, 27–30
    ovary, 23–30, 34
    oviduct, 34
    pregnancy, 31, 33–34
    prolactin release, 8–9
    synthesis of, 23–27
    uterus, 30–33
  in females, ovarian hormone regulation of LH and PRL release
    brain, 3–10
    pituitary, 10–14
  in males, 48, see also Renin-angiotensin system in male reproductive structures
    ductus deferens, 54–55
    epididymis, 49
    prostate, 56
    seminal fluid, 56–57
    seminal vessels, 56
    testis, 46, 47
  NPY and, 122
  OXY and AVP neurons, 64–65, 71
  renin-angiotensin system, 48
Angiotensinogen, 23, see also Angiotensin II
  amniotic fluid, 36
  in glial cells, 3
  in males, 48
  uterus, 30
Anterior hypothalamic preoptic area, 6
Anterior pituitary gland, see Pituitary, anterior
Antral fluid, 3
Arachidonate, 183
Arcuate nucleus, 107, 123, 131
  galanin, 196, 197
  naloxone effects on LH release, 169, 170
  neurotensin, 181
  NPY fibers, 116, 171
  opioids, 164, 171

A receptors, CCK, 214
Argyriphil cells, motilin in, 243
AVP, see Vasopressin

B

B-cells, 254, 263
Behavioral effects, CCK, 222–228
β-endorphins, 71, 158, 163, see also
     Opioid peptides
Bicarbonate secretion, motilin and, 248
Birth
  galanin expression, 202
  motilin effects, 249
Blood flow, angiotensin II and, 30
Blood pressure, 4, 25
Bovine luteinizing hormone (bLH), 46
Brain, see also Central nervous system;
     specific structures
  CCK receptors (B receptors), 214, 220,
     229–230
  insulin secretion, 265
Brainstem
  NPY in, 113, 116
  PP-like activity, 108
  PYY-like immunoreactivity, 110
B receptors, CCK, 214, 220, 229–230
BROCA region, 126

C

Calcitonin, and GIP, 254
Calcitonin gene-related peptide, 257
Calcium, 127, 128, 183
Captopril, 46
Cardiovascular system
  blood pressure, 4, 25
  NPY and, 112
Castration, and NPY, 131, 132
Catecholaminergic neurons, PP-like
     activity, 108
Catecholamines
  and GIP, 254–255
  neurotensin regulation of prolactin
     release, 182–183
  NPY colocalization, 116
  opioid-LHRH neuron functional links,
     164–165
Caudate nucleus, CCK and, 224
CAV-259, 96
CCK, see Cholecystokinin
Central administration, see also Intracere-
     broventricular administration

CCK, and oxytocin and vasopressin,
     220–221
  galanin, 200
  OXY, 68–69
  OXY antisera, 69–71
  VIP, 89–90
Central nervous system, see also Brain;
     specific structures
  motilin effects, 248
  NPY in, 113–114
  OXY- and AVP-containing neurons,
     63–64
  VIP neurons in, 86–87
Cerebral cortex
  motilin in, 244
  NPY binding, 111
  NPY in, 114
  PP-like activity, 108
  VIP neurons in, 86–87
Cervix, 34
Cholecystokinin
  effects, 214–222
    on CRH and ACTH, 222
    on gonadotropins, 214–215
    on oxytocin and vasopressin, 215–221
    on PRL, 221–222
  functions, 222–228
    grooming behavior, 223
    ingestive behavior, 224–228
    reproductive behavior, 223–224
  future directions, 229–231
    functional mapping, 230–231
    receptor antagonists, 229–230
  and GIP, 254, 257
  neuroanatomy and neurophysiology,
     211–214
Cholinergic pathways, motilin secretion,
     246–247
Cholinomimetics, and GIP effects on
     insulin secretion, 257
Chylomicrons, 253
Circumventricular organs, 265
  GIP and, 259
  motilin access to, 249–250
Clonidine, 165
Colocalization
  of angiotensin II in LH secretory gran-
     ules, 11
  CCK, 211–213, 223–224
  galanin, 196, 197, 202–203
  motilin, 244

neurotensin, 181
NPY, 116, 121
VIP, 87
Corpora lutea, 24, 30
Cortex, CCK in, 211
Cortical neurons, VIP neurons, 87
Corticosterone, 112, 122
Corticotropin releasing factor (CRF), 212, 213, 222, 226
Corticotropin releasing factor (CRF)-neurons, NPY and, 112
C-peptide, 254

D

Development
  galanin expression, 202
  motilin during, 244
Diacylglycerol, 183
Diestrus, see also Estrous cycle
  galanin gene expression, 202
  OXY effects, 66–67, 72
  NPY, 130, 169
Diet, see also Food intake, 262, 263
Dihydrotestosterone, 161
5,7-Dihydroxytryptamine, 122
Domperidone, 96, 97
Dopamine
  CCK effects on gonadotropins, 215, 224
  galanin and, 200
  neurotensin and, 181, 183
  and prolactin release, 7, 8, 97, 183
Dopamine β-hydroxylase, 196
Dopamine receptors, and PRL response, 96–97
Dopaminergic neurons, 116
  CCK colocalization, 213
  galanin and, 200
  neurotensin and, 188
  ovarian steroids and, 15
Dorsomedial hypothalamus
  CCK in, 212
  VIP neurons, 87
Dorsomedial nuclei, 131
Ductus deferens, 54–55
Dynorphin, see Opioid peptides

E

EC$_2$ cells, motilin in, 243
Endogenous opioid peptides, see Opioid peptides

Endometrium, angiotensin II binding in, 31–33
β-Endorphins, 71, 158, 163, see also Opioid peptides
Enterchromaffin cells, motilin in, 243
Enterogastrone, 253, 256
Epididymis, 49–54
Epinephrine
  opioid-LHRH neuron functional links, 164–165
  uterus, 30
Estradiol, 157
  GIP and, 258–259
  and LHRH release, 71
  neuropeptide Y and, 167–173
  opioid peptides and, 161–163
  OXY and AVP neurons, 65
  renin-angiotensin system
    and angiotensin II receptor binding, 4
    in males, 48
Estradiol-primed ovariectomized animals, see Ovariectomy
Estrogens, 129, see also Gonadal steroids; Ovarian hormones
  angiotensin II and, 29, 30
  insulin and, 260
  and renin production in testis, 47
Estrous cycle
  and angiotensin II, see also Renin-angiotensin system in female reproductive structures
    levels of, 4
    prolactin regulation, 14
    uterus, 31
  galanin gene expression, 201–202
  and NPY, 135
  oxytocin effects, 72
  opioid peptides and, see also Opioid peptides
  galanin gene expression, 202
  in obese (fa/fa) rat, 264
Exocrine secretion, motilin and, 248
Exogenous hormone infusion, see Central administration; Intracerebroventricular administration
Eye, NPY in, 113

F

Fat metabolism, GIP and, 257
Fatty acids, 253

Feeding, see Food intake
Female reproductive tract, see also Renin-
    angiotensin system in female
    reproductive structures
  angiotensinergic activity, 23–38
  hyperandrogenism, 261–263
  NPY effects, 120–127
  NPY in, 115, 118
Fetal development, galanin expression, 202
Fluid and electrolyte balance
  angiotensin II and, 4, 25
  and galanin, 196
Fluid secretion, motilin and, 248
Fluoxetine, 182
Follicle stimulating hormone (FSH), 29,
    73–74
  GIP and, 258–260
  insulin and, 261, 264
  neurotensin and, 184–186, 189, 190
  NPY and, 121, 122, 131, 133–135
  in obese (fa/fa) rat, 264
  OXY, 73
  vasopressin and, 74
Follicle stimulating hormone receptors,
    insulin and, 261
Follicle stimulating hormone-releasing
    factor, 73, 259
Follicular fluid renin activity, 24
Food intake
  CCK and, 224–228, 230
  and GIP, 252–254
  and hyperandrogenism in women, 262,
    263
  motilin effects, 248
  and motilin secretion, 245
FSH, see Follicle stimulating hormone

G

GABA neurons, galanin and, 197
Galanin, 195–205
  expression of, 201–203
  and GH, 197–199
  and GIP effects on insulin secretion, 257
  and LH, 200–201
  localization of, 196–197
  and PRL, 199–200
  regulation of, 203–204
Gallbladder, 258
Gastric inhibitory peptide (GIP)
  cellular localization and species distribu-

tion, 251–252
  isolation and characterization, 250–251
  metabolic effects, 257
  other actions, 258
  as releasing hormone, 255–257
  and reproductive hormone secretion,
    258–265
    anterior pituitary hormones, 258–259
    insulin-mediated, 259–265
  secretion of, 252–255
Gastric inhibitory polypeptide, 85, 86
Gastric motility
  CCK and, 226, 230
  GIP and, 258
Gastrin
  CCK receptors and, 214
  and GIP, 254
Gastrointestinal system, see also specific
    hormones
  NPY in, 114
  PYY-like immunoreactivity, 110
GH, see Growth hormone
GHRH, see Growth hormone releasing
    hormone
GIP, see Gastric inhibitory peptide
Glial cells, 3, 4
Glicentin, 250
Glucagon, 85, 86, 250, 254, 257, 264
Glucagon-like peptides (GlP), 250
Glutamic acid decarboxylase, 244
Gonadal steroids, see also Ovarian
    hormones
  galanin gene expression, 201
  and hypothalamic levels of neuropeptide
    Y, 130–132
  and LH release, OXY and AVP influ-
    ence, 65–66
  motilin effects, 249
  and NPY, 134–135
  and NPY effect, 126–130
  and PP effects, 109
Gonadotropins, 29, see also Follicle
    stimulating hormone; Luteinizing
    hormone
  CCK and, 214–215
  and galanin, 202
  male, renin-angiotensin system and,
    45–46
  OXY and AVP effects, 64
  and renin activity, follicular fluid, 24
Granulosa cells, 29, 30

Grooming behavior, CCK and, 223
Growth hormone (GH)
  galanin and, 197–199, 202–203
  GIP and, 260
  motilin colocalization, 244
  motilin effects, 249
  neurotensin and, 187, 189, 190
  NPY and, 121, 133–134
  PP and, 108
Growth hormone releasing factor (GRF),
    85, 86
Growth hormone releasing hormone
    (GHRH), 116
  galanin and, 197–199
  GIP and, 259
  GIP homology, 250
  neurotensin and, 187
  PP and, 108

H

Haloperidol, 123
hCG, 29, 46
Hippocampus
  CCK in, 211
  insulin and, 265
  NPY binding, 111
  NPY in, 114
  VIP neurons, 87
Human chorionic gonadotropin, 29, 46
Hydration state, see Fluid and electrolyte
    balance
5-Hydroxytryptamine, 71, see also
    Serotonin
5-Hydroxytryptophan, 74, 94, 182
Hyperandrogenism, 261–263
Hyperglucagonemia, 264
Hyperglycemia, 263–264
Hyperinsulinemia, 261–262
Hypophysis, see Pituitary
Hypothalamus
  CCK, 211, 212
  galanin, 197
  insulin, 265
  motilin, 244, 249
  neurotensin, 181, 189–190
  NPY binding, 111
  NPY effects, 112, 121–132
    gonadal steroids and, 130–131
    hormone release *in vitro,* 127–130
    hormone release *in vivo,* 121–127

  secretion of NPY, 131–132
  NPY in, 114–116
  opioid-LHRH neuron functional link, 164
  opioid peptides, 157–158
  OXY- and AVP-containing neurons, 63
  PP-like immunoreactivity, 107–108
  renin-angiotensin system, 4, 28
  VIP neurons, 87

I

I-cells, 252
Infants, motilin in, 245
Ingestive behavior, see Food intake
Inhibin, 259
Inositol triphosphate, 183
Insulin, 112, 253, 259–265
Intravenous administration, see Peripheral
    administration
Intracerebroventricular administration, see
    also Central administration
  angiotensin II, 5–10
  CCK
    and gonadotropins, 214–215
    and PRL, 220
  cupric acetate, and NPY, 131–132
  neurotensin, 183
    and FSH and LH levels, 184–185
    and GH, 187
    and prolactin release, 183–184
    and TSH, 186–187
  NPY, 122
  OXY and AVP antisera, 71, 73
  PP, 108

K

K-cells, 252, 263

L

Lactation, 64, 75
Lateral hypothalamus, 181
Lateral vestibular nucleus, 248
Lepidoptera, GIP in, 251
Leukotrienes, 183
Leydig cells, 48
Leydig tumor cells, 46
LH, see Luteinizing hormone
Limbic system
  and angiotensin II levels, 4
  galanin, 196

neurotensin, 181
OXY- and AVP-containing neurons, 63
VIP neurons, 87
Lipids, GIP and, 257
Lipoproteins, insulin and, 260
Lordosis, 223–234
Luteinizing hormone (LH)
  CCK and, 214–215
  galanin and, 200–201
  GIP and, 258–260
  insulin and, 261
  motilin effects, 249
  neurotensin and, 185–186, 189, 190
  NPY and, 120–127, 131, 133–134
  opioid peptides and, 158–163
  PP and, 108–109
  release of, 65–73
    AVP, 72
    OXY, 66–72
  renin-angiotensin system
    angiotensin II colocalization, 11
    angiotensin II regulation of, 6–7
    in males, 48
  testis effects, 48
  and testosterone production, 47
Luteinizing hormone-releasing hormone
    (LHRH), 71
  and LH surge in NAL-infused rat, 158
  NPY and, 121–127, 131, 167–173
    anterior pituitary effects, 132–135
    gonadal steroids and, 130–132
    in vitro, 127–128
  OXY and AVP effects, 64
  PP and, 108
  and prolactin secretin, 11–14
  renin-angiotensin system
    angiotensin II effects on LH, 7
    in males, 48
  VIP and, 90–91
Luteinizing hormone-releasing hormone
    neurons
  neurotensin and, 189
  opioid peptides and, 164

M

Magnocellular neurons
  CCK in, 211–212, 226
  galanin, 196
Males, see also Renin-angiotensin system
    in male reproductive structures

angiotensinergic activity, 45–57
galanin
  and GH, 197–198
  and LH, 200–201
  gene expression, 202
  neurotensin, 182
  regulation of GH, 187
  regulation of prolactin release, 183–184
  NPY effects, 122–123, 131
  NPY in, 115
  prolactin
    angiotensin II and, 8
    AVP and, 75
    neurotensin and, 183–184
Mamillary body, motilin in, 244
Mapping, CCK, 230–231
M-cells, motilin in, 243
Medial basal hypothalamus, opioid-LHRH
    neuron functional link, 164
Medial preoptic area (MPOA)
  CCK, 215, 223–224
  naloxone effects on LH release, 169
  neurotensin, 181, 185
  NPY-opioid peptide interactions, 171
Medial preoptic nucleus, 107, 116, 126,
    131
Median eminence
  CCK effects, 211
  CCK in, 223–224
  galanin, 196, 197, 202
  GIP and, 259
  motilin in, 244
  naloxone effects on LH release, 169
  neuropeptide Y-opioid peptide regulation
    of LH surge, 167–173
  neurotensin, 181, 188
  NPY in, 118, 130–131, 171
  opioid peptides, 158, 164, 169, 171
  OXY- and AVP-containing neurons, 63
  renin-angiotensin system, in males, 48
  VIP, 87, 90
Medulla, motilin in, 244
Mesenteric blood flow, 258
Metestrus, 66–67, 72
α-Methylparatyrosine (α-MT), 122, 123,
    182–183, 200
Midbrain
  CCK in, 211
  motilin in, 244
Migrating motor complex (MMC), 245,
    247

Milk ejection, 64
MK 329, 229–230
Monosodium glutamate, 197
Morphine, 165
Motilin
  actions of, 247–250
  in brain, 243–245
  cellular localization and species distribution, 242–243
  isolation and structure, 241–242
  secretion of, 245–247
Motor activity, GIP and, 258
Motor fibers, VIP, 86
Myometrium, angiotensin II binding in, 31–33

N

Naloxone, 158–162
  and catecholamine release, 165
  and LH surge in estrogen treated OVX rats, 163
  and NPY, 169, 170
Neural clock (NC), 157, 158
Neural lobe, CCK in, 212, 213
Neural pathways, see also specific pathways
  and GIP, 254
  opioid-LHRH neuron functional links, 164–165
Neurohypophyseal hormones, 76, see also Oxytocin; Vasopressin
Neurohypophysis, see Pituitary, posterior
Neuropeptide Y, 64
  amino acid sequence, 109
  distribution of
    CNS, 113–114
    GI system, 114
    research references, 115–116
  effects of
    on hypothalamus, 121–132
    on pituitary, anterior, 132–136
  opioid-LHRH neuron functional links, 164–165
  opioid peptides and, 165–173
  OXY neuron interactions, 71
  pancreatic polypeptide, 107–111
  proteolytic processing, 111
Neurotensin
  and FSH and LH, 184–186, 189, 190
  galanin colocalization, 197

and GH, 187, 189, 190
  intrahypothalamic action, 189–190
  and prolactin, 182–184, 188, 190
  and TSH, 186–187
Neurotransmitters, see also specific transmitters and systems
  and GIP, 254
  neurotensin action, 182
Non-adrenergic non-cholinergic pathways
  and GIP, 254–256
  motilin secretion, 247
Nonagrentaffin cells, motilin in, 243
Noradrenergic neurons
  NPY in, 116, 121–122
  ovarian steroids and, 15
  PP-like activity, 108
Norepinephrine
  angiotensin II and, 7, 55
  opioid-LHRH neuron functional links, 164–165
  OXY and AVP neurons, 64–65
  OXY neuron interactions, 71
NPY, see Neuropeptide Y
Nucleus accumbens, NPY in, 113
Nucleus interstitalis stria terminalis, 181
Nutrients
  and GIP, 252–254
  and motilin secretion, 245

O

Obesity, insulin in, 263–264
Olfactory bulb, NPY in, 113
Opioid peptides, 157–173
  adrenergic connection, 164–165
  LHRH neuron communications, 164
  LH surge induction, 158–163
    in OVX rats, 161–163
    preovulatory, 158–161
  neuropeptide Y and, 165–173
  ovarian steroids and, 15
Organum vasculosum of lamina terminalis (OVLT), 244, 259
Ovarian hormones, see also Gonadal steroids
  and angiotensin II, pituitary, 11
  and LH release, 5, 6
  and prolactin release, 7, 8–10
Ovariectomy
  and antiogensinogen levels, 4
  CCK and, 215

galanin
  and GH, 197–198
  and LH, 200–201
  and PRL, 199
growth hormone, 187, 197–198
LH
  AVP and, 72
  galanin and, 200–201
  OXY effects, 66–67
  opioid peptides and, 158, 161–163
neuropeptide Y and, 167–168
neurotensin, 182
  and FSH and LH levels, 184–186
  and GH, 187
  and prolactin release, 183–184
  and TSH, 186–187
and NPY, 120–122, 126–132, 134–135
opioid peptides, 158, 161–163
oxytocin and, 66–67, 75, 95, 96
PP effects, 108
and prolactin release, 7–8, 95, 96,
      183–184, 199
VIP effects, 95, 96
Ovary
  androgen biosynthesis, 261
  angiotensinergic activity, 23–30
  hyperandrogenism, 261–263
  insulin and insulin-like receptors, 260
  NPY fibers, 118, 120
  renin-angiotensin system localization, 37
Oviduct, 34–35, 37
Ovulation, see also Estrous cycle
  angiotensin II and, 29, 30
  opioid peptides and, see Opioid peptides
Oxytocin (OXY)
  anatomy, 63–65
  CCK and, 211, 215–221, 230
    colocalization, 213
    feeding behavior effects, 225–228, 230
  and FSH release, 73
  galanin, 196
  and LH release, 66–72
  and PRL secretion, 74–75, 93–94

P

Pancreas, VIP, 86
Pancreatic islets, 112
Pancreatic peptide, 165, 166
Pancreatic polypeptide, 107–110
Paraspinal ganglia, PP-like activity, 108

Parasympathomimetics, and GIP action,
      255
Paraventricular nucleus, NPY, 107, 112,
      116, 131
Paraventricular region
  neurotensin, 181
  VIP neurons, 87
Pars distalis, 244
Pars intermedialis, 244
Parvocellular neurons
  CCK, feeding behavior effects, 226
  CCK in, 211–212
  galanin in, 196
Pentagastrin, 214, 255, 258
Peptide histidine isoleucine-27 (PHI-27),
      85, 86, 250
Peptidergic system, galanin studies, 197
Peripheral administration
  of angiotensin II, and prolactin release,
      11
  of CCK
    and CRH and ACTH, 222
    feeding behavior effects, 226
    and gonadotropins, 215
    and oxytocin and vasopressin, 216–220
    and PRL, 220
  of LHRH, and prolactin regulation by
      angiotensin II, 14
Periventricular nucleus, NPY, 107, 116
Periventricular region, neurotensin, 181
PHI-27, 90
Phlorizidin, 252
Phosphoinositides, VIP and, 87
Phospholipase A2, 183
Pimozide, 215
Pineal gland, 244
Pituitary, anterior
  AVP neurons and, 72–76
  FSH release, 73–74, see also Follicle
      stimulating hormone
  galanin, 195–205
  LH release, 65–73, see also Luteinizing
      hormone
  motilin effects, 249–250
  motilin in, 244
  neuropeptide Y and, 122–127, 132–136
  neurotensin and, 181–190
  NPY binding, 111
  NPY-opioid peptide interactions, 171
  OXY neurons and, 66–75
  PP and, 108

prolactin release, 74–76
VIP receptors, 88
Pituitary, posterior, see also Oxytocin;
    Vasopressin
CCK effects, 211
NPY binding, 111
NPY in, 115, 118, 120
Pituitary portal system
NPY in, 115, 135
VIP in, 84, 88
Placenta, 35–37
Pons, 63, 244
Potassium-stimulated release of NPY, 132
Pregnancy
  galanin and, 202
  motilin effects, 249
  renin in, 30
Pregnant mare's serum gonadotropin
    (PMSG)
  and angiotensin II receptors, 28–29
  and galanin, 202
Preoptic area, motilin in, 244
Preoptic tuberal pathway, 164, 171
Preventricular nucleus
  CCK, feeding behavior effects, 226, 227
  CCK in, 212, 230–231
  galanin, 196, 197
Proestrus
  angiotensin II and, 6–7
  galanin gene expression, 201–202
  opioid peptides and, 158–163, 172
  OXY effects, 66–67, 72, 73, 75
  NPY in, 130, 169, 172
Progesterone, see also Ovarian steroids
  insulin and, 260
  and LH surge, 161, 162
  in proestrus, 129
  progesterone-primed ovariectomized
      animals, neurotensin in, 182
    and FSH and LH levels, 185–186
    and GH, 187
    and prolactin release, 183–184
    and TSH, 186–187
Proglumide, 229
Prolactin
  angiotensin II and, 10–11
    brain system, 9–10
    pituitary, 11–14
  CCK and, 230
    feeding behavior effects, 225
    oxytocin and, 221–222

galanin and, 199–200, 202–203
motilin and, 249
neurotensin and, 182–184, 188, 190
oxytocin and, 74–75, 95, 221–222
in obese (fa/fa) rat, 264
PP and, 108
vasopressin and, 75–76
Prolactin-releasing factor, VIP and, 88–94
Prorenin, 24
  female reproductive organs, 37
  placenta and amniotic fluid, 35–36
Prostaglandins, 31, 128
Prostate, 56
Protein kinase C, 183, 220
Purkinje cells, motilin in, 244
Push-pull perfusion, 125, 132, 190

R

Raphe nucleus, galanin in, 196
Releasing factor activity, VIP, 88–94
Renin, 23, see also Angiotensin II
  brain, 3
  central administration, 8
  in females, 35–37
  in males, 45–46, 48
Renin-angiotensin system in female
    reproductive structure
  localization, 37
  other structures, 34–37
  ovary, 23–30
  uterus, 30–34
Renin-angiotensin system in male repro-
    ductive structures
  ductus deferens, 54–55
  epididymis, 49–54
  prostate, 56
  seminal fluid, 56–57
  seminal vesicles, 56
  testis, 45–48
Reproductive behavior, CCK and, 223–224

S

Saralasin, 14
Sarthran, 13
Second messenger systems, 87, 183, 220
Secretin, 85, 86, 90, 250
Seminal fluid, 56–57
Seminal vesicles, 56
Sequence homologies

GIP, 250
  motilin, 245
  NPY and PP, 109–110
Serotonergic neurons, galanin in, 196
Serotonin, 122
  galanin and, 200
  neurotensin regulation of prolactin
    release, 182
  OXY neuron interactions, 71
  and prolactin release, 97
Sexual dimorphism, 212–213, 223–224
Somatostatin
  GIP and, 254, 256–257, 259
  motilin and, 246, 249
  neurotensin and, 187
  PP and, 108
Spermatozoa, 33, 53, 57
Spinal cord
  CCK in, 213
  motilin in, 244
  NPY in, 113
  PYY-like immunoreactivity, 110
Staurosporine, 220
Steroidogenesis, 260
Stress
  CCK and, 221, 222, 225
  and LH and prolactin levels, 13
Stria terminalis, 116
Striatum, 108, 113
Subcortical areas, motilin in, 244
Subfornical organ, 4, 259
Substance P, 213
Substantia innominata, 113
Substantia nigra, 213
Suprachiasmatic nucleus, 107
  NPY fibers, 116
  VIP neurons, 87
Supraoptic nucleus, 196
Sympathetic neurons, and GIP, 254–255
Synaptogenesis, motilin and, 244, 245

T

Testis
  NPY fibers, 120
  renin-angiotensin system, 45–48
Testosterone
  angiotensin II and, 47
  insulin and, 260, 261
  LH and, 48
  and NPY, 131, 132
Thalamus, NPY in, 114

Third ventricle, see Intracerebroventricular
    administration
Thyroid stimulating hormone (TSH)
  galanin and, 199
  neurotensin and, 186–187, 189, 190
  NPY and, 134, 135
  PP and, 108
Thyrotropin releasing hormone (TRH), 93,
    199
Triacylglycerol, 253
Trophic factor, motilin as, 244
Trophoblasts, 35
Tropomyosin, 245
Tuberoinfundibular dopaminergic system,
    183
Tyrosine hydroxylase, 196, 197, 213

U

Uterus
  angiotensinergic activity, 30–34
  motilin effects, 249
  oxytocin and, 64
  renin-angiotensin system localization, 37

V

Vagotomy, 217, 222
Vagus
  CCK and, 222, 226, 228
  and GIP, 254
  and motilin secretion, 246–247
Vasoactive intestinal peptide
  actions of, 87–94, 221
    presence in hypophyseal portal plasma,
      88
    receptors in the anterior pituitary gland,
      88
    releasing factor activity *in vitro*, 88–89
    releasing factor activity *in vivo*, 89–94
  characterization, 85
  dual site hypothesis for effects of, 94–98
  galanin and, 199, 200
  GIP homology, 250
  OXY and, 75
  and prolactin response, 96
  sites of production and release, 85–87
Vasoactive intestinal peptidergic neurons,
    199
Vasopressin
  anatomy, 63–65
  CCK and, 211–213, 215–221

CCK colocalization, 212, 213
and FSH release, 74
and galanin, 196–197
and LH release, 72
NPY and, in males, 122
and prolactin release from anterior
    pituitary, 75–76
Vasotocin, 72
Ventral tegmental area, CCK in, 213
Ventricular cannulation, 122, see also
    Intracerebroventricular administra-
    tion
Ventromedial hypothalamus, 131

CCK in, 212, 223
VIP neurons, 87

W

Water intake, angiotensin II and, 4

Y

Yohimbine, 121

Z

Zucker rat, 264